GRAMOPHONE, FILM, TYPEWRITER

WRITING SCIENCE

EDITORS Timothy Lenoir and Hans Ulrich Gumbrecht

GRAMOPHONE, FILM, TYPEWRITER

FRIEDRICH A. KITTLER

Translated, with an Introduction, by

GEOFFREY WINTHROP-YOUNG AND MICHAEL WUTZ

STANFORD UNIVERSITY PRESS

STANFORD, CALIFORNIA

The publication of this work was assisted by a subsidy from Inter Nationes, Bonn

Gramophone, Film, Typewriter was originally published in German in 1986 as *Grammophon Film Typewriter*,

Stanford University Press
Stanford, California

Printed in the United States of America

CIP data appear at the end of the book

TRANSLATORS' ACKNOWLEDGMENTS

A translation by Dorothea von Mücke of Kittler's Introduction was first published in *October* 41 (1987): 101–18. The decision to produce our own version does not imply any criticism of the *October* translation (which was of great help to us) but merely reflects our decision to bring the Introduction in line with the bulk of the book to produce a stylistically coherent text.

All translations of the primary texts interpolated by Kittler are our own, with the exception of the following: Rilke, "Primal Sound," has been reprinted from Rainer Maria Rilke, *Selected Works*, vol. 1, *Prose*, trans. G. Craig Houston (New York: New Directions, 1961), 51–56. © 1961 by New Directions Publishing Corporation; used with permission. The translation of Heidegger's lecture on the typewriter originally appeared in Martin Heidegger, *Parmenides*, trans. André Schuwer and Richard Rojcewicz (Bloomington: Indiana Univ. Press, 1992), 80–81, 85–86.

We would like to acknowledge the help we have received from June K. Phillips, Stefan Scherer, Candadai Seshachari, Shirin Shenassa, Steven Taubeneck, David Tompson, The Hemingway Trust and the Research and Professional Growth Committee of Weber State University, and the Interlibrary Loan Divisions at the University of British Columbia and Weber State University.

G.W.-Y.
M.W.

CONTENTS

TRANSLATORS' INTRODUCTION: *Friedrich Kittler and Media Discourse Analysis*

It was the Germans, those disastrous people, who first discovered that slag heaps and by-products might also count as learning, but I doubt if we can blame any one race or nation in particular for setting dumps and dustbins above the treasure cabinets of scholarship.

— H. G. WELLS, *The Camford Visitation*

MEDIA AWAKENINGS: THE USUAL SUSPECTS

In October 1939, in the first fall of the war, students and instructors at the University of Toronto abandoned their classes to listen to the enemy. A loudspeaker installed on a street close to Victoria College was broadcasting a speech by Adolf Hitler, who in the wake of Germany's victory over Poland was exhorting those still deluded enough to resist him to call it quits. Among the audience was a mesmerized classicist:

> The strident, vehement, staccato sentences clanged out and reverberated and chased each other along, series after series, flooding over us, battering us, half drowning us, and yet kept us rooted there listening to a foreign tongue which we somehow could nevertheless imagine that we understood. This oral spell had been transmitted in the twinkling of an eye, across thousands of miles, had been automatically picked up and amplified and poured over us.[1]

Half a century later, Eric Havelock—whose work on the Hellenic shift from orality to early literacy had become required reading for media and communication historians—recounted his wireless rapture in an attempt to explain why the early 1960s witnessed a sudden interest in the hitherto neglected topic of orality. In 1962–63, five prominent texts shedding light on the role of oral communication appeared within twelve months: *La Pensée sauvage* (Claude Lévi-Strauss), *The Gutenberg Galaxy* (Marshall McLuhan), *Animal Species and Evolution* (Ernst Mayr), "The Consequences of Literacy" (Jack Goody and Ian Watt), and Havelock's own *Preface to Plato*. What united these publications, Havelock argued, was the fact that their authors belonged to the first generation to be shaped by a world in which a print-biased media ecology had been altered by new ways of recording, storing, and transmitting sounds and voices, including the radiogenic Austrian dialect of a German dictator. Indeed, how could a generation of listeners acoustically nurtured on short-wave broadcasts of fireside chats, burning airships, Martian invasions, and calls for total war not grow up to ponder the changing relationship between speech and writing? "Here was the moving mouth, the resonant ear, and nothing more, our servants, or our masters; never the quiet hand, the reflective eye. Here was orality indeed reborn."[2]

"Media," the opening line of Friedrich Kittler's *Gramophone, Film, Typewriter* states with military briskness, "determine our situation" (xxxix). They certainly determine our appreciation of them. The media of the present influence how we think about the media of the past or, for that matter, those of the future. Without phonography and its new ability to faithfully manipulate the spoken word in ways that no longer require that speech be translated into writing, there would be no academic enterprises aimed at understanding the communicative household of cultures with few or no symbol-based external storage capacities. Our "reborn" or, to use Walter Ong's better-known phrase, "secondary" orality retroactively created the bygone word-of-mouth world that was not yet at the mercy of the quiet hand and the reflective eye.[3] Not surprisingly, many media histories adhere to a tripartite structure that uses these two oralities to bracket an interim period known as the "Gutenberg Galaxy" or the "Age of Print." Such framing, however, implies that the (re)discovery of a past orality will affect the perception of our present literacy, since every exploration of the dynamics of orality is a renegotiation of the limits and boundaries of literacy and its associated media networks. Why, then, separate the quantum leap in the research into orality from the emergence of the more comprehensive attention toward mediality in general? We need only add to Havelock's list a couple of equally divergent and influential contemporary titles—most prominently, André Leroi-Gourhan's *Geste et Parole* (1964–65) and McLuhan's *Understanding Media* (1964)—to realize that the watershed Havelock had in mind concerned more than questions of orality versus literacy. A widespread interest cutting across all disciplinary boundaries started to focus on the materialities of communication. At a time when the term "media" either was still missing from many dictionaries or conjured up visions of spiritualism, numerous scholars were attempting to bring into focus the material and technological aspects of communication and to assess the psychogenetic and sociogenetic impact of changing media ecologies. Such attempts set themselves the tasks of establishing criteria for the examination of storage and communication technologies, pondering the relationships among media, probing their social, cultural, and political roles, and, if possible, providing guidelines for future use.

Of course there were predecessors, and some are still being quoted. Of the many learned clichés circulating in the widening gyre of media studies, the most persistent may be the assurance that all the nasty things we can say about computers were already spelled out in Plato's critique of writing in *Phaedrus*.[4] In this century, Walter Benjamin's famous essay "The Work of Art in the Age of Mechanical Reproduction" was first published in 1936, and Harold Innis's *Empire and Communications* and *Bias of Communication*, the first attempts to conjugate world history according to the workings of different media technologies, appeared in 1950 and 1951, respectively. The list of works published before 1960 could be expanded, especially if one were to include the many single-medium theorists and commentators—such as Münsterberg, Arnheim, Balázs, and Kracauer on film, or Brecht and Lazarsfeld on radio—as well as the growth of North American communication studies, but media theory as we know it today first emerged in the 1960s.

Much of this work tends to go by generic names such as "media theory" or "media studies." Such terms are so hospitable as to be ridiculous, as if the combined trades, skills, and disciplines of paper production, book binding, bibliography, textual criticism, literary analysis, and the economics of publishing were to be labeled "paper theory." But their vagueness reflects a genuine diversity of possible approaches, for at the end of the twentieth century the study of media is roughly where the study of literature was at its beginning. When Boris Eichenbaum, one of the proponents of Russian formalism, tried to defend the "formal method" against the growing encroachment of state-sponsored Socialist Realism, he quoted the impatient comments of his fellow critic Roman Jakobson to underline the specificity and appropriateness of their new approach:

> The object of the science of literature is not literature, but literariness—that is, that which makes a given work a work of literature. Until now literary historians

have preferred to act like the policeman who, intending to arrest a certain person, would, at any opportunity, seize any and all persons who chanced into the apartment, as well as those who passed along the street. The literary historian used everything—anthropology, psychology, politics, philosophy. Instead of a science of literature, they created a conglomeration of homespun disciplines. They seemed to have forgotten that their essays strayed into related disciplines . . . and that these could rightly use literary masterpieces only as defective, secondary documents.[5]

The same impatience underlies Friedrich Kittler's comment that "media science" (*Medienwissenschaft*) will remain mere "media history" as long as the practitioners of cultural studies "know higher mathematics only from hearsay."[6] Just as the formalist study of literature should be the study of "literariness," the study of media should concern itself primarily with mediality and not resort to the usual suspects—history, sociology, philosophy, anthropology, and literary and cultural studies—to explain how and why media do what they do. It is necessary to rethink media with a new and uncompromising degree of scientific rigor, focusing on the intrinsic technological logic, the changing links between body and medium, the procedures for data processing, rather than evaluate them from the point of view of their social usage.

This centering upon media is reminiscent of the work of Marshall McLuhan, and, not surprisingly, the growing interest in the media-related work of Kittler, Vilém Flusser, Paul Virilio, Arthur Kroker, and Régis Debray coincides with McLuhan's resurrection as a critic of modernity worthy of being mentioned in the same breath as Adorno, Foucault, or Heidegger.[7] During McLuhan's lifetime this respectability would have amazed many a critic, since he appeared to be second to none when it came to making life easy for his detractors: his questionable politics, his casual and at times cynical dismissal of social issues, his delight in hobnobbing with the corporate and political elite, not to mention the breezy shallowness of his work following *Understanding Media*, all conspired to make him and his "Summa Popologica"[8] a well-placed punching bag, especially for the learned Left. McLuhan's focus on technologies, media formats, and materialities of communication did not fit easily within an intellectual landscape shaped more by questions of media ownership, audience manipulation, and strategies for communicative emancipation.

The intellectual Left's dismissal of McLuhan was equally pronounced in Germany. In a well-known media essay of 1971, Hans Magnus Enzensberger rejected him as a reactionary "ventriloquist" for the apolitical avant-garde, a "charlatan" ignorant of social processes "whose confused books serve as a quarry of undigested observations for the media industry."[9] Building on Brecht and Benjamin, Enzensberger attempted to formulate a "socialist strategy" for the emancipatory use of media. Anticipating a theme of great importance in *Gramophone, Film, Typewriter* (although stripped of its political overtones in Kittler's book), he pointed out that in principle, technologies such as the transistor radio recognize no contradiction between transmitter and receiver. Rather, these technical distinctions reflect the social division of labor into producers and consumers and therefore are ultimately predicated on the contradiction between the ruling and ruled classes. If passive consumers were to become active citizens and producers, they would have to take charge of this untapped technological potential, install themselves as producers, and thereby "bring the communications media, which up to now have not deserved the name, into their own."[10]

This notion of liberating media "into their own" provoked a vociferous response from Jean Baudrillard, who in his essay "Requiem for the Media" charged Enzensberger with regurgitating the old Marxist delusion that underneath the capitalist veneer of exchange value resides a more natural use value waiting to be uncovered.[11] It was erroneous to believe, Baudrillard argued, that media are neutral technological systems whose social impact depended upon who uses them to say what; rather, it was "in their form and very operation" that they induced social relations. In other words, media are "not *coefficients* but *effectors* of ideology"[12]—which was Baudrillard's way of terminologically updating McLuhan's mantra that the medium is the message. In short, media do not mediate; they are anti-mediatory and intransitive. The "revolutionary" events of May '68, Baudrillard claimed, could not survive their mediation because "transgression and subversion never get 'on the air' without being subtly negated as they are; transformed into models, neutralized into signs, they are eviscerated of their meaning."[13]

In his attempt to show that media destroy the aura of an event, Baudrillard was, in essence, transferring structuralist and semiotic explanations of the production and maintenance of meaning and ideology from texts and signs to media. To him, writing in France in the early 1970s, it was clear that "ideology" could no longer be constructed as an essence of social interests or manipulative intents fabricated at a hidden center and then channeled through the media. Just as recent scholarship had analyzed ideology and meaning as the result of an interplay of signs, a media theory inspired by structuralism and semiotics saw them to be inherent in the ways media operated. "'The medium is the message' operates a transfer of meaning onto the medium itself qua technological structure."[14]

However little they otherwise may have in common, the work of Kittler and Baudrillard is located on the same intellectual trajectory. Both reconceptualize the media issue in terms of recent theoretical developments commonly grouped together as "French theory." Superficially, Kittler's work can be seen as a merger of Foucault, Lacan, and McLuhan, that is, a combination of discourse analysis, structuralist psychoanalysis, and first-generation media theory. To distinguish it from the more generic terms "media studies" and "media theory," we will call it "media discourse analysis"[15] and present it in the following discussion as a distinctly German offshoot of poststructuralism that can only be understood against the German reception in the 1970s of the French triumvirate of Derrida, Foucault, and Lacan (with Virilio to be added later).

"LACANCAN AND DERRIDADA": THE FRENCH ACROSS THE RHINE

When poststructuralist theorizing crossed the Rhine from France into Germany in the late 1970s, it was not received with open arms. It is perhaps unsurprising that the harshest attacks against it were directed not at the *maître penseurs* themselves but at their German adepts. One outspoken critic chastised the work of the latter as "Lacancan and Derridada," an "unconditional and frequently uncritical adaptation to French theories" afflicted by a "congestion of linguistic expressiveness" that "above all desires one thing—not to be understood."[16] One no doubt can find similar sentiments in reaction to North American appropriations of poststructuralism, but to understand what Kittler says—and why he chooses to say it with a certain panache—it is necessary to describe briefly what distinguishes the German reception of poststructuralism from its North American counterparts.[17]

In Germany there was no signature event such as Derrida's presentation of "Structure, Sign, and Play in the Discourse of the Human Sciences" at Johns Hopkins, no "Yale School," and no "deconstruction" to speak of. There was instead, in Robert Holub's words, "a coterie of scholars"—among them Kittler, the philosopher Norbert Bolz, and the Germanist Jochen Hörisch—who had no "spiritual father" or "intellectual center" and at some point became intrigued with French theory.[18] Whereas in North America theory profited from a form of intellectual Reaganomics, a trickle-down effect by which the work of reputable scholars at allegedly superior institutions percolated downward and outward, its German reception tended to start at the academic margins—with students, junior faculty, reading groups, small publishing houses—and then gradually, and against notable resistance, move inward and upward. To a large extent resistance came from the Left, since what Derrida, Lacan, Foucault, and their disciples had to say seemed at first incompatible with positions inspired by Marx or the Frankfurt School (unlike today, where so much research goes into showing how like-minded they are). At times the struggle that ensued was motivated less by theory than by power. As had already happened in France in the wake of the events of 1968, the established Left was in danger of losing ground with one of its most important constituencies. If structuralist and poststructuralist criticism of Marx and his progeny prevailed, then disenchanted students, artists, and intellectuals might no longer be at the more-or-less exclusive disposal of the Left. Who, for example, could take Herbert Marcuse's sanguine Freudian-Marxist brew seriously after reading Lacan?[19] Faced with this challenge, the Left, which itself had faced stiff opposition during its fight for recognition, was quick to resort to the ubiquitous *Irrationalismusvorwurf*—that is, it accused French-inspired theorizing of downplaying history, eradicating the subject, and conjuring up impersonal, determinist symbolic chains and networks of irrationalism. Given National Socialism's mobilization and exploitation of the strong antirational tradition in German thought, this reproach carries considerable weight in Germany. Kittler has acknowledged the challenge: in a recent interview he described his magnum opus, *Discourse Networks*, as "written in black in every sense."[20] This phrase not only refers to the book's typographical appearance or to the fact that it was written in and for the black academic market (that is, outside established schools and trends) but also alludes to the German political color coding that associates black with conservatism.

Not that the Right and Center were any more welcoming, despite the fact that several of the German poststructuralists who later rose to prominence began their careers under the tutelage of well-known traditional literary scholars. (Kittler, for instance, started as an assistant to Gerhard Kaiser, one of the more prominent representatives of the hermeneutic tradition.)[21] Once again, conflict was probably unavoidable, and once again, it took on a certain edge because the opposing parties, despite their widely differing approaches and terminologies, were not that far removed from one another and were frequently concerned with identical issues. German critics of Derrida, especially those steeped in the hermeneutic tradition, have repeatedly claimed that he is not particularly original if read closely. His indebtedness to Heidegger is well known, and yet an assumption persists—explored in great detail in Manfred Frank's study *What Is Neostructuralism?*—that questions regarding the mediation of reference and subjectivity by and through language were already addressed, and at least partly solved, in the writings of Schleiermacher and several post-Kantian German idealist and Romantic philosophers.[22] In short, what was good about French poststructuralism was not new, and what was new was not good.

The poststructuralists responded with a threefold approach. First, leaving aside the purported inferiority of French philosophers of 1950–80 to their German counterparts of 1790–1820, they argued that the very fact that French poststructuralism was posing the same questions and dealing with related issues urged for its increased reception rather than its dismissal. Second, instead of neutralizing the French poststructuralists by referring them back to their German antecedents, they proposed that the latter be radicalized by focusing on those instances where they anticipated or came close to the solutions put forward by French theorists. This strategy was adopted, for example, by Hörisch, who plays off the brash, young (as it were, proto-French), antihermeneutic Schleiermacher against the elderly, cryptohermeneutical Schleiermacher so dear to the established German tradition.[23] It also helps to explain why, since the 1977 publication of the collection *Urszenen*, German poststructuralism has been so drawn to "difficult" texts and writers of that era.[24] If Hölderlin, Kleist, or even the long novels of Goethe are seen as inspired by, playing with, and taking apart the proto-French aesthetic and philosophic axioms of their day, then discourse analysis, Lacanian theorizing, and Derridean deconstruction become the more appropriate tools for dealing with them.[25]

The third and most straightforward approach consisted in informing traditional hermeneutic scholars that they were unable to face the true dimensions of the French theory offerings, an objection that sometimes took the shape of gleefully or defiantly confirming their worst suspicions of what poststructuralism is up to. In his critique of *What Is Neostructuralism?* Kittler honed in on Frank's fearful assumption that French theorists were promoting the "dream of a subjectless machine."[26] Discussing Lacan's famous account of human consciousness as a camera that captures and stores images even when nobody is around,[27] Frank had argued that Lacan, in the final analysis, could not do without some kind of subject endowed with self-reflective consciousness. Not so, Kittler responded: this mechanical Polaroid consciousness was all Lacan had in mind because his technological materialism, just like Freud's, "reasoned only as far as the information machines of his era—no more and no less."[28] By emphasizing Lacan's frequent references to circuits and feedback (not to mention Lacan's refusal to discuss the subject of language with anybody not versed in cybernetics), Kittler moved Lacan out of the hermeneutically soiled realms of old-style psychoanalysis, philosophy, and literary scholarship and into the far more appropriate posthermeneutic domain of information theory. Nowadays, Kittler noted disapprovingly, even newspapers regurgitate Lacan's famous dictum that the unconscious is the discourse of the other, "but that this discourse of the other is the discourse of the circuit is cited by no one."[29]

To associate French poststructuralism with modern media technology has become a commonplace in current North American literary theory. George Landow's *Hypertext*, with its programmatic subtitle, *The Convergence of Contemporary Critical Theory and Technology*, asserts that hypertext presents an "almost embarrassingly literal embodiment" of Derrida's emphasis on de-centering and Barthes's conception of the readerly versus the writerly text.[30] Gregory Ulmer claims that the grammatological works of Derrida "already reflect an internalization of the electronic media, thus marking what is really at stake in the debate surrounding Western metaphysics."[31] Eugene Provenzo and Mark Poster, in turn, link Foucault's analysis of surveillance techniques to databases and electronic control procedures.[32] It now appears that these links, analogies, and correspondences also can be projected back in time. What hypertext and hypermedia are to poststructuralism, cybernetics was to structuralism and semiotics, and in both instances the human implication has been profound:

> Without passing through linguistics at all, Norbert Wiener (inventor of cybernetics) had already as early as 1948 defined man without reference to interiority as a communication machine, a machine for exchanging information with his environment. The idea that all reality must be broken up in the final analysis into a set of relations between elements came together by an entirely different angle with the structural postulate, imputing every effect of meaning to a combination of minimal units or pertinent traits of a determinate code. While resolutely aware of it, French semiology was metaphorizing and "culturalizing" the American mechanist paradigm.[33]

In a chapter entitled "Structures—Discourses—Media" in his book *Philosophie nach ihrem Ende* (Philosophy after its end), Bolz describes the "clear paradigm sequence" that has ruled French theory production since Saussure. First, Saussure's insight that the meaning of signs is an effect of differential articulation reappears in Lévi-Strauss to describe the human mind as a set of matrices for the emergence of structures, while Lacan, combining structural linguistics with cybernetic theory, "transforms structural psychoanalysis into a media theory of the unconscious."[34] In the second stage, Foucault builds on this link to describe the relays and circuits of discursive practices. Finally, Paul Virilio's "dromological" and "chronopolitical" analyses—which will be of great importance to the "Film" section of *Gramophone, Film, Typewriter*—link the mutation of human perception to changes in military media technology. Step 1: We recognize that we are spoken by language. Step 2: We understand that language is not some nebulous entity but appears in the shape of historically limited discursive practices. Step 3: We finally perceive that these practices depend on media. In short, structuralism begot discourse analysis, and discourse analysis begot media theory.

Media, then, are (at) the end of theory because in practice they were already there to begin with. Accordingly, Kittler ties the emergence of structuralism to the introduction of the typewriter, and he criticizes Foucault for neither reflecting on the mediality of the discursive practices he analyzed nor going beyond the confines of the Gutenberg Galaxy. Thus, whereas Foucault's archives are based on the hegemony of written language, on the silent assumption that print is the primary (if not the only) carrier of signification, Kittler's archeology of the present seeks to include the technological storage and communication media of the post-print age(s). "Even writing itself, before it ends up in libraries, is a communication medium, the technology of which the archeologist [Foucault] simply forgot. It is for this reason that all his analyses end immediately before that point in time at which other media penetrated the library's stacks. Discourse analysis cannot be applied to sound archives and towers of film rolls" (5).

Media are the alpha and omega of theory. If media do indeed "determine our situation," then they no doubt also determine, and hence configure, our intellectual operations. One could easily reappropriate Derrida's much-deferred pronouncement *il n'y a pas de hors-texte* and suggest that the fundamental premise of media discourse analysis is *il n'y a pas de hors-media*.

DISCOURSE NETWORKS: FROM MOTHER TONGUES TO MATTERS OF INSCRIPTION

Kittler's intellectual career can be broken down into three parts, each roughly covering one decade. In the 1970s, his focus was on discourse analysis; in the 1980s, he turned his attention to the technologizing of discourse by electric media; and in the 1990s, to its subsequent digitization.

Beginning as a *Privatdozent* in Freiburg, he dealt with the so-called Age of Goethe (1770–1830) in most of his early work, concentrating on canonical authors like Lessing, Schiller, and Goethe himself. The influence of Foucault and Lacan is obvious—his highly demanding reading of E. T. A. Hoffmann's "The Sandman" ranks as "the most compressed and programmatic of all applications of Lacan"[35]—as is the attempt to fuse the two. One of his principal goals is to relate Lacanian notions of subject (de)formation, specifically within the framework of the nuclear family that emerged in the second half of the eighteenth century, to the discursive practices that came to regulate the new roles and relationships of mothers, fathers, and children on the one hand and authorities and subjects on the other. Looking back at this early work, Kittler emphasized that the nuclear family between the ages of the Enlightenment and Romanticism was "not a fact of social history" but a "code," a "veritable discourse machine" that produced all the secrets and intimacies that were subsequently mistaken as essential components of an equally essential human nature. Hence, texts such as Lessing's family dramas or Goethe's *Bildungsromane* have to be read as instances of a cultural inscription program: German literature around 1800, so often hailed as the apex of Germanic cultural output culminating in the twin peaks of Goethe and Schiller, becomes a means of programming people, part of the overall recoding enterprise that ushered in an age that saw not only the spread of the nuclear family but also the growth of literacy, the notion of authorship as the expression of ineffable individuality and *Innerlichkeit*, and the preindustrial mobilization of the modern nation state on all ideological, administrative, and military levels.[36] "The official locus of production for German Poetry was the nuclear family; scholars saw to its multiplication; and a science that claimed the title Science provided its justification."[37]

The 1980s (during which Kittler moved from Freiburg to Bochum) brought a considerable broadening of his interests and increasing forays into non-German, and non-Germanist, areas. Always a prolific scholar, he produced essays on (among others) Nietzsche, Pink Floyd, Peter Handke, Dashiell Hammett, Bram Stoker, Richard Wagner, and Thomas Pynchon.[38] More importantly, "media"—a word rarely used in the previous decade—made a grand entry, and with good reason. If literature is programming, how exactly does it proceed? Obviously, it involves the production, circulation, and consumption of texts. Interpreting those texts, that is, isolating and forcing them to reveal something beyond the materialities and orders of communication that produced them in the first place, will be of little help. Instead, discourse analysis begins by simply registering them as material communicative events in historically contingent, interdiscursive networks that link writers, archivists, addresses, and interpreters.[39] In so doing, discourse analysis does not deny interpretation; it merely concentrates on something more interesting. First of all, it focuses on the brute fact that certain texts *were* produced—rather than not, and rather than others. Second, it shows that these texts, regardless of the variegated social practices to which they may be related, exhibit certain regularities that point to specific rules programming what people can say and write.

Third and perhaps most surprising, discourse analysis highlights the fact that, given the growing social complexity and expanding communicative networks of the early 1800s, standardized interpretation appears to have been possible and, indeed, was ever more desirable. The hermeneutic master plan seems to have been to offset increasing social complexity with interpretative homogenization. This plan can only work, however, if people are trained to work with language in standardized ways that downplay its changing materiality. For instance—to choose one example of importance to Kittler—people have to be trained to read the smooth and continuous flow of ink on paper as the manifestation of an equally smooth and continuous flow of personality. In Hegel's words, the essence of individuality has its "appearance and externality" in handwriting. But people also have to be trained to disregard the change from handwriting to print.[40] This point, then, is crucial: beginning in the Age of Goethe—not coincidentally one of the formative periods of German history—stable cultural references such as authorship, originality, individuality, and *Geist*, all accessible by way of standardized interpretation practices, cut through and homogenized increasing social complexity; this could only occur, however, because a naturalized language now seen as a lucid carrier of meaning cut through and homogenized the different media. In short, people were programmed to operate upon media in ways that enabled them to elide the materialities of communication. But if there is any truth to what media theory, following Innis, Ong, and McLuhan, has been claiming for decades, media have their own "biases" and "messages" that must be taken into account. The question of how people operate upon media thus has to be complemented by the equally important question of how media operate upon people. Subsequently, discourse analysis has to be expanded as well as supplemented by media theory. Scholars such as Kittler, Bolz, and Hörisch, as it were, played Marx to Foucault's Hegel: they pulled discourse analysis off its textual and discursive head and set it on its media-technological feet.

The new dimensions of Kittler's analysis are contained in a nutshell in the important essay "Autorschaft und Liebe" (Authorship and love), first published in 1980 as part of a volume polemically and programmatically entitled *Austreibung des Geistes aus den Geisteswissenschaften: Programme des Poststrukturalismus* (Expulsion of the Spirit from the humanities: programs of poststructuralism). The essay is organized around the sharp contrast between two very different body-medium links that represent two very different ways that writers evoked and readers experienced love. First, Kittler presents Paolo and Francesca, Dante's infernal couple, whose doomed love drastically short-circuits texts and bodies, leading them to physically (re)enact the adulterous love story they had been reading out loud. (Their narrative, in turn, manages to physically knock out their spellbound listener.) Against this Kittler sets the equally ill-fated love recorded by Goethe of Werther and Lotte, who celebrate a far less physical but no less delirious communion by allowing their souls to share the spirit of Klopstock's beloved poetry.[41] Impassioned bodies cede to yearning souls, nameless desires communicated by an anonymous text make way for the spirit of authorship, and manuscripts to be read aloud in the company of others are replaced by printed books to be devoured in solitary silence: the contrastive technique employed here is reminiscent of Foucault, whose presence is equally evident in the structural macrolevel of *Discourse Networks*, first published in German in 1985 (and now in its third, revised edition).

Indeed, in discussing *Discourse Networks* Kittler confirmed that Foucault, as "the most historical" of the French triumvirate, is the most important to him—more important than Lacan and far more than Derrida.[42] As David Wellbery points out in his excellent foreword to the English translation, there are substantial affinities. In *The Order of Things*, Foucault periodizes European conceptions of life, labor, and language on the basis of three generalized "epistemes": the "Renaissance," the "classical," and the "modern." Kittler, in turn, presents three historical moments corresponding more or less to Foucault's: the "Republic of Scholars" is the approximate equivalent to Foucault's "Renaissance" and "classical" epistemes; the historical datum "1800" correlates roughly to Foucault's "modern" period; and "1900" designates a discourse network that matches Foucault's emergent postmodernism.[43] In Kittler's usage, "discourse network" designates "the network of technologies and institutions that allow a given culture to select, store, and produce relevant data."[44] The term is very extensive: it attempts to link physical, technological, discursive, and social systems in order to provide epistemic snapshots of a culture's administration of power and knowledge. Not unlike the approach taken in Jonathan Goldberg's acclaimed study *Writing Matter*, the aim is to combine a "Foucauldian" analysis of historically contingent rules and regulations, which allow or force people to speak in certain ways, with the examination of equally contingent physical and mental training programs and the analysis of the contemporary media technologies that link the two.

Although Kittler leaves his "Republic of Scholars" largely undeveloped, the discursive field of "1800"—the period known as German Classicism, Romanticism, or the Age of Goethe—is described in terms of the spiritualized oralization of language. Kittler argues that the process of alphabetization came to be associated with the Mother as an embodiment of Nature—more specifically, with "the Mother's mouth," now reconceptualized as an erotic orifice linking sound, letter, and meaning into a primary linguistic unit charged with pleasure. German children learned to read through both the physical and sexual immediacy of and proximity to the *Muttermund* (which in German signifies both the literal mouth of the mother as well as the opening of the uterus). By associating erotic pleasure with the act of composition and rereading, and with Mother Nature more generally, writers of the Classical and Romantic periods understood language as a form of originary orality, a transcendental inner voice superior and anterior to any form of written language. In the same way, Woman was constructed as the primordial site of linguistic origin and inspiration, which urged male writers such as Goethe both to serve as state bureaucrats and to produce texts for a predominantly female audience. And prominent educators addressed mothers as the primary targets of children's socialization into language, initiating pedagogical reforms that centered on the pronunciation-based acquisition of reading and writing. Originary orality, in that sense, was the effect of a feedback loop involving didactic techniques, media reform, and a peculiar surcharge of the maternal imago.

The discourse network of 1800 depended upon writing as the sole, linear channel for processing and storing information. For sights, sounds, and other data outside the traditional purview of language to be recorded, they had to be squeezed through the symbolic bottleneck of letters, and to be processed in meaningful ways they had to rely on the eyes and ears of hermeneutically conditioned readers. Reading, in that sense, was an exercise in scriptographically or typographically induced verbal hallucinations, whereby linguistic signs were commuted into sounds and images. With the advent of phonography and film, however, sounds and pictures were given their own, far more appropriate channels, resulting in a differentiation of data streams and the virtual abolition of the Gutenberg Galaxy. Language's erstwhile hegemony was divided among media that were specific to the type of information they processed. Writing, a technology of symbolic encoding, was subverted by new technologies of storing physical effects in the shape of light and sound waves. "Two of Edison's developments—the phonograph and the kinetoscope—broke the monopoly of writing, started a non-literary (but equally serial) data processing, established an industry of human engineering, and placed literature in the ecological niche which (and not by chance) Remington's contemporaneous typewriter had conquered."[45]

But if, in the discourse network of 1800, Woman is constructed as the source of poetic language, how is this construct affected by the new differentiation of data processing? The discourse network of 1900, Kittler argues, demystifies the animating function of Woman and the conception of language as naturalized inner voice. No longer reducible to "the One Woman or Nature," the women of the discourse network of 1900 are "enumerable singulars,"[46] released from their supplemental function to the male creative process. No longer destined to engender poetic activity in male writers and subsequently to validate the (male) author-function by making sense of the texts written for their consumption, women now become producers themselves. While male writers, deprived of a female decoding network, devolved from inspired poets to simple word processors, women began to process texts themselves. The sexually closed circuits of the Gutenberg Galaxy's old boys' network are severed. Exchanging needlework for typewriters and motherhood for a university education, women commenced to fabricate textures of a different cloth and thus asserted equal access to the production of discourse. Yet, while the typewriter did away with either sex's need for a writing stylus (and in the process giving women control over a writing machine–qua-phallus), it reinscribed women's subordination to men: women not only became writers but also became secretaries taking dictation on typewriters, frequently without comprehending what was being dictated.

As a correlate to the Edisonian specification of inscription technologies, writers became increasingly aware of the materiality of language and communication. Thought of around 1800 as a mysterious medium encoding prelinguistic truth, writing in the Age of Edison began to be understood as only one of several media possessed of an irreducible facticity. In Mallarmé's succinct phrase, "one does not make poetry with ideas, but with *words*," bare signifiers that inverted the logic of print as a vehicle of linguistic communication and instead emphasized "textuality as such, turning words from means to ends-in-themselves."[47] Fundamentally, these words were nothing but marks against a background that allowed meaning to occur on the basis of difference. What the typewriter had instituted, namely, the inscription of (standardized) black letters on white paper, was replicated in the processing modes of both the gramophone and film. The gramophone recorded on a cylinder covered with wax or tinfoil, and eventually on a graphite disk, whereas film recorded on celluloid; but both recorded indiscriminately what was within the range of microphones or camera lenses, and both thereby shifted the boundaries that distinguished noise from meaningful sounds, random visual data from meaningful picture sequences, unconscious and unintentional inscriptions from their conscious and intentional counterparts. This alternation between foreground and background, and the corresponding oscillation between sense and nonsense on a basis of medial otherness, a logic of pure differentiality—which on a theoretical level was to emerge in the shape of Saussure's structural linguistics—typifies the discourse network of 1900. The transcendental signified of Classical and Romantic poets has ceded to the material signifier of modernism.

Bewundert viel und viel gescholten (much admired and much admonished): Helen's iambic self-diagnosis in the second part of Goethe's *Faust* comes to mind when assessing the reception of *Discourse Networks*. To some, it is more than a book of genius and inspiring breadth; it is a watershed beyond which the study of literature and culture must follow a different course. In a discussion of Nietzsche, the mechanized philosopher who more than any other heralded the posthermeneutic age of the new media, Kittler quotes the poet-doctor Gottfried Benn: "Nietzsche led us out of the educated and erudite, the scientific, the familiar and good-natured that in so many ways distinguished German literature in the nineteenth century." Almost exactly one hundred years later, Kittler's work appears to some, particularly among the younger generation, as what is leading us out of the similarly stagnant pools of erudition and familiarity that have come to distinguish German, and not only German, literary scholarship. To others it is a sloppy mosaic that runs roughshod over more nuanced, contextualized, and academically acceptable research undertaken in cultural studies, literary history, and the history of science, not to mention feminism. Critics might instead be tempted to apply the second half of Benn's statement (not quoted by Kittler) to Kittler's role in contemporary scholarship: "Nietzsche led us . . . into intellectual refinement, into formulation for the sake of expression; he introduced a conception of artistry into Germany that he had taken over from France."[48] And finally, there is a third reaction, one Helen could not complain of: the book is much ignored. This is, no doubt, partly due to the difficulties involved; to an audience outside of German studies, the exclusively German focus of the first part, describing the discourse network of 1800, poses considerable problems. *Gramophone, Film, Typewriter*, however, is far more accessible by virtue of its focus on the *Mediengründerzeit*—a coinage derived from the historiographical term *Gründerzeit*, which denotes the first decades of the Second German Empire founded in 1871, and which Kittler reappropriates to refer to the "founding age" of new technological media pioneered by Edison and others during the same time period.

MARSHALL MCNIETZSCHE: THE ADVENT OF THE ELECTRIC TRINITY

At first glance, *Gramophone, Film, Typewriter* appears to be a lengthy addendum to the second part of *Discourse Networks* ("1900"), providing further and more detailed accounts of the ruptures brought about by the differentiation of media and communication technologies. The book could be understood as a relay station that mediates—Kittler uses the more technical term *verschalten* (to wire)—various forgotten or little-known texts on the new electric media and the condition of print in the age of its technological obsolescence. Kittler reprints, in their entirety, Rilke's essay "Primal Sound," the vignettes "Goethe Speaks into the Phonograph" and "Fata Morgana Machine" by Salomo Friedlaender (a.k.a. Mynona), Heidegger's meditation on the typewriter, and Carl Schmitt's quasiphilosophical essay "The Buribunks," among others, passing from one to another through his own textual passages. In that sense, *Gramophone, Film, Typewriter* is engineered to function as a kind of intertextual archive, rescuing unread texts from oblivion. Because these texts were written between the 1890s and the 1940s, that is, in the immediate presence of a changing media ecology, they registered with particular acuity the cultural effects of the new recording technologies, including the erosion of print's former monopoly. Print reflects, within the limits of its own medium, on its own marginalization.

The overall arrangement is simple. As the title indicates, the book comprises three parts, each dedicated to one of the new information channels. What distinguishes the post-Gutenberg methods of data processing from the old alphabetic storage and transmission monopoly is the fact that they no longer rely on *symbolic mediation* but instead record, in the shape of light and sound waves, visual and acoustic *effects of the real*. "Gramophone" addresses the impact and implications of phonography, "Film" concentrates on early cinematography, and "Typewriter" addresses the new, technologically implemented materiality of writing that no longer lends itself to metaphysical soul building. For those more interested in theoretical issues, and technological extensions of poststructuralism in particular, it will be important to keep in mind that Kittler relates phonography, cinematography, and typing to Lacan's axiomatic registers of the real, the imaginary, and the symbolic. In brief, writing in a postprint environment is associated with the symbolic, with linguistic signs that have been reduced to their bare "materiality and technicity" and comprise a "finite set without taking into account philosophical dreams of infinity" (15). The imaginary, by contrast, is linked with the technology of film, because the sequential processing of single frames into a projected continuity and wholeness corresponds to Lacan's mirror stage—that is, the child's experience of its imperfect body (in terms of motor control and digestive function) as a perfect reflection, an imagined and imagistic composition in the mirror. The real is in turn identified with phonography, which, regardless of meaning or intent, records all the voices and utterances produced by bodies, thus separating the signifying function of words (the domain of the imaginary in the discourse network of 1800) as well as their materiality (the graphic traces corresponding to the symbolic) from unseeable and unwritable noises. The real "forms the waste or residue that neither the mirror of the imaginary nor the grid of the symbolic can catch: the physiological accidents and stochastic disorder of bodies" (16). Hence, the distinctions of Lacanian psychoanalysis, what Bolz calls a "media theory of the unconscious," appear as the "theory" or "historical effect" of the possibilities of information processing existent since the beginning of this century.[49]

Readers will find much that is familiar from *Discourse Networks*: Kittler continues to pay sustained attention to the coincidence of psychoanalysis and Edisonian technology, and includes a suggestive discussion of "psychoanalytic case studies, in spite of their written format, as media technologies" (89), since they adhere to the new, technological media logic positing that consciousness and memory are mutually exclusive. He further develops the contradictory and complicated relays between gender and media technology, including a "register" of this century's "literary desk couples" (214)—couples who, according to Kittler, have exchanged lovemaking for text processing. And once again, Kittler questions a motley crew of friendly and unfriendly witnesses—among them Nietzsche, Freud, Kafka, Rilke, Ernst Jünger, Roger Waters, and William Burroughs—to ascertain what exactly happened when the intimate and stately (that is, increasingly quaint and cumbersome) processing technology called writing was challenged, checked, modified, and demoted by new storage and communication technologies. Nietzsche in particular takes on a key role as the first philosopher to use a typewriter and thus as the first thinker to fully recognize that theoretical and philosophical speculations are the effects of the commerce between bodies and media technologies. Nietzsche had this recognition in mind, Kittler suggests, when he observed in one of his few typed letters that "Our writing tools are also working on our thoughts" (*Unser Schreibzeug arbeitet mit an unseren Gedanken*). When the progressively myopic retired philologist began using a typewriter—a Danish writing ball by Malling Hansen that did not allow him to see the letter imprinted at the moment of inscription—he not only anticipated *écriture automatique* but also began to change his way of writing and thinking from sustained argument and prolonged reflection to aphorisms, puns, and "telegram style." After abandoning his malfunctioning machine, Nietzsche elevated the typewriter itself to the "status of a philosophy," suggesting in *On the Genealogy of Morals* that humanity has shifted away from its inborn faculties (such as knowledge, speech, and virtuous action) in favor of a memory machine. Crouched over his mechanically defective writing ball, the physiologically defective philosopher realizes that "writing . . . is no longer a natural extension of humans who bring forth their voice, soul, individuality through their handwriting. On the contrary, . . . humans change their position—they turn from the agency of writing to become an inscription surface" (210).

Nietzsche—or, better, this technologically informed, poststructuralist reading of Nietzsche—points to an elementary trope governing Kittler's narrative. Regardless of its convictions or ideological direction, poststructuralism claims to reveal many key concepts (such as the Subject, Authorship, Truth, Presence, "so-called Man," and the Soul) to be a kind of conceptual vapor or effect that arises from, and proceeds to cover up, underlying discursive operations and materialities. In posthermeneutic scholarship such as Kittler's, these effects are not so much denied as bracketed through a shift of focus toward certain external points—in particular, bodies, "margins," power structures, and, increasingly, media technologies—in the interstices of which those phantasms had come to life in the first place. Thus, both Nietzsche's and Kittler's intellectual careers consist in pushing the brackets together, until everything that had frolicked between them is squeezed out of existence. When a camera (as in Lacan's example) does all the registering, storing, and developing on its own, there is no need for an intervening Subject and its celebrated Consciousness; when the inspiring maternal imago of Woman turns into a secretary, there is no need for binding Love; when the phonograph mercilessly stores all that people have to say and then some, there might be an unconscious but no meditating Soul. The sad spectacle of the allegedly insane Nietzsche in the last ten years of his life, "screaming inarticulately," mindlessly filling notebooks with simple "writing exercises," and "'happy in his element' as long as he had pencils,"[50] is where the converging brackets meet. It is, as it were, the ground zero of all hermeneutically inclined theorizing: on the one hand, a body in all its vulnerable nakedness; on the other, media technologies in all their mindless impartiality; and between them nothing but the exchange of noise that only a certain amount of focused delusion can arrange into deeper meanings.

But as we know only too well, the switch from the Gutenberg Galaxy to Edison's Universe has been followed by the more recent move into the Turing World. With obedience to this succession, *Gramophone, Film, Typewriter* begins with Edison's phonograph and ends with Turing's COLOSSUS, a move already hinted at in the first paragraph of "Gramophone." Shifting from tinfoil and paraffin paper to charge-coupled devices, surface-wave filters, and digital signal processors, the book moves away from "technological media" such as the gramophone and kinetoscope to the computer, and it thus signals the beginning of the third stage in Kittler's intellectual career (during which he was installed as Professor of Aesthetics and Media History at Berlin's Humboldt University). If Kittler's passage from the 1970s to the 1980s, with his progressive grounding of discourse in the materialities of communication, is analogous to the switch from the symbol-based discourse network of 1800 to the technology-based discourse network of 1900, then his passage from the 1980s to the 1990s approximates the switch from the electric discourse network of 1900 to an electronic "systems network 2000," with its reintegration of formerly differentiated media technologies and communication channels by the computer, the medium to end all media. Once again, his essays signal an increasing movement of interest toward computer hardware and software, the archeology of the digital takeover (Kittler edited and introduced the German translation of Alan Turing's works), and military technology and strategy.[51] All of this first appears, fully orchestrated, in the concluding passages of *Gramophone, Film, Typewriter*.

Finally, a word about style. A book on the materialities of communication can hardly be oblivious to its own materialities and historical situatedness, so it comes as no surprise that *Gramophone, Film, Typewriter* itself carries the imprint of the media of which it speaks. The mosaic-like qualities of much of the text, for instance, the sometimes sudden shifts from one passage or paragraph to another and, alternately, the gradual fade-outs from Kittler's own texts to those of his predecessors, derives, in both theory and practice, from the jump-cutting and splicing techniques fundamental to cinema. But media technologies could also be invoked to explain Kittler's idiosyncratic stylistics on the micro-level of the individual sentence or paragraph. Long stretches are characterized by a quality of free association—not to say, automatic writing—that once again could be labeled cinematic, with one idea succeeding the other, strung together by a series of leitmotifs. One such leitmotif is the aforementioned dictum by Nietzsche, "Our writing tools are also working on our thoughts," which Kittler quotes repeatedly, suggesting certain stylistic and intellectual affinities with his mechanized predecessor. (And who could question their similarities? Nietzsche was the first German professor of philology to use a typewriter; Kittler is the first German professor of literature to teach computer programming.) Certainly, Kittler's prose is somewhat Nietzschean in that syntactic coherence frequently yields to apodictic aperçus, sustained argument to aphoristic impression, and reasoned logic to sexy sound bites. This enigmatic prose is further exacerbated by stylistic peculiarities all Kittler's own. Most noticeable among these is the frequent use of adverbs or adverbial constructions such as *einfach*, *einfach nur*, *bekanntlich*, *selbstredend*, or *nichts als* (variously translated as "merely," "simply," "only," "as is known," and "nothing but"), as in this explanation of the computerized recording of phonemes: "The analog signal is simply digitized, processed through a recursive filter, and its autocorrelation coefficients calculated and electronically stored" (75). Such sentences (call them Kittler's Just So Stories) are, with casual hyperbole, meant to suggest the obvious, bits of common knowledge that don't require any elaboration, even though (or precisely because) their difficult subjects would urge the opposite. Similarly, Kittler is fond of separating consecutive clauses (in the German original, they tend to lead off with *weswegen*) from their main clauses, as in this explanation of the physiological bases of the typewriter: "Blindness and deafness, precisely when they affect speech or writing, yield what would otherwise be beyond each: information on the human information machine. Whereupon its replacement by mechanics can begin" (189). Despite their casual, ostensibly unpolished, conversational qualities, these clauses almost always refer to important points. Which is why sentences like this simply deserve special attention.

Not surprisingly, Kittler's rhetorical bravado has drawn sharp criticism. One critic attributed the paradox that Kittler confidently employs writing to ferret out superior and more advanced media technologies to "stylistic means consciously used for the production of theoretical fantasy literature."[52] To Robert Holub, the

> single most disturbing factor of Kittler's prose [is] the style in which it is written. Too often arguments seem obscure and private. One frequently has the impression that its author is writing not to communicate, but to amuse himself. His text consists of a tapestry of leitmotifs, puns, and cryptic pronouncements, which at times makes for fascinating reading, but too often resembles free association as much as it does serious scholarship.[53]

As with McLuhan, Kittler's prose carries a flashy dexterity that makes many claims seem invulnerable to substantive critique precisely because of their snappy and elegant phrasing. To this litany one could add Kittler's penchant for maneuvering between engineering parlance and medical jargon, as well as his use of a whole register of specialized terminologies that, in Holub's estimation, suggest "a semblance of profundity"[54] but do not ultimately contribute to a sustained argument. To top it off, a growing number of younger scholars have modeled their writing on Kittler's very personal style: to the delight of connoisseurs of German academese, *Kittlerdeutsch* is already as distinct an idiom as the equally unmistakable *Adornodeutsch*.

Rather than take Kittler to task for his virtuoso play on the keyboard of poststructuralist rhetoric, we would urge consideration of his writing style in the larger context of the tradition he writes in—and, more important, against. Clearly, he cultivates a cool, flippant, and playful style to subvert the academic ductus of German university prose, a tongue-in-cheek rhetoric to thumb his nose at the academic establishment. If style, as Derrida reminds us (not coincidentally, in his analysis of Nietzsche's writing) is always "the question of a pointed object . . . sometimes only a pen, but just as well a stylet, or even a dagger,"[55] then Kittler is certainly twisting his own stylus into the body of German intellectual discourse, which has kept alive for far too long what he feels to be the obsolete hermeneutic tradition. To counteract the widespread use of stiff and lugubrious academic prose, he indulges in stylistic *jouissance*, a spirited playfulness meant to assault and shock conventional scholarly sensibilities. And indeed, what better way is there to debunk highfalutin theories than a wry recourse to the materialities of comunication?[56] No less than the philosopher with a hammer of a century ago, who smashed notions of selfhood and forged a style of his own by hammering on the keys of his writing ball, Kittler plays the *enfant terrible* of the German humanities who pummels literary-critical traditions with a rhetorical freestyle all his own. Indeed, to paraphrase Nietzsche, the inscription technologies of the present have contributed to Kittler's thinking.

ONLY CONNECT: THEORY IN THE AGE OF INTELLIGENT MACHINES

But Friedrich Nietzsche is not the real hero of *Gramophone, Film, Typewriter*. That part goes to Thomas Alva Edison, a casting decision that Kittler believes will appeal to a North American audience: "Edison . . . is an important figure for American culture, like Goethe for German culture. But between Goethe and myself there is Edison."[57] Indeed, Kittler credits his sojourns in California—in particular, the requirement that he furnish Stanford undergraduates with updated, shorthand summaries of German history—with providing the impetus to focus on technological issues. Much could be said about the history behind this alleged dichotomy between the United States and Germany, or of the implied distinction between technology and culture, but there can be no doubt that North American readers will find much of interest in *Gramophone, Film, Typewriter*. They will, however, also find cause for irritation beyond the question of style. In conclusion, we will briefly point to five particularly promising or problematic issues for the North American reception of Kittler.

1. Back to the ends of Man. After years of "antihumanist" rhetoric, a lull appears to be settling in. A spirit of compromise is afoot in the humanities, and "subjects" are being readmitted into scholarly discourse, provided they behave themselves and do not suffer any self-aggrandizing Cartesian or Kantian relapse. In the face of such imminent harmony, Kittler's rhetoric may seem like a throwback to the heady days of militant antihumanism. His work no doubt invites the plotting of a historical graph in which the human being is reduced from its original function as *homo faber* to an accessory in a scenario of technological apocalypse, in which the "omnipotence of integrated circuits" will lead to a fine-tuning of the self-replicating Turing machine that relegates human ingenuity and idealism to the junkyard of history. Implicit in much of *Gramophone, Film, Typewriter* is the belief that "so-called Man" (*der sogenannte Mensch*—a mocking phrase repeated like a mantra throughout the book) is about to disappear as a cognitive and self-determining agent (if such an agent ever existed) and be subsumed by the march of technological auto-sophistication. We are faced with the *Aufhebung* of human processes into silicon microprocessors, the dissolution of human software into computer hardware, for if computer technologies, beginning with the earliest storage facilities, ultimately substitute for physiological impairments and extend the sensory apparatus, then technology's prosthetic function could allow for the complete replacement of the human. Heidegger's notion of technology as *Gestell*, a supportive framing of human being, turns out to be an entire *Ersatz* for human being. Furthermore, it is not only a question of so-called Man disappearing *now*; He was never there to begin with, except as a figment of cultural imagination based on media-specific historical underpinnings. To appropriate Max Weber's famous term, Kittler's work contributes in radical fashion to the ongoing process of *Entzauberung*, or disenchantment.

As we have already indicated, some of Kittler's rhetoric of *épater l'humaniste bourgeois* must be seen against the background of specifically German poststructuralist debates, but we would nonetheless invite readers to consider the possibility that Kittler, especially when viewed in conjunction with North American discussions of subject formation under electronic conditions, is highlighting a crucial point: that the question of the subject has not been answered yet, for as long as we are not addressing it in its media-technological context, we are not even able to come up with the right question.

2. *The stop and go of history.* Not surprisingly, Kittler has been charged with a cavalier attitude toward the vicissitudes of historical change. Instead of tracing and assigning value to the agencies and contingencies that explain the unfolding transformation from one historical moment to another, his broad typologies tend "to obscure those subterranean disturbances that can build into a paradigm shift."[58] His descriptive and nonevolutionary model favoring sudden ruptures and transformations at the expense of genetic causalities is derived from Foucault, but it takes on a certain edge because epistemological breaks are tied to technological ruptures. The emphasis on discontinuity, however, is less problematic than the obvious technological determinism. As Timothy Lenoir has noted, Kittler explicitly rejects any characterization of his work as "'new historicism' or sociology of literature," opting instead to describe his project in terms that "frequently invoke McLuhan's deterministic media theories."[59]

Certainly, Kittler's emphasis on technological breakthroughs to the exclusion of other causative factors is indicative of a sometimes facile neglect of the dynamic complexities of development and evolution—technological or otherwise. But there are important exceptions, most notably his ingenious description of the discourse network of 1800 as the confluence of social practices, such as the role of speaking mothers in the socialization of children, the publicly mandated methodologies of language acquisition, the training of civil servants, and the beginning of hermeneutic literary criticism, among others. The media environment of 1800, therefore, particularly in the forms of writing and interpretation, is clearly seen as a historically specific contingency; it is not, as McLuhanites would have it, part of the makeup of the Gutenberg Galaxy by default. Media determine our situation, but it appears that our situation, in turn, can do its share to determine our media. In some of his more recent essays, Kittler argues that the discourse network of 1800 itself prepared the ground for the technological developments associated with its successor: "Romantic literature as a virtual media technology, as it was supported by the complicity between author, reader, and hero, contributed itself to the subversion of the unchallenged monopoly of print in Europe and to the change of guards from image-based literature to the mass media of photography and film."[60] Here Kittler appears to retrace the well-known theoretical footsteps of Walter Benjamin, who observed that every historical era "shows critical epochs in which a certain art form aspires to effects which could be fully obtained only with a changed technical standard."[61] At the risk of oversimplifying matters, we could say that Kittler espouses a type of technomaterialism that, albeit only on a formal level, bears some resemblance to Marxism's historical and dialectical materialism. Out of the dialectical exchange between the media-technological "base" and the discursive "superstructure" arise conflicts and tensions that sooner or later result in transformations at the level of media. At a given point in time, that is, during the discourse network of 1800, a widely used storage technology—the printed book—forms the material basis for new, hermeneutically programmed reading techniques that enable readers to experience an "inner movie"; subsequently, a desire arises in these readers to invent, or at least immediately select, the new cinematographic technology that provides images for real.

3. *Arms and no Man.* One element that may strike some readers as disturbing is Kittler's virtual fetishism of technological innovations produced by military applications, spin-offs that owe their existence to mil-

itary combat. Along with Paul Virilio and Norbert Bolz, Kittler derives a veritable genealogy of media in which war functions as the father of all things technical. In *Gramophone, Film, Typewriter* and related essays, he argues that the history of film coincides with the history of automatic weapons technology, that the development of early telegraphy was the result of a military need for the quick transmission of commands and intelligence, that television is a by-product of radar technology, and that the computer evolved in the context of the Second World War and the need both to encrypt and decode military intelligence and to compute missile trajectories. Modern media are suffused with war, and the history of communication technologies turns out to be "a series of strategic escalations."[62] Needless to say, humans as the subjects of technological innovations are as important as the individual soldier in the mass carnage of the First World War or the high-tech video wars of the present. If we had to name the book that comes closest to Kittler in this respect, it would be Manuel De Landa's eminently readable *War in the Age of Intelligent Machines*, a history of war technology written from the point of view of a future robot who, for obvious reasons, has little interest in what this or that human has contributed to the evolution of the machinic phylum.[63]

But such a unilateral war-based history of media technology would not meet with the approval of all historians and theorists of communication. James Beniger, for example, has argued that the science of cybernetics and its attendant technologies—the genesis of which Kittler locates in the communicative vicissitudes of the Second World War—is ultimately the result of the crisis of control and information processing experienced in the early heyday of the Industrial Revolution. In the wake of capitalist expansion of productivity and the distribution of goods, engineers had to invent ever-more refined feedback loops and control mechanisms to ensure the smooth flow of products to their consumers, and more generally to regulate the flow of data between market needs and demands (what cybernetics would call output and input). "Microprocessors and computer technologies, contrary to currently fashionable opinion, are not new forces only recently unleashed upon an unprepared society." On the contrary, "many of the computer's major contributions were anticipated along with the first signs of a control crisis in the mid-nineteenth century."[64] Building upon Beniger, Jochen Schulte-Sasse for one has taken Kittler to task for conflating the history of communication technologies with the history of warfare while ignoring the network of enabling conditions responsible for breakthroughs in technological innovations.[65]

4. *Hail the conquering engineer.* Kittler's work tends to champion a special class of technologists that made both the founding age and the digital age of modern media possible: the engineer. Edison, Muybridge, Marey, the Lumière brothers, Turing, and von Neumann have left behind a world—or rather, have *made* a world—in which technology, in more senses than one, reigns supreme. And one of their fictional counterparts, Mynona's ingenious Professor Pschorr, even manages to "beat" Goethe and get the girl in the short story "Goethe Speaks into the Phonograph." As we have mentioned, Kittler contrasts his "American" attitude to the purported technophobia of German academics, but it may serve readers well to point out that Kittler is speaking from a long German tradition of engineer worship reaching as far back as the second part of Goethe's *Faust* and including immensely successful science fiction novels by Dominik and Kellermann, the construction of the engineer as a leader into a new world in late-nineteenth- and early-twentieth-century technocratic utopias (including Thea von Harbou's *Metropolis*), and, above all, the apotheosis of the engineer at the conclusion of Oswald Spengler's *Decline of the West*.[66] In turn, Kittler's somewhat quaint portrayal of the United States as a haven of technophilia also has easily recognizable German roots: it harks back to the boisterous "Americanism" of the Weimar Republic that saw a Fordist and Taylorized United States as a model for overcoming the backwardness of the Old World.[67]

5. *Reactionary postmodernism?* The Fordism of the Weimar Republic was related to a cultural current that was to have considerable influence on conservative and, subsequently, Nazi ideology. Labeled "reactionary modernism" by Jeffrey Herf, it was an attempt to reject Enlightenment values while embracing technology in order to reconcile the strong antimodernist German tradition with technological progress. In spite of all the unrest and disorientation caused by the rapid modernization of late nineteenth-century Germany, the reactionary modernists claimed that "Germany could be *both* technologically advanced *and* true to its soul."[68] One of reactionary modernism's key components was to sever the traditional—and traditionally unquestioned—link between social and technological progress. No longer ensnared by the humanist ideology of the Enlightenment, the technological achievements of the modern age could be made to enter a mutually beneficial union with premodern societal structures. Among the most important thinkers to contribute to this distinctly German reaction to the travails of modernization were Oswald Spengler, Carl Schmitt, Ernst Jünger, Werner Sombart, and Mar-

tin Heidegger, some of whom figure prominently in the writings of Kittler and Bolz. To be sure, writing about the likes of Jünger, Benn, and Heidegger is anything but synonymous with endorsing the extremist political ideologies they may have held at one time or another. Nevertheless, readers of *Gramophone, Film, Typewriter* and Kittler's related essays might be left with the impression that in spite of all distancing maneuvers, Kittler seems to feel a certain reverence, if not for the writers themselves, then certainly for their largely unquestioning admiration of (media-)technological innovations. Jünger—who features prominently in "Film"—is a case in point: the way in which the workers and soldiers of his early novels and essays are dwarfed by productions and weapons technologies that dissolve their *Innerlichkeit*, or inner experience of being, into a spray of media effects is distinctly reminiscent of Kittler's poststructuralist erasure of the subject.

Of course there is a major difference: Kittler is as far removed as one can be from the traditional right-wing rhetoric of "soul," "*Volk*" and the "national body"; if these or related terms appear, they do so only as examples of the crude historical conceptualizations of the growing connectivity and communication spaces established by modern media technologies. But the question remains whether certain affinities exist that might suggest that some of Kittler's work be labeled a "postmodern" variant of the old reactionary modernism—most prominently, the determination to sever the connection between technological and social advancement, to jettison the latter in favor of the former and install, as it were, Technology as the new, authentic subject of history. What gives this approach an additional edge, however, is the growing awareness of the degree to which the French poststructuralists from whom Kittler takes his cue were themselves influenced by these right-wing German thinkers.[69] (Naturally, Heidegger comes to mind, but one should not underestimate Jünger.) But if it is true that the "antihumanists" of French poststructuralism owe a lasting debt to Nietzsche as well as to the Weimar thinkers of the Right, then Kittler's media discourse analysis, with its insistence that media determine our situation and that our situation changed decisively during the *Mediengründerzeit*, exposes their intellectual origins as well as technological matrix that shaped them.

PREFACE

Tap my head and mike my brain,
Stick that needle in my vein.

— THOMAS PYNCHON

Media determine our situation, which—in spite or because of it—deserves a description.

Situation conferences were held by the German General Staff, great ones around noon and smaller ones in the evening: in front of sand tables and maps, in war and so-called peace. Until Dr. Gottfried Benn, writer and senior army doctor, charged literature and literary criticism as well with the task of taking stock of the situation. His rationale (in a letter to a friend): "As you know, I sign: On behalf of the Chief of the Army High Command: Dr. Benn."[1]

Indeed: in 1941, with the knowledge of files and technologies, enemy positions and deployment plans, and located at the center of the Army High Command in Berlin's Bendlerstraße, it may still have been possible to take stock of the situation.[2]

The present situation is more obscure. First, the pertinent files are kept in archives that will all remain classified for exactly as many years as there remains a difference between files and facts, between planned objectives and their realization. Second, even secret files suffer a loss of power when real streams of data, bypassing writing and writers, turn out merely to be unreadable series of numbers circulating between networked computers. Technologies that not only subvert writing, but engulf it and carry it off along with so-called Man, render their own description impossible. Increasingly, data flows once confined to books and later to records and films are disappearing into black holes and boxes that, as artificial intelligences, are bidding us farewell on their way to nameless high commands. In this situation we are left only with reminiscences, that is to say, with stories. How that which is written in no book came to pass may still be for books to record. Pushed to their margins even obsolete media become sensitive enough to register the signs and clues of a situation. Then, as in the case of the sectional plane of two optical media, patterns and moirés emerge: myths, fictions of science, oracles . . .

This book is a story made up of such stories. It collects, comments upon, and relays passages and texts that show how the novelty of technological media inscribed itself into the old paper of books. Many of these papers are old or perhaps even forgotten, but in the founding age of technological media the terror of their novelty was so overwhelming that literature registered it more acutely than in today's alleged media pluralism, in which anything goes provided it does not disturb the assumption of global dominance by Silicon Valley. An information technology whose monopoly is now coming to an end, however, registers this very information: an aesthetics of terror. What writers astonished by gramophones, films, and typewriters—the first technological media—committed to paper between 1880 and 1920 amounts, therefore, to a ghostly image of our present as future.[3] Those early and seemingly harmless machines capable of storing and therefore separating sounds, sights, and writing ushered in a technologizing of information that, in retrospect, paved the way for today's self-recursive stream of numbers.

Obviously, stories of this kind cannot replace a history of technology. Even if they were countless they would remain numberless and thus would fail to capture the real upon which all innovations are based. Conversely, number series, blueprints, and diagrams never turn back into writing, only into machines.[4] Heidegger said as much with his fine statement that technology itself prevents any experience of its essence.[5] However, Heidegger's textbook-like confusion of writing and experience need not be; in lieu of philosophical inquiries into essence, simple knowledge will do.

We can provide the technological and historical data upon which fictional media texts, too, are based. Only then will the old and the new, books and their technological successors, arrive as the information they are. Understanding media—despite McLuhan's title—remains an impossibility precisely because the dominant information technologies of the day control all understanding and its illusions. But blueprints and diagrams, regardless of whether they control printing presses or mainframe computers, may yield historical traces of the unknown called the body. What remains of people is what media can store and communicate. What counts are not the messages or the content with which they equip so-called souls for the duration of a technological era, but rather (and in strict accordance with McLuhan) their circuits, the very schematism of perceptibility.

Whosoever is able to hear or see the circuits in the synthesized sound of CDs or in the laser storms of a disco finds happiness. A happiness beyond the ice, as Nietzsche would have said. At the moment of merciless submission to laws whose cases we are, the phantasm of man as the creator of media vanishes. And it becomes possible to take stock of the situation.

In 1945, in the half-burned, typed minutes of the Army High Command's final conferences, war was already named the father of all things: in a very free paraphrase of Heraclitus, it spawns most technological inventions.[6] And since 1973, when Thomas Pynchon's *Gravity's Rainbow* was published, it has become clear that real wars are not fought for people or fatherlands, but take place between different media, information technologies, data flows.[7] Patterns and moirés of a situation that has forgotten us . . .

But no matter what: without the research and contributions of Roland Baumann this book would not have been written. And it would have not have come about without Heidi Beck, Norbert Bolz, Rüdiger Campe, Charles Grivel, Anton (Tony) Kaes, Wolf Kittler, Thorsten Lorenz, Jann Matlock, Michael Müller, Clemens Pornschlegel, Friedhelm Rong, Wolfgang Scherer, Manfred Schneider, Bernhard Siegert, Georg Christoph (Stoffel) Tholen, Isolde Tröndle-Azri, Antje Weiner, David E. Wellbery, Raimar Zons, and Agia Galini.

F.K.

SEPTEMBER 1985

GRAMOPHONE, FILM, TYPEWRITER

INTRODUCTION

Optical fiber networks. People will be hooked to an information channel that can be used for any medium—for the first time in history, or for its end. Once movies and music, phone calls and texts reach households via optical fiber cables, the formerly distinct media of television, radio, telephone, and mail converge, standardized by transmission frequencies and bit format. The optoelectronic channel in particular will be immune to disturbances that might randomize the pretty bit patterns behind the images and sounds. Immune, that is, to the bomb. As is well known, nuclear blasts send an electromagnetic pulse (EMP) through the usual copper cables, which would infect all connected computers.

The Pentagon is engaged in farsighted planning: only the substitution of optical fibers for metal cables can accommodate the enormous rates and volumes of bits required, spent, and celebrated by electronic warfare. All early warning systems, radar installations, missile bases, and army staffs in Europe, the opposite coast,[1] finally will be connected to computers safe from EMP and thus will remain operational in wartime. In the meantime, pleasure is produced as a by-product: people are free to channel-surf among entertainment media. After all, fiber optics transmit all messages imaginable save for the one that counts—the bomb.

Before the end, something is coming to an end. The general digitization of channels and information erases the differences among individual media. Sound and image, voice and text are reduced to surface effects, known to consumers as interface. Sense and the senses turn into eyewash. Their media-produced glamor will survive for an interim as a by-product of strategic programs. Inside the computers themselves everything becomes a number: quantity without image, sound, or voice. And once optical fiber networks turn formerly distinct data flows into a standardized series of digitized numbers, any medium can be translated into any other. With numbers, everything goes. Modulation, transformation, synchronization; delay, storage, transposition; scrambling, scanning, mapping—a total media link on a digital base will erase the very concept of medium. Instead of wiring people and technologies, absolute knowledge will run as an endless loop.

But there still are media; there still is entertainment.

Today's standard comprises partially connected media links that are still comprehensible in McLuhan's terms. According to him, one medium's content is always other media: film and radio constitute the content of television; records and tapes the content of radio; silent films and audiotape that of cinema; text, telephone, and telegram that of the semi–media monopoly of the postal system. Since the beginning of the century, when the electronic tube was developed by von Lieben in Germany and De Forest in California, it has been possible to amplify and transmit signals. Accordingly, the large media networks, which have been in existence since the thirties, have been able to fall back on all three storage media—writing, film, and photography—to link up and send their signals at will.

But these links are separated by incompatible data channels and differing data formats. Electrics does not equal electronics. Within the spectrum of the general data flow, television, radio, cinema, and the postal service constitute individual and limited windows for people's sense perceptions. Infrared radiations or the radio echoes of approaching missiles are still transmitted through other channels, unlike the optical fiber networks of the future. Our media systems merely distribute the words, noises, and images people can transmit and receive. But they do not compute these data. They do not produce an output that, under computer control, transforms any algorithm into any interface effect, to the point where people take leave of their senses. At this point, the only thing being computed is the transmission quality of storage media, which appear in the media links as the content of the media. A compromise between engineers and salespeople regulates how poor the sound from a TV set can be, how fuzzy movie images can be, or how much a beloved voice on the telephone can be filtered. Our sense perceptions are the dependent variable of this compromise.

A composite of face and voice that remains calm, even when faced during a televised debate by an opponent named Richard M. Nixon, is deemed telegenic and may win a presidential election, as in Kennedy's case. Voices that an optical close-up would reveal as treacherous, however, are called radiogenic and rule over the VE 301, the *Volksempfänger* of the Second World War. For, as the Heidegger disciple among Germany's early radio experts realized, "death is primarily a radio topic."[2]

But these sense perceptions had to be fabricated first. For media to link up and achieve dominance, we need a coincidence in the Lacanian sense: that something ceases not to write itself. Prior to the electrification of media, and well before their electronic end, there were modest, merely mechanical apparatuses. Unable to amplify or transmit, they nevertheless were the first to store sensory data: silent movies stored sights, and Edison's phonograph (which, unlike Berliner's later gramophone, was capable both of recording and reproducing) stored sounds.

On December 6, 1877, Edison, lord of the first research laboratory in the history of technology, presented the prototype of the phonograph to the public. On February 20, 1892, the same lab in Menlo Park (near New York) added the so-called kinetoscope. Three years later, the Lumière brothers in France and the Skladanowsky brothers in Germany merely had to add a means of projection to turn Edison's invention into cinema.

Ever since that epochal change we have been in possession of storage technologies that can record and reproduce the very time flow of acoustic and optical data. Ears and eyes have become autonomous. And that changed the state of reality more than lithography and photography, which (according to Benjamin's thesis) in the first third of the nineteenth century merely propelled the work of art into the age of its technical reproducibility. Media "define what really is";[3] they are always already beyond aesthetics.

What phonographs and cinematographs, whose names not coincidentally derive from writing, were able to store was time: time as a mixture of audio frequencies in the acoustic realm and as the movement of single-image sequences in the optical. Time determines the limit of all art, which first has to arrest the daily data flow in order to turn it into images or signs. What is called style in art is merely the switchboard of these scannings and selections. That same switchboard also controls those arts that use writing as a serial, that is, temporally transposed, data flow. To record the sound sequences of speech, literature has to arrest them in a system of 26 letters, thereby categorically excluding all noise sequences. Not coincidentally, this system also contains as a subsystem the seven notes, whose diatonics—from A to G—form the basis of occidental music. Following a suggestion made by the musicologist von Hornbostel, it is possible to fix the chaos of exotic music assailing European ears by first interpolating a phonograph, which is able to record this chaos in real time and then replay it in slow motion. As the rhythms begin to flag and "individual measures, even individual notes resound on their own," occidental alphabetism with its staffs can proceed to an "exact notation."[4]

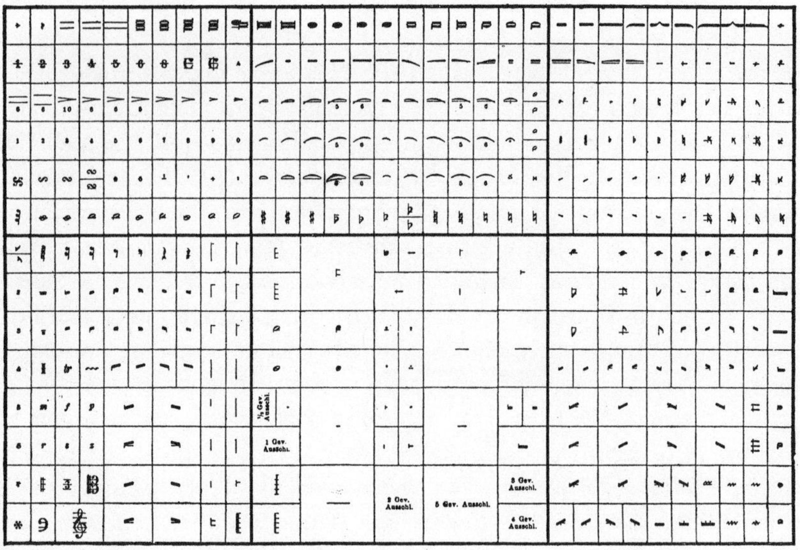

Texts and scores—Europe had no other means of storing time. Both are based on a writing system whose time is (in Lacan's term) symbolic. Using projections and retrievals, this time memorizes itself—like a chain of chains. Nevertheless, whatever ran as time on a physical or (again in Lacan's terms) real level, blindly and unpredictably, could by no means be encoded. Therefore, all data flows, provided they really were streams of data, had to pass through the bottleneck of the signifier. Alphabetic monopoly, grammatology.

If the film called history rewinds itself, it turns into an endless loop. What will soon end in the monopoly of bits and fiber optics began with the monopoly of writing. History was the homogenized field that, as an academic subject, only took account of literate cultures. Mouths and graphisms were relegated to prehistory. Otherwise, stories and histories (both deriving from *historia*) could not have been linked. All the orders and judgments, announcements and prescriptions (military and legal, religious and medical) that produced mountains of corpses were communicated along the very same channel that monopolized the descriptions of

The oldest depiction of a print shop, 1499—as a dance of death.

those mountains of corpses. Which is why anything that ever happened ended up in libraries.

And Foucault, the last historian or first archeologist, merely had to look things up. The suspicion that all power emanates from and returns to archives could be brilliantly confirmed, at least within the realms of law, medicine, and theology. A tautology of history, or its calvary. For the libraries in which the archeologist found so much rich material collected and catalogued papers that had been extremely diverse in terms of addressee, distribution technique, degree of secrecy, and writing technique—Foucault's archive as the entropy of a post office.[5] Even writing itself, before it ends up in libraries, is a communication medium, the technology of which the archeologist simply forgot. It is for this reason that all his analyses end immediately before that point in time at which other media penetrated the library's stacks. Discourse analysis cannot be applied to sound archives or towers of film rolls.

As long as it was moving along, history was indeed Foucault's "wavelike succession of words."[6] More simply, but no less technically than tomorrow's fiber optic cables, writing functioned as a universal medium—

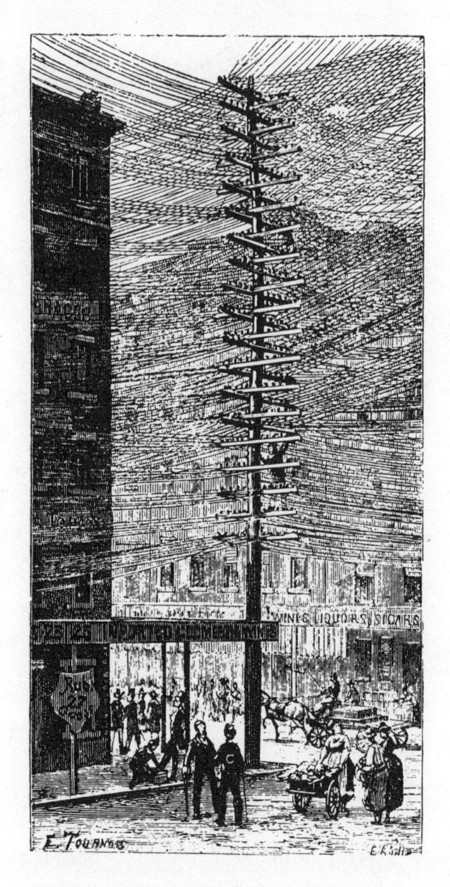

Telephone lines, New York, 1888.

in times when there was no concept of medium. Whatever else was going on dropped through the filter of letters or ideograms.

"Literature," Goethe wrote, "is a fragment of fragments; only the smallest proportion of what took place and what was said was written down, while only the smallest proportion of what was written down has survived."[7]

Accordingly, oral history today confronts the historians' writing monopoly; accordingly, a media theoretician like the Jesuit priest Walter J. Ong, who must have been concerned with the spirit of the Pentecostal mystery, could celebrate a primary orality of tribal cultures as opposed to the secondary orality of our media acoustics. Such research remained unthinkable as long as the opposite of "history" was simply termed (again following Goethe) "legend."[8] Prehistory was subsumed by its mythical name; Goethe's definition of literature did not even have to mention optical or acoustic data flows. And even legends, those oralized segments of bygone events, only survived in written format; that is, under pretechnological but literary conditions. However, since it has become possible to record the epics of the last Homeric bards, who until recently were wandering through Serbia and Croatia, oral mnemotechnics or cultures have become reconstructible in a completely different way.[9] Even Homer's rosy-fingered Eos changes from a Goddess into a piece of chromium dioxide that was stored in the memory of the bard and could be combined with other pieces into whole epics. "Primary orality" and "oral history" came into existence only after the end of the writing monopoly, as the technological shadows of the apparatuses that document them.

Writing, however, stored writing—no more and no less. The holy books attest to this. Exodus, chapter 20, contains a copy of what Yahweh's own finger originally had written on two stone tablets: the law. But of the thunder and lightning, of the thick cloud and the mighty trumpet which, according to scripture, surrounded this first act of writing on Mount Sinai, that same Bible could store nothing but mere words.[10]

Even less is handed down of the nightmares and temptations that afflicted a nomad called Mohammed following his flight to the holy mountain of Hira. The Koran does not begin until the one God takes the place of the many demons. The archangel Gabriel descends from the seventh heaven with a roll of scripture and the command to decipher the scroll. "Rejoice in the name of the Lord who created—created man from clots of blood. Recite! Your Lord is the Most Bountiful One, who by pen taught man what he did not know."[11]

Mohammed, however, answers that he, the nomad, can't read; not even the divine message about the origin of reading and writing. The archangel has to repeat his command before an illiterate can turn into the founder of a book-based religion. For soon, or all too soon, the illegible scroll makes sense and presents to Mohammed's miraculously alphabetized eyes the very same text that Gabriel had already uttered twice as an oral command. Mohammed's illuminations began, according to tradition, with this 96th sura—in order then to be "memorized by the faithful and written down on primitive surfaces such as palm leaves, stones, wood, bones, and pieces of leather, and to be recited, again and again, by Mohammed and select believers, especially during Ramadan."[12]

Writing therefore merely stores the fact of its authorization. It celebrates the storage monopoly of the God who invented it. And since the realm of this God consists of signs that only nonreaders can't make sense of, all books are books of the dead, like the Egyptian ones with which literature began.[13] The book itself coincides with the realm of the dead beyond all senses into which it lures us. When the Stoic philosopher Zeno asked the oracle at Delphi how he should best lead his life, he was given the answer "that he should mate with the dead. He understood this to mean that he should *read* the *ancients*."[14]

The story of how the divine instructions to use quills extended beyond Moses and Mohammed and reached simpler and simpler people is a lengthy one that nobody can write, because it would be history itself. In much the same way, the storage capacities of our computers will soon coincide with electronic warfare and, gigabyte upon gigabyte, exceed all the processing capacities of historians.

Suffice it to say that one day—in Germany, this may have already been the case during the age of Goethe—the homogenous medium of writing also became homogenous in the social sphere. Compulsory education engulfed people in paper. They learned a way of writing that, as an "abuse of language" (according to Goethe), no longer had to struggle with cramped muscles and individual letters, but rather proceeded in rapture or darkness. They learned to read "silently to one's self," a "sorry substitute for speech"[15] that consumed letters without effort by bypassing oral organs. Whatever they emitted and received was writing. And because only that exists which can be posted, bodies themselves fell under the regime of the symbolic. What is unthinkable today was once reality: no film stored the movements they made or saw, no phonograph, the noise they made or heard. For whatever existed failed before time. Silhouettes or pastel drawings fixed facial expressions, and scores were unable to store noise. But once a hand took hold of a pen, something miraculous occurred: the body, which did not cease not to write itself, left strangely unavoidable traces.

I'm ashamed to tell of it. I'm ashamed of my handwriting. It exposes me in all my spiritual nakedness. My handwriting shows me more naked than I am with my clothes off. No leg, no breath, no clothes, no sound. Neither voice nor reflection. All cleaned out. Instead, a whole man's being, shriveled and misshapen, like his scribble-scrabble. His lines are all that's left of him, as well as his self-propagation. The uneven tracings of his pencil on paper, so minimal that a blind man's fingertips would hardly detect them, become the measure of the whole fellow.[16]

Today, this shame, which overcomes the hero of Botho Strauss's last love story, *Dedication*, whenever he sees his handwriting, is no more than an anachronism. The fact that the minimal unevenness between stroke and paper can store neither a voice nor an image of a body presupposes in its exclusion the invention of phonography and cinema. Before their invention, however, handwriting alone could guarantee the perfect securing of traces. It wrote and wrote, in an energetic and ideally uninterrupted flow. As Hegel so correctly observed, the alphabetized individual had his "appearance and externality"[17] in this continuous flow of ink or letters.

And what applied to writing also applied to reading. Even if the alphabetized individual known as the "author" finally had to fall from the private exteriority of handwriting into the anonymous exteriority of print in order to secure "all that's left of him, as well as his self-propagation"—alphabetized individuals known as "readers" were able to reverse this exteriorization. "If one reads in the right way," Novalis wrote, "the words will unfold in us a real, visible world."[18] And his friend Schlegel added that "one believes to hear what one merely reads."[19] Perfect alphabetization was to supplement precisely those optical and acoustic data flows that, under the monopoly of writing, did not cease not to write themselves. Effort had been removed from writing, and sound from reading, in order to naturalize writing. The letters that educated readers skimmed over provided people with sights and sounds.

Aided by compulsory education and new alphabetization techniques, the book became both film and record around 1800—not as a media-technological reality, but in the imaginary of readers' souls. As a surrogate of unstorable data flows, books came to power and glory.[20]

In 1774 an editor by the name of Goethe committed handwritten letters or *Sorrows of Young Werther* to print. The "nameless throng" (to quote the dedication of *Faust*), too, was to hear an "early song" that, like "some old half-faded song," revived "old griefs" and "old friends."[21] This was the new literary recipe for success: to surreptitiously turn the voice or handwriting of a soul into Gutenbergiana. In the last letter he wrote and sealed but did not send off before committing suicide, Werther gave his beloved the very promise of poetry: during her lifetime she would have to remain with Albert, her unloved husband, but afterwards she would be united with her lover "in the sight of the Infinite One in eternal embraces."[22] Indeed: the addressee of handwritten love letters, which were then published by a mere editor, was to be rewarded with an immortality in the shape of the novel itself. It alone was able to create the "beautiful realm"[23] in which the lovers of Goethe's *Elective Affinities*, according to the hope of their narrator, "will waken together once more."[24] Strangely enough, Eduard and Ottilie had one and the same handwriting during their lifetime. Their death elevated them to a paradise that under the storage monopoly of writing was called poetry.

And maybe that paradise was more real than our media-controlled senses can imagine. Reading intently, Werther's suicidal readers may well have perceived their hero in a real, visible world. And the lovers among Goethe's female readers, like Bettina Brentano, may well have died with the heroine of his *Elective Affinities* only to be "reborn in a more beautiful youth" through Goethe's "genius."[25] Maybe the perfectly alphabetized readers of 1800 were a living answer to the question with which Chris Marker concludes his film essay *Sans Soleil*:

> Lost at the end of the world on my island, Sal, in the company of my dogs strutting around, I remember that January in Tokyo, or rather I remember the images I filmed in Tokyo in January. They have now put themselves in place of my memory, they *are* my memory. I wonder how people who do not film, take photos, or record tapes remember, how humankind used to go about remembering.[26]

It is the same with language, which only leaves us the choice of either retaining words while losing their meaning or, vice versa, retaining meaning while losing the words.[27] Once storage media can accommodate optical and acoustic data, human memory capacity is bound to dwindle. Its "liberation"[28] is its end. As long as the book was responsible for all serial data flows, words quivered with sensuality and memory. It was the passion of all reading to hallucinate meaning between lines and letters: the visible and audible world of Romantic poetics. And the passion of all writing was (in the words of E. T. A. Hoffmann) the poet's desire to "describe" the hallucinated "picture in one's mind with all its vivid colors, the light and the shade," in order to "strike [the] gentle reader like an electric shock."[29]

Electricity itself put an end to this. Once memories and dreams, the dead and ghosts, become technically reproducible, readers and writers no longer need the powers of hallucination. Our realm of the dead has withdrawn from the books in which it resided for so long. As Diodor of Sicily once wrote, "it is no longer only through writing that the dead remain in the memory of the living."

The writer Balzac was already overcome by fear when faced with photography, as he confessed to Nadar, the great pioneer of photography. If (according to Balzac) the human body consists of many infinitely thin

Spirit photography, 1904.

layers of "specters," and if the human spirit cannot be created from nothingness, then the daguerreotype must be a sinister trick: it fixes, that is, steals, one layer after the other, until nothing remains of the specters and the photographed body.[30] Photo albums establish a realm of the dead infinitely more precise than Balzac's competing literary enterprise, the *Comédie humaine*, could ever hope to create. In contrast to the arts, media do not have to make do with the grid of the symbolic. That is to say, they reconstruct bodies not only in a system of words or colors or sound intervals. Media and media only fulfill the "high standards" that (according to Rudolf Arnheim) we expect from "reproductions" since the invention of photography: "They are not only supposed to resemble the object, but rather guarantee this resemblance by being, as it were, a product of the object in question, that is, by being mechanically produced by it—just as the illuminated objects of reality imprint their image on the photographic layer,"[31] or the frequency curves of noises inscribe their wavelike shapes onto the phonographic plate.

A reproduction authenticated by the object itself is one of physical precision. It refers to the bodily real, which of necessity escapes all symbolic grids. Media always already provide the appearances of specters. For, according to Lacan, even the word "corpse" is a euphemism in reference to the real.[32]

Accordingly, the invention of the Morse alphabet in 1837 was promptly followed by the tapping specters of spiritistic seances sending their messages from the realm of the dead. Promptly as well, photographic plates—even and especially those taken with the camera shutter closed—furnished reproductions of ghosts or specters, whose black-and-white fuzziness only served to underscore the promise of resemblance. Finally, one of the ten applications Edison envisioned for his newly invented phonograph in the *North American Review* (1878) was to record "the last words of dying persons."

It was only a small step from such a "family record,"[33] with its special consideration of revenants, to fantasies that had telephone cables linking the living and the dead. What Leopold Bloom in *Ulysses* could only wish for in his Dublin graveyard meditations had already been turned into science fiction by Walter Rathenau, the AEG chairman of the board and futurist writer.[34] In Rathenau's story "Resurrection Co.," the cemetery administration of Necropolis, Dacota/USA, following a series of scandalous premature burials in 1898, founds a daughter company entitled "Dacota and Central Resurrection Telephone Bell Co." with a capital stock of $750,000. Its sole purpose is to make certain that the inhabitants of graves, too, are connected to the public telephone network. Whereupon the dead avail themselves of the opportunity to prove, long before McLuhan, that the content of one medium is always another medium—in this concrete case, a *déformation professionelle*.[35]

These days, paranormal voices on tape or radio, the likes of which have been spiritistically researched since 1959 and preserved in rock music since Laurie Anderson's 1982 release *Big Science*,[36] inform their researchers of their preferred radio wavelength. This already occurred in 1898, in the case of Senate President Schreber: when a paranormal, beautifully autonomous "base or nerve language" revealed its code as well as its channels,[37] message and channel became one. "You just have to choose a middle-, short-, or long-wave talk-show station, or the 'white noise' between two stations, or the 'Jürgenson wave,' which, depending on where you are, is located around 1450 to 1600 kHz between Vienna and Moscow."[38] If you replay a tape that has been recorded off the radio, you will hear all kinds of ghost voices that do not originate from any known radio station, but that, like all official newscasters, indulge in radio self-advertisement. Indeed, the location and existence of that "Jürgenson wave" was pinpointed by none other than "Friedrich Jürgenson, the Nestor of vocal research."[39]

The realm of the dead is as extensive as the storage and transmission capabilities of a given culture. As Klaus Theweleit noted, media are always flight apparatuses into the great beyond. If gravestones stood as symbols at the beginning of culture itself, our media technology can retrieve all gods. The old written laments about ephemerality, which measured no more than distance between writing and sensuality, suddenly fall silent. In our mediascape, immortals have come to exist again.

War on the Mind is the title of an account of the psychological strategies hatched by the Pentagon. It reports that the staffs planning the electronic war, which merely continues the Battle of the Atlantic,[40] have already compiled a list of the propitious and unpropitious days in other cultures. This list enables the U.S. Air Force "to time [its] bombing campaigns to coincide with unpropitious days, thus 'confirming' the forecasts of local gods." As well, the voices of these gods have been recorded on tape to be broadcast from helicopters "to keep tribes in their villages." And finally, the Pentagon has developed special film projectors capable of projecting those gods onto low-hanging clouds.[41] A technologically implemented beyond . . .

Of course the Pentagon does not keep a handwritten list of good and bad days. Office technology keeps up with media technology. Cinema and the phonograph, Edison's two great achievements that ushered in the present, are complemented by the typewriter. Since 1865 (according to European accounts) or 1868 (according to American ones), writing has no longer been the ink or pencil trace of a body whose optical and acoustic signals were irretrievably lost, only to reappear (in readers' minds) in the surrogate sensuality of handwriting. In order to store series of sights and sounds, Old Europe's only storage technology first had to be mechanized. Hans Magnus Malling Hansen in Copenhagen and Christopher Latham Sholes in Milwaukee developed mass-producible typewriters. Edison commented positively on the invention's potential when Sholes visited him in

Newark to demonstrate his newly patented model and to invite the man who had invented invention to enter a joint venture.[42]

But Edison declined the offer—as if, already in 1868, the phonograph and kinetoscope preoccupied their future inventor. Instead, the offer was grabbed by an arms manufacturer suffering from dwindling revenues in the post–Civil War slump. Remington, not Edison, took over Sholes's discourse machine gun.

Thus, there was no Marvelous One from whose brow sprang all three media technologies of the modern age. On the contrary, the beginning of our age was marked by separation or differentiation.[43] On the one hand, we have two technological media that, for the first time, fix unwritable data flows; on the other, an "'intermediate' thing between a tool and a machine," as Heidegger wrote so precisely about the typewriter.[44] On the one hand, we have the entertainment industry with its new sensualities; on the other, a writing that already separates paper and body during textual production, not first during reproduction (as Gutenberg's movable types had done). From the beginning, the letters and their arrangement were standardized in the shapes of type and keyboard, while media were engulfed by the noise of the real—the fuzziness of cinematic pictures, the hissing of tape recordings.

In standardized texts, paper and body, writing and soul fall apart. Typewriters do not store individuals; their letters do not communicate a beyond that perfectly alphabetized readers can subsequently hallucinate as meaning. Everything that has been taken over by technological media since Edison's inventions disappears from typescripts. The dream of a real visible or audible world arising from words has come to an end. The historical synchronicity of cinema, phonography, and typewriting separated optical, acoustic, and written data flows, thereby rendering them autonomous. That electric or electronic media can recombine them does not change the fact of their differentiation.

In 1860, five years before Malling Hansen's mechanical writing ball (the first mass-produced typewriter), Gottfried Keller's "Misused Love Letters" still proclaimed the illusion of poetry itself: love is left with the impossible alternatives of speaking either with "black ink" or with "red blood."[45] But once typing, filming, and recording became equally valid options, writing lost such surrogate sensualities. Around 1880 poetry turned into literature. Standardized letters were no longer to transmit Keller's red blood or Hoffmann's inner forms, but rather a new and elegant tautology of technicians. According to Mallarmé's instant insight, literature is made up of no more and no less than twenty-six letters.[46]

Lacan's "methodological distinction"[47] among the real, the imaginary, and the symbolic is the theory (or merely a historical effect) of that differentiation. The symbolic now encompasses linguistic signs in their materiality and technicity. That is to say, letters and ciphers form a finite set without taking into account philosophical dreams of infinity. What counts are differences, or, in the language of the typewriter, the spaces between the elements of a system. For that reason, Lacan designates "the world of the symbolic [as] the world of the machine."[48]

The imaginary, however, comes about as the mirror image of a body that appears to be, in terms of motor control, more perfect than the infant's own body, for in the real everything begins with coldness, dizziness, and shortness of breath.[49] Thus, the imaginary implements precisely those optical illusions that were being researched in the early days of cinema. A dismembered or (in the case of film) cut-up body is faced with the illusionary continuity of movements in the mirror or on screen. It is no coincidence that Lacan recorded infants' jubilant reactions to their mirror images in the form of documentary footage.

Finally, of the real nothing more can be brought to light than what Lacan presupposed—that is, nothing. It forms the waste or residue that neither the mirror of the imaginary nor the grid of the symbolic can catch: the physiological accidents and stochastic disorder of bodies.

The methodological distinctions of modern psychoanalysis clearly coincide with the distinctions of media technology. Every theory has its historical a priori. And structuralist theory simply spells out what, since the turn of the century, has been coming over the information channels.

Only the typewriter provides writing as a selection from the finite and arranged stock of its keyboard. It literally embodies what Lacan illustrated using the antiquated letter box. In contrast to the flow of handwriting, we now have discrete elements separated by spaces. Thus, the symbolic has the status of block letters. Film was the first to store those mobile doubles that humans, unlike other primates, were able to (mis)perceive as their own body. Thus, the imaginary has the status of cinema. And only the phonograph can record all the noise produced by the larynx prior to any semiotic order and linguistic meaning. To experience pleasure, Freud's patients no longer have to desire what philosophers consider good. Rather, they are free to babble.[50] Thus, the real—especially in the talking cure known as psychoanalysis—has the status of phonography.

Once the technological differentiation of optics, acoustics, and writing exploded Gutenberg's writing monopoly around 1880, the fabrication of so-called Man became possible. His essence escapes into apparatuses. Machines take over functions of the central nervous system, and no longer, as in times past, merely those of muscles. And with this differentiation—and not with steam engines and railroads—a clear division occurs between matter and information, the real and the symbolic. When it comes to inventing phonography and cinema, the age-old dreams of humankind are no longer sufficient. The physiology of eyes, ears, and brains have to become objects of scientific research. For mechanized writing to be optimized, one can no longer dream of writing as the expression of individuals or the trace of bodies. The very forms, differences, and frequencies of its letters have to be reduced to formulas. So-called Man is split up into physiology and information technology.

When Hegel summed up the perfect alphabetism of his age, he called it Spirit. The readability of all history and all discourses turned humans or philosophers into God. The media revolution of 1880, however, laid the groundwork for theories and practices that no longer mistake information for spirit. Thought is replaced by a Boolean algebra, and consciousness by the unconscious, which (at least since Lacan's reading) makes of Poe's "Purloined Letter" a Markoff chain.[51] And that the symbolic is called the world of the machine undermines Man's delusion of possessing a "quality" called "consciousness," which identifies him as something other and better than a "calculating machine." For both people and computers are "subject to the appeal of the signifier"; that is, they are both run by programs. "Are these humans," Nietzsche already asked himself in 1874, eight years before buying a typewriter, "or perhaps only thinking, writing, and speaking machines?"[52]

In 1950 Alan Turing, the practitioner among England's mathematicians, gave the answer to Nietzsche's question. He observed, with formal elegance, that there is no question to begin with. To clarify the issue, Turing's essay "Computing Machinery and Intelligence"—appearing in, of all places, the philosophical periodical *Mind*—proposed an experiment, the so-called Turing game: A computer *A* and human *B* exchange data via some kind of telewriter interface. The exchange of texts is monitored by a censor *C*, who also only receives written information. *A* and *B* both pretend to be human, and *C* has to decide which of the two is simulating and which merely is Nietzsche's thinking, writing, and speaking machine. But the game remains open-ended, because each time the machine gives itself away—be it by making a mistake or, more likely, by not making any—it will refine its program by learning.[53] In the Turing game, Man coincides with his simulation.

And this is, obviously, already so because the censor *C* receives plotter printouts and typescripts rather than handwritten texts. Of course, computer programs could simulate the "individuality" of the human hand, with its routines and mistakes, but Turing, as the inventor of the universal discrete machine, was a typist. Though he wasn't much better or skilled at it than his tomcat Timothy, who was allowed to jump across the keyboard in Turing's chaotic secret service office,[54] it was at least somewhat less catastrophic than his handwriting. The teachers at the honorable Sherborne School could hardly "forgive" their pupil's chaotic lifestyle and messy writing. He got lousy grades for brilliant exams in mathematics only because his handwriting was "the worst . . . ever seen."[55] Faithfully, schools cling to their old duty of fabricating individuals (in the literal sense of the word) by drilling them in a beautiful, continuous, and individual handwriting. Turing, a master in subverting all education, however, dodged the system; he made plans for an "exceedingly crude" typewriter.[56]

Nothing came of these plans. But when, on the meadows of Grantchester, the meadows of all English poetry from the Romantics to Pink

Floyd, he hit upon the idea of the universal discrete machine, his early dreams were realized and transformed. Sholes's typewriter, reduced to its fundamental principle, has supported us to this day. Turing merely got rid of the people and typists that Remington & Son needed for reading and writing.

And this is possible because a Turing machine is even more exceedingly crude than the Sherborne plan for a typewriter. All it works with is a paper strip that is both its program and its data material, its input and its output. Turing slimmed down the common typewriter page to this little strip. But there are even more economizations: his machine doesn't need the many redundant letters, ciphers, and signs of a typewriter keyboard; it can do with one sign and its absence, 1 and 0. This binary information can be read or (in Turing's technospeak) scanned by the machine. It can then move the paper strip one space to the right, one to the left, or not at all, moving in a jerky (i.e., discrete) fashion like a typewriter, which in contrast to handwriting has block caps, a back spacer, and a space bar. (From a letter to Turing: "Pardon the use of the typewriter: I have come to prefer discrete machines to continuous ones.")[57] The mathematical model of 1936 is no longer a hermaphrodite of a machine and a mere tool. As a feedback system it beats all the Remingtons, because each step is controlled by scanning the paper strip for the sign or its absence, which amounts to a kind of writing: it depends on this reading whether the machine keeps the sign or erases it, or, vice versa, whether it keeps a space blank or replaces it with a sign, and so on and so forth.

That's all. But no computer that has been built or ever will be built can do more. Even the most advanced Von Neumann machines (with program storage and computing units), though they operate much faster, are in principle no different from Turing's infinitely slow model. Also, while not all computers have to be Von Neumann machines, all conceivable data processing machines are merely a state *n* of the universal discrete machine. This was proved mathematically by Alan Turing in 1936, two years before Konrad Zuse in Berlin built the first programmable computer from simple relays. And with that the world of the symbolic really turned into the world of the machine.[58]

Unlike the history to which it put an end, the media age proceeds in jerks, just like Turing's paper strip. From the Remington via the Turing machine to microelectronics, from mechanization and automatization to the implementation of a writing that is only cipher, not meaning—one century was enough to transfer the age-old monopoly of writing into the omnipotence of integrated circuits. Not unlike Turing's correspondents, everyone is deserting analog machines in favor of discrete ones. The CD digitizes the gramophone, the video camera digitizes the movies. All data streams flow into a state *n* of Turing's universal machine; Romanticism notwithstanding, numbers and figures become the key to all creatures.

GRAMOPHONE

"Hullo!" Edison screamed into the telephone mouthpiece. The vibrating diaphragm set in motion a stylus that wrote onto a moving strip of paraffin paper. In July 1877, 81 years before Turing's moving paper strip, the recording was still analog. Upon replaying the strip and its vibrations, which in turn set in motion the diaphragm, a barely audible "Hullo!" could be heard.[1]

Edison understood. A month later he coined a new term for his telephone addition: phonograph.[2] On the basis of this experiment, the mechanic Kruesi was given the assignment to build an apparatus that would etch acoustic vibrations onto a rotating cylinder covered with tinfoil. While he or Kruesi was turning the handle, Edison once again screamed into the mouthpiece—this time the nursery rhyme "Mary Had a Little Lamb." Then they moved the needle back, let the cylinder run a second time—and the first phonograph replayed the screams. The exhausted genius, in whose phrase genius is 1 percent inspiration and 99 percent perspiration, slumped back. Mechanical sound recording had been invented. "Speech has become, as it were, immortal."[3]

It was December 6, 1877. Eight months earlier, Charles Cros, a Parisian writer, bohemian, inventor, and absinthe drinker, had deposited a sealed envelope with the Academy of Sciences. It contained an essay on the "Procedure for the Recording and Reproduction of Phenomena of Acoustic Perception" (*Procédé d'enregistrement et de reproduction des phénomènes perçus par l'ouïe*). With great technological elegance this text formulated all the principles of the phonograph, but owing to a lack of funds Cros had not yet been able to bring about its "practical realization." "To reproduce" the traces of "the sounds and noises" that the "to and fro" of an acoustically "vibrating diaphragm" leaves on a rotating disk—that was also the program of Charles Cros.[4]

But once he had been preceded by Edison, who was aware of rumors of the invention, things sounded different. "Inscription" is the title of the poem with which Cros erected a belated monument to honor his inventions, which included an automatic telephone, color photography, and, above all, the phonograph:

> Comme les traits dans les camées
> J'ai voulu que les voix aimées
> Soient un bien qu'on garde à jamais,
> Et puissent répéter le rêve
> Musical de l'heure trop brève;
> Le temps veut fuir, je le soumets.
>
> Like the faces in cameos
> I wanted beloved voices
> To be a fortune which one keeps forever,
> And which can repeat the musical
> Dream of the too short hour;
> Time would flee, I subdue it.[5]

The program of the poet Cros, in his capacity as the inventor of the phonograph, was to store beloved voices and all-too-brief musical reveries. The wondrously resistant power of writing ensures that the poem has no words for the truth about competing technologies. Certainly, phonographs can store articulate voices and musical intervals, but they are capable of more and different things. Cros the poet forgets the noises mentioned in his precise prose text. An invention that subverts both literature and music (because it reproduces the unimaginable real they are both based on) must have struck even its inventor as something unheard of.

Hence, it was not coincidental that Edison, not Cros, actually built the phonograph. His "Hullo!" was no beloved voice and "Mary Had a Little Lamb" no musical reverie. And he screamed into the bell-mouth not only because phonographs have no amplifiers but also because Edison, following a youthful adventure involving some conductor's fists, was half-deaf. A physical impairment was at the beginning of mechanical sound recording—just as the first typewriters had been made by the blind for the blind, and Charles Cros had taught at a school for the deaf and mute.[6]

Whereas (according to Derrida) it is characteristic of so-called Man and his consciousness to hear himself speak[7] and see himself write, media

The first talking machine, built by Kruesi.

dissolve such feedback loops. They await inventors like Edison whom chance has equipped with a similar dissolution. Handicaps isolate and thematize sensory data streams. The phonograph does not hear as do ears that have been trained immediately to filter voices, words, and sounds out of noise; it registers acoustic events as such. Articulateness becomes a second-order exception in a spectrum of noise. In the first phonograph letter of postal history, Edison wrote that "the articulation" of his baby "was loud enough, just a bit indistinct . . . not bad for a first experiment."[8]

Wagner's *Gesamtkunstwerk*, that monomaniacal anticipation of modern media technologies,[9] had already transgressed the traditional boundaries of words and music to do justice to the unarticulated. In *Tristan*, Brangäne was allowed to utter a scream whose notation cut straight through the score.[10] Not to mention *Parsifal*'s Kundry, who suffered from a hysterical speech impairment such as those which were soon to occupy the psychoanalyst Freud: she "gives a loud wail of misery, that sinks gradually into low accents of fear," "utters a dreadful cry," and is reduced to "hoarse and broken," though nonetheless fully composed, garbling.[11] This labored inception of language has nothing to do with operas and dramas that take it for granted that their figures can speak. Composers of 1880, however, are allied with engineers. The undermining of articulation becomes the order of the day.

In Wagner's case this applies to both text and music. The *Rhinegold* prelude, with its infinite swelling of a single chord, dissolves the E-flat major triad in the first horn melody as if it were not a matter of musical harmony but of demonstrating the physical overtone series. All the harmonics of E-flat appear one after the other, as if in a Fourier analysis; only the seventh is missing, because it cannot be played by European instruments.[12] Of course, each of the horn sounds is an unavoidable overtone mixture of the kind only the sine tones of contemporary synthesizers can avoid. Nevertheless, Wagner's musico-physiological dream[13] at the outset of the tetralogy sounds like a historical transition from intervals to frequencies, from a logic to a physics of sound. By the time Schoenberg, in 1910, produced the last analysis of harmony in the history of music, chords had turned into pure acoustics: "For Schoenberg as well as for science, the physical basis in which he is trying to ground all phenomena is the overtone series."[14]

Overtones are frequencies, that is, vibrations per second. And the grooves of Edison's phonograph recorded nothing but vibrations. Intervals and chords, by contrast, were ratios, that is, fractions made up of integers. The length of a string (especially on a monochord) was subdivided, and the fractions, to which Pythagoras gave the proud name *logoi*, resulted in octaves, fifths, fourths, and so on. Such was the logic upon which was founded everything that, in Old Europe, went by the name of music: first, there was a notation system that enabled the transcription of clear sounds separated from the world's noise; and second, a harmony of the spheres that established that the ratios between planetary orbits (later human souls) equaled those between sounds.

The nineteenth century's concept of frequency breaks with all this.[15] The measure of length is replaced by time as an independent variable. It is a physical time removed from the meters and rhythms of music. It quantifies movements that are too fast for the human eye, ranging from 20 to 16,000 vibrations per second. The real takes the place of the symbolic. Certainly, references can also be established to link musical intervals and acoustic frequencies, but they only testify to the distance between two discourses. In frequency curves the simple proportions of Pythagorean music turn into irrational, that is, logarithmic, functions. Conversely, overtone series—which in frequency curves are simply inte-

gral multiples of vibrations and the determining elements of each sound—soon explode the diatonic music system. That is the depth of the gulf separating Old European alphabetism from mathematical-physical notation.

Which is why the first frequency notations were developed outside of music. First noise itself had to become an object of scientific research, and discourses "a privileged category of noises."[16] A competition sponsored by the Saint Petersburg Academy of Sciences in 1780 made voiced sounds, and vowels in particular, an object of research,[17] and inaugurated not only speech physiology but also all the experiments involving mechanical language reproduction. Inventors like Kempelen, Maelzel, and Mical built the first automata that, by stimulating and filtering certain frequency bands, could simulate the very sounds that Romanticism was simultaneously celebrating as the language of the soul: their dolls said "Mama" and "Papa" or "Oh," like Hoffmann's beloved automaton, Olympia. Even Edison's 1878 article on phonography intended such toy mouths voicing

the parents' names as Christmas presents.[18] Removed from all Romanticism, a practical knowledge of vowel frequencies emerged.

Continuing these experiments, Willis made a decisive discovery in 1829. He connected elastic tongues to a cogwheel whose cogs set them vibrating. According to the speed of its rotation, high or low sounds were produced that sounded like the different vowels, thus proving their frequency. For the first time pitch no longer depended on length, as with string or brass instruments; it became a variable dependent on speed and, therefore, time. Willis had invented the prototype of all square-curve generators, ranging from the bold verse-rhythm experiments of the turn of the century[19] to *Kontakte*, Stockhausen's first electronic composition.

The synthetic production of frequencies is followed by their analysis. Fourier had already provided the mathematical theory, but that theory had yet to be implemented technologically. In 1830, Wilhelm Weber in Göttingen had a tuning fork record its own vibrations. He attached a pig's bristle to one of the tongues, which etched its frequency curves into sooty glass. Such were the humble, or animal, origins of our gramophone needles.

From Weber's writing tuning fork Edouard Léon Scott, who as a Parisian printer was, not coincidentally, an inhabitant of the Gutenberg Galaxy, developed his phonautograph, patented in 1857. A bell-mouth amplified incoming sounds and transmitted them onto a membrane, which in turn used a coarse bristle to transcribe them onto a soot-covered cylinder. Thus came into being autographs or handwritings of a data stream that heretofore had not ceased not to write itself. (Instead, there was handwriting.) Scott's phonautograph, however, made visible what, up to this point, had only been audible and had been much too fast for ill-equipped human eyes: hundreds of vibrations per second. A triumph of the concept of frequency: all the whispered or screamed noises people emitted from their larynxes, with or without dialects, appeared on paper. Phonetics and speech physiology became a reality.[20]

They were especially real in the case of Henry Sweet, whose perfect English made him the prototype of all experimental phonetics as well as the hero of a play. Recorded by Professor F. C. Donders of Utrecht,[21] Sweet was also dramatized by George Bernard Shaw, who turned him into a modern Pygmalion out to conquer all mouths that, however beautiful, were marred by dialect. To record and discipline the dreadful dialect of the flower girl Eliza Doolittle, "Higgins's laboratory" boasts "a phonograph, a laryngoscope, [and] a row of tiny organ pipes with a bellows."[22] In the world of the modern *Pygmalion*, mirrors and statues are unnecessary; sound storage makes it possible "to inspect one's own speech or discourse as in a mirror, thus enabling us to adopt a critical stance toward our products."[23] To the great delight of Shaw, who saw his medium or his readability technologically guaranteed to all English speakers,[24] machines easily solve a problem that literature had not been able to tackle on its own, or had only been able to tackle through the mediation of pedagogy:[25] to drill people in general, and flower girls in particular, to adopt a pronunciation purified by written language.

It comes as no surprise that Eliza Doolittle, all of her love notwithstanding, abandons her Pygmalion (Sweet, a.k.a. Higgins) at the end of the play in order to learn "bookkeeping and typewriting" at "shorthand schools and polytechnic classes."[26] Women who have been subjected to phonographs and typewriters are souls no longer; they can only end up in musicals. Renaming the drama *My Fair Lady*, Rodgers and Hammerstein will throw Shaw's *Pygmalion* among Broadway tourists and record labels. "On the Street Where You Live" is sound.

In any event, Edison, ancestor of the record industry, only needed to combine, as is so often the case with inventions. A Willis-type machine gave him the idea for the phonograph; a Scott-type machine pushed him toward its realization. The synthetic production of frequencies combined with their analysis resulted in the new medium.

Edison's phonograph was a by-product of the attempt to optimize telephony and telegraphy by saving expensive copper cables. First, Menlo Park developed a telegraph that indented a paraffin paper strip with Morse signs, thus allowing them to be replayed faster than they had been transmitted by human hands. The effect was exactly the same as in Willis's case: pitch became a variable dependent on speed. Second, Menlo Park developed a telephone receiver with a needle attached to the diaphragm. By touching the needle, the hearing-impaired Edison could check the amplitude of the telephone signal. Legend has it that one day the needle drew blood—and Edison "recognized how the force of a membrane moved by a magnetic system could be put to work." "In effect, he had found a way to transfer the functions of his ear to his sense of touch."[27]

A telegraph as an artificial mouth, a telephone as an artificial ear—the stage was set for the phonograph. Functions of the central nervous system had been technologically implemented. When, after a 72-hour shift ending early in the morning of July 16, 1888, Edison had finally completed a talking machine ready for serial production, he posed for the hastily summoned photographer in the pose of his great idol. The French emperor, after all, is said to have observed that the progress of national welfare (or military technology) can be measured by transportation costs.[28] And no means of transportation are more economical than those which convey information rather than goods and people. Artificial mouths and ears, as technological implementations of the central nervous system, cut down on mailmen and concert halls. What Ong calls our secondary orality has the elegance of brain functions. Technological sound storage provides a first model for data streams, which are simultaneously becoming objects of neurophysiological research. Helmholtz, as the perfecter of vowel theory, is allied with Edison, the perfecter of measuring instruments. Which is why sound storage, initially a mechanically primitive affair on the level of Weber's pig bristle, could not be invented until the soul fell prey to science. "O my head, my head, my head," groans the phonograph in the prose poem Alfred Jarry dedicated to it. "All white underneath the silk sky: They have taken my head, my head—and put me into a tea tin!"[29]

Which is why Villiers de l'Isle-Adam, the symbolist poet and author of the first of many Edison novels, is mistaken when, in *Tomorrow's Eve*, he has the great inventor ponder his delay.

> What is most surprising in history, almost unimaginable, is that among all the great inventors across the centuries, not one thought of the Phonograph! And yet most of them invented machines a thousand times more complicated. The Phonograph is so simple that its construction owes nothing to materials of scientific composition. Abraham might have built it, and made a recording of his

calling from on high. A steel stylus, a leaf of silver foil or something like it, a cylinder of copper, and one could fill a storehouse with all the voices of Heaven and Earth.[30]

This certainly applies to materials and their processing, but it misses the historical a priori of sound recording. There are also immaterials of scientific origin, which are not so easy to come by and have to be supplied by a science of the soul. They cannot be delivered by any of the post-Abraham candidates whom Villiers de l'Isle-Adam suspects of being able to invent the phonograph: neither Aristotle, Euclid, nor Archimedes could have underwritten the statement that "The soul is a notebook of phonographic recordings" (but rather, if at all, a tabula rasa for written signs, which in turn signify acts of the soul). Only when the soul has become the nervous system, and the nervous system (according to Sigmund Exner, the great Viennese neurophysiologist) so many facilitations (*Bahnungen*), can Delboeuf's statement cease to be scandalous. In 1880, the philosopher Guyau devoted a commentary to it. And this first theory of the phonograph attests like no other to the interactions between science and technology. Thanks to the invention of the phonograph, the very theories that were its historical a priori can now optimize their analogous models of the brain.

JEAN-MARIE GUYAU, "MEMORY AND PHONOGRAPH" (1880)

Reasoning by analogy is of considerable importance to science; indeed, in as far as it is the principle of induction it may well form the basis of all physical and psychophysical sciences. Discoveries frequently start with metaphors. The light of thinking could hardly fall in a new direction and illuminate dark corners were it not reflected by spaces already illuminated. Only that which reminds us of something else makes an impression, although and precisely because it differs from it. To understand is to remember, at least in part.

Many similes and metaphors have been used in the attempt to understand mental abilities or functions. Here, in the as yet imperfect state of science, metaphors are absolutely necessary: before we *know* we have to start by *imagining* something. Thus, the human brain has been compared to all kinds of objects. According to Spencer it shows a certain analogy to the mechanical pianos that can reproduce an infinite number of melodies. Taine makes of the brain a kind of print shop that incessantly produces and stores innumerable clichés. Yet all these similes appear somewhat sketchy. One normally deals with the brain at rest; its images are perceived to be fixed, *stereotyped*; and that is imprecise. There is nothing finished in the brain, no real images; instead, we see only virtual, potential images waiting for a sign to be transformed into actuality. How this transformation into reality is really achieved is a matter of speculation. The greatest mystery of brain mechanics has to do with dynamics—not with statics. We are in need of a comparative term that will allow us to see not only how an object receives and stores an imprint, but also how this imprint at a given time is reactivated and produces new vibrations within the object. With this in mind, the most refined instrument (both receiver and motor in one) with which the human brain may be compared is perhaps Edison's recently invented phonograph. For some time now I have been wanting to draw attention to this comparison, ever since I came across a casual observation in Delboeuf's last article on memory that confirmed my intentions: "The soul is a notebook of phonographic recordings."

Upon speaking into a phonograph, the vibrations of one's voice are transferred to a point that engraves lines onto a metal plate that correspond to the uttered sounds—uneven furrows, more or less deep, depending on the nature of the sounds. It is quite probable that in analogous ways, invisible lines are incessantly carved into the brain cells, which provide a channel for nerve streams. If, after some time, the stream encounters a channel it has already passed through, it will once again proceed along the same path. The cells vibrate in the same way they vibrated the first time; psychologically, these similar vibrations correspond to an emotion or a thought analogous to the forgotten emotion or thought.

This is precisely the phenomenon that occurs when the phonograph's small copper disk, held against the point that runs through the grooves it has etched, starts to reproduce the vibrations: to our ears, these vibrations turn back into a voice, into words, sounds, and melodies.

If the phonographic disk had self-consciousness, it could point out while replaying a song that it remembers this particular song. And what appears to us as the effect of a rather simple mechanism would, quite probably, strike the disk as a miraculous ability: memory.

Let us add that it could distinguish new songs from those already played, as well as new impressions from simple memories. Indeed, a certain effort is necessary for first impressions to etch themselves into metal or brain; they encounter more resistance and, correspondingly, have to exert more force; and when they reappear, they vibrate all the stronger. But when the point traces already existing grooves instead of making new ones, it will do so with greater ease and glide along without applying any pressure. The *inclination* of a memory or reverie has been spoken of; to pursue a memory, in fact: to smoothly glide down a slope, to wait for a certain number of complete memories, which appear one after the other, all in a row and without shock. There is, therefore, a significant difference between impressions in the real sense and memory. Impressions tend to belong to either of two classes: they either possess greater intensity, a unique sharpness of outline and fixity of line, or they are weaker, more blurred and imprecise, but nevertheless arranged in a certain order that imposes itself on us. To *recognize* an image means to assign it to the second class. One *feels* in a less forceful way and is aware of this emotion. A memory consists in the awareness, first, of the diminished intensity of an impression, second, of its increased ease, and third, of the connections it entertains with other impressions. Just as a trained eye can see the difference between a copy and the original, we learn to distinguish memories from impressions and are thus able to recognize a memory even before it has been located in time and space. We project this or that impression back into the past without knowing which part of the past it belongs to. This is because a memory retains a unique and distinguishing character, much like a sensation coming from the stomach differs from an acoustic or visual impression. In a similar manner, the phonograph is incapable of reproducing the human voice in all its strength and warmth. The voice of the apparatus will remain shrill and cold; it has something imperfect and abstract about it that sets it apart. If the phonograph could hear itself, it would learn to recognize the difference between the voice that came from the outside and forced itself onto it and the voice that it itself is broadcasting and which is a simple echo of the first, following an already grooved path.

A further analogy between the phonograph and our brain exists in that the speed of the vibrations impressed on the apparatus can noticeably change the character of the reproduced sounds or recalled images. Depending on whether you increase or decrease the rotation of the phonographic disk, a melody will be transposed from one octave to another. If you turn the handle faster, a song will rise from the deepest and most indistinct notes to the highest and most piercing. Does not a similar effect occur in the brain when we focus our attention on an initially blurred image, increasing its clarity step by step and thereby moving it, as it were, up the scale? And could this phenomenon not be explained by the increased or decreased speed and strength of the vibrations of our cells? We have within us a kind of scale of images along which the images we conjure up and dismiss incessantly rise and fall. At times they vibrate in the depths of our being like a blurred "pedal"; at times their sonic fullness radiates above all others. As they dominate or recede, they appear to be closer or farther away from us, and sometimes the length of time separating them from the present moment seems to be waning or waxing. I know of impressions I received ten years ago that, under the influence of an association of ideas or simply owing to my attention or some change of emotion, suddenly seem to date from yesterday. In the same way singers create the impression of distance by lowering their voice; they merely need to raise it again to suggest the impression of approaching.

These analogies could be multiplied. The principal difference between the brain and the phonograph is that the metal disk of Edison's still rather primitive machine remains deaf to itself; there is no transition from movement to consciousness. It is precisely this wondrous transition that keeps occurring in the brain. It remains an eternal mystery that is less astonishing than it appears, however. Were the phonograph able to hear itself, it would be far less mystifying in the final analysis than the idea of our hearing it. But indeed we do: its vibrations really turn into impressions and thoughts. We therefore have to concede the transformation of movement into thought that is always possible—a transformation that appears more likely when it is a matter of internal brain movement than when it comes from the outside. From this point of view it would be neither very imprecise nor very disconcerting to define the brain as an infinitely perfected phonograph—a conscious phonograph.

It doesn't get any clearer than that. The psychophysical sciences, to which the philosopher Guyau has absconded, embrace the phonograph as the only suitable model for visualizing the brain or memory. All questions concerning thought as thought have been abandoned, for it is now a matter of implementation and hardware. Thus memory, around 1800 a wholly "subordinate inner power,"[31] moves to the fore eighty years later. And because Hegel's spirit is thereby ousted from the start, the recently invented phonograph, not yet even ready for serial production, is superior to all other media. Unlike Gutenberg's printing press or Ehrlich's automatic pianos in the brain metaphors of Taine and Spencer, it alone can combine the two actions indispensable to any universal machine, discrete or not: writing and reading, storing and scanning, recording and replaying. In principle, even though Edison for practical reasons later separated recording units from replaying ones, it is one and the same stylus that engraves and later traces the phonographic groove.

Which is why all concepts of trace, up to and including Derrida's grammatological *ur*-writing, are based on Edison's simple idea. The trace preceding all writing, the trace of pure difference still open between reading and writing, is simply a gramophone needle. Paving a way and retracing a path coincide. Guyau understood that the phonograph implements memory and thereby makes it unconscious.

It is only because no philosopher, not even one who has abandoned philosophy for psychophysics, can rid himself of his professional delusions that Guyau attempts to crown or surpass the unconscious mnemonic capabilities of the phonograph at the end of his essay by contrasting them with conscious human abilities. But consciousness, the quality that Guyau ascribes to the brain in order to celebrate the latter as an infinitely perfected phonograph, would result in an infinitely inferior one. Rather than hearing the random acoustic events forcing their way into the bell-mouth in all their real-time entropy, Guyau's conscious phonograph would attempt to understand[32] and thus corrupt them. Once again, alleged identities or meaning or even functions of consciousness would come into play. Phonographs do not think, therefore they are possible.

Trademark, "Writing Angel."

Guyau's own, possibly unconscious example alludes to the imputation of consciousness and inner life: if a phonograph really possessed the consciousness attributed to it and were able to point out that it remembered a song, it would consider this a miraculous ability. But impartial and external observers would continue to see it as the result of a fairly simple mechanism. When Guyau, who had observed the brain simply as a technical apparatus, turns his experimental gaze inward, he falls short of his own standards. It was, after all, an external gaze that had suggested the beautiful comparison between attention and playback speed. If the focusing of blurred mental images by way of attention amounts to nothing more or less than changing the time axis of acoustic events by increasing playback speed or indulging in time axis manipulation (TAM), then there is no reason to celebrate attention or memory as miraculous abilities. Neither gramophone needles nor brain neurons need any self-consciousness to retrace a groove faster than it was engraved. In both cases it boils down to programming. For that reason alone the diligent hand of the phonograph user, who in Edison's time had difficulties sticking to the correct time while turning the handle, could be replaced by clockworks and electronic motors with adjustable speed. The sales catalogues of American record companies warned their customers of the friend who "comes to you and claims that your machine is too slow or too fast. Don't listen to him! He doesn't know what he is talking about."[33]

But standardization is always upper management's escape from technological possibilities. In serious matters such as test procedures or mass entertainment, TAM remains triumphant. The Edison Speaking Phonograph Company, founded two months after Edison's primitive prototype of December 1877, did its first business with time axis manipulation: with his own hand the inventor turned the handle faster than he had during the recording in order to treat New York to the sensational pleasure of frequency-modulated musical pieces. Even the modest cornet of a certain Levy acquired brilliance and temperament.[34] Had he been among the delighted New Yorkers, Guyau would have found empirical proof that frequency modulation is indeed the technological correlative of attention.

Of course Europe's written music had already been able to move tones upward or downward, as the term "scale" itself implies. But transposition doesn't equal TAM. If the phonographic playback speed differs from its recording speed, there is a shift not only in clear sounds but in entire noise spectra. What is manipulated is the real rather than the symbolic. Long-term acoustic events such as meter and word length are affected as well. This is precisely what von Hornbostel, albeit without recognizing what distinguished it from transposition, praised as the "special advantage" of the phonograph: "It can be played at faster and slower speeds, allowing us to listen to musical pieces whose original speed was too fast at a more settled pace, and accordingly transposed, in order to analyze them."[35]

The phonograph is thus incapable of achieving real-time frequency shifts. For this we need rock bands with harmonizers that are able to reverse—with considerable electronic effort—the inevitable speed changes, at least to deceivable human ears. Only then are people able to return simultaneously and in real time from their breaking voices, and women can be men and men can be women again.

Time axis reversal, which the phonograph makes possible, allows ears to hear the unheard-of: the steep attack of instrumental sounds or spoken syllables moves to the end, while the much longer decay moves to the front. The Beatles are said to have used this trick on "Revolution 9" to whisper the secret of their global success to the tape freaks among their fans:[36] that Paul McCartney had been dead for a long time, replaced on album covers, stage, and in songs by a multimedia double. As the Columbia Phonograph Company recognized in 1890, the phonograph can be used as machine for composing music simply by allowing consumers to play their favorite songs backwards: "A musician could get one popular melody every day by experimenting in that way."[37]

TAM as poetry—but poetry that transgresses its customary boundaries. The phonograph cannot deny its telegraphic origin. Technological media turn magic into a daily routine. Voices that start to migrate through frequency spectra and time axes do not simply continue old literary word-game techniques such as palindromes or anagrams. This letter-bending had become possible only once the primary code, the alphabet itself, had taken effect. Time axis manipulation, however, affects the raw material of poetry, where manipulation had hitherto been impossible. Hegel had referred to "the *sound*" as "a disappearing of being in the act of being," subsequently celebrating it as a "saturated expression of the manifestation of inwardness."[38] What was impossible to store could not be manipulated. Ridding itself of its materiality or clothes, it disappeared and presented inwardness as a seal of authenticity.

But once storage and manipulation coincide in principle, Guyau's thesis linking phonography and memory may be insufficient. Storage facilities, which according to his own insight are capable of altering the character of the replayed sounds (thanks to time manipulation), shatter the very concept of memory. Reproduction is demoted once the past in all its sensuous detail is transmitted by technical devices. Certainly, hi-fi means "high fidelity" and is supposed to convince consumers that record companies remain loyal to musical deities. But it is a term of appeasement. More precise than the poetic imagination of 1800, whose alphabetism or creativity confronted an exclusively reproductive memory, technology literally makes the unheard-of possible. An old Pink Floyd song spells it out:

> When that old fat sun in the sky's falling
> Summer ev'ning birds are calling
> Summer Sunday and a year
> The sound of music in my ear
> Distant bells
> New mown grass smells
> Songs sweet
> By the river holding hands.
> And if you see, don't make a sound
> Pick your feet up off the ground
> And if you hear as the wall night falls
> The silver sound of a tongue so strange,
> Sing to me sing to me.[39]

The literally unheard-of is the site where information technology and brain physiology coincide. To make no sound, to pick your feet up off the ground, and to listen to the sound of a voice when night is falling—we all do it when we put on a record that commands such magic.

And what transpires then is indeed a strange and unheard-of silver noise. Nobody knows who is singing—the voice called David Gilmour that sings the song, the voice referred to by the song, or maybe the voice of the listener who makes no sound and is nonetheless supposed to sing once all the conditions of magic have been met. An unimaginable closeness of sound technology and self-awareness, a simulacrum of a feedback loop relaying sender and receiver. A song sings to a listening ear, telling it to sing. As if the music were originating in the brain itself, rather than emanating from stereo speakers or headphones.

That is the whole difference between arts and media. Songs, arias, and operas do not rely on neurophysiology. Voices hardly implode in our ears, not even under the technical conditions of a concert hall, when singers are visible and therefore discernible. For that reason their voices have been trained to overcome distances and spaces. The "sound of music in my ear" can exist only once mouthpieces and microphones are capable of recording any whisper. As if there were no distance between the recorded voice and listening ears, as if voices traveled along the transmitting bones of acoustic self-perception directly from the mouth into the ear's labyrinth, hallucinations become real.

And even the distant bells that the song listens to are not merely signifiers or referents of speech. As a form of literature, lyric had been able to provide as much and no more. Countless verses used words to conjure up acoustic events as lyrical as they were indescribable. As rock songs, lyric poetry can add the bells themselves in order to fill attentive brains with something that, as long as it had been confined to words, had remained a mere promise.

In 1898, the Columbia Phonograph Company Orchestra offered the song "Down on the Swanee River" as one of its 80 cylinders. Advertisements promised Negro songs and dances, as well as the song's location and subject: pulling in the gangplank, the sounds of the steam engine, and, 80 years before Pink Floyd, the chiming of a steamboat bell[40]—all for 50 cents. Songs became part of their acoustic environment. And lyrics fulfilled what psychoanalysis—originating not coincidentally at the same time—saw as the essence of desire: hallucinatory wish fulfillment.

Freud's "Project for a Scientific Psychology" (1895) saw the state of "being hallucinated in a backward flow of Q to ϕ and also to ω."[41] In other words: impermeable brain neurons occupied by memory traces rid themselves of their charge or quantity by transferring them onto permeable neurons designed for sensory perception. As a result, data already stored appear as fresh input, and the psychic apparatus becomes its own simulacrum. Backflow or feedback comes as close to perfect hallucinatory wish fulfillment as Freud's "Project for a Scientific Psychology" does to technological media. "The intention is to furnish a psychology that shall be a natural science: that is, to represent psychical processes as quantitatively determinate states of specifiable material particles, thus making those processes perspicuous and free from contradiction."[42] That is psychophysics at its best. All of Freud's elaborations on neurons and their cathexes and on facilitations and their resistance are based on the "views on localization held by [the] cerebral anatomy"[43] of his time. That the psychic apparatus (already technified by its name) can transmit and store data while remaining both permeable and impermeable would remain an insoluble paradox were its analogy modeled upon writing. (At best, Freud's famous "Mystic Writing-Pad," commented upon by Derrida,[44] might be able to carry out both functions.) A brain physiology that followed Broca and Wernicke's subdivision of discourse into numerous subroutines and located speaking, hearing, writing, and reading in various parts of the brain (because it exclusively focused on the states of specifiable material particles) had to model itself on the phonograph—an insight anticipated by Guyau. It comes as no surprise, then, that Sigmund Exner, whose research formed the basis for Freud's notion of facilitation in "Scientific Project," also "provided the basis for the construction of a scientific phonographic museum" at the University of Vienna.[45]

"When it comes to molecules and cranial pathways, we"—that is, the brain researchers and art physiologists of the turn of the century—"automatically think of a process similar to that of Edison's *phonograph*."[46] These are the words of Georg Hirth, author of the first German treatise on art physiology. Twenty years later, they were written into art itself. In 1919, Rilke completed a prose "essay" that, using the modest means of bricolage or literature, translated all the discoveries of brain physiology into modern poetry.

RAINER MARIA RILKE, "PRIMAL SOUND" (1919)

It must have been when I was a boy at school that the phonograph was invented. At any rate it was at that time a chief object of public wonder; this was probably the reason why our science master, a man given to busying himself with all kinds of handiwork, encouraged us to try our skill in making one of these instruments from the material that lay nearest to hand. Nothing more was needed than a piece of pliable cardboard bent to the shape of a funnel, on the narrower orifice of which was stuck a piece of impermeable paper of the kind used to bottle fruit. This provided a vibrating membrane, in the middle of which we stuck a bristle from a coarse clothes brush at right angles to its surface. With these few things one part of the mysterious machine was made, receiver and reproducer were complete. It now only remained to construct the receiving cylinder, which could be moved close to the needle marking the sounds by means of a small rotating handle. I do not remember what we made it of; there was some kind of cylinder which we covered with a thin coating of candle wax to the best of our ability. Our impatience, brought to a pitch by the excitement of sticking and fitting the parts as we jostled one another over it, was such that the wax had scarcely cooled and hardened before we put our work to the test.

How now this was done can easily be imagined. When someone spoke or sang into the funnel, the needle in the parchment transferred the sound waves to the receptive surface of the roll slowly turning beneath it, and then, when the moving needle was made to retrace its path (which had been fixed in the meantime with a coat of varnish), the sound which had been ours came back to us tremblingly, haltingly from the paper funnel, uncertain, infinitely soft and hesitating and fading out altogether in places. Each time the effect was complete. Our class was not exactly one of the quietest, and there can have been few moments in its history when it had been able as a body to achieve such a degree of silence. The phenomenon, on every reception of it, remained astonishing, indeed positively staggering. We were confronting, as it were, a new and infinitely delicate point in the texture of reality, from which something far greater than ourselves, yet indescribably immature, seemed to be appealing to us as if seeking help. At the time and all through the intervening years I believed that that independent sound, taken from us and preserved outside of us, would be unforgettable. That it turned out otherwise is the cause of my writing the present account. As will be seen, what impressed itself on my memory most deeply was not the sound from the funnel but the markings traced on the cylinder; these made a most definite impression.

I first became aware of this some fourteen or fifteen years after my school days were past. It was during my first stay in Paris. At that time I was attending the anatomy lectures in the Ecole des Beaux-Arts with considerable enthusiasm. It was not so much the manifold interlacing of the muscles and sinews nor the complete inner agreement of the inner organs with another that appealed to me, but rather the bare skeleton, the restrained energy and elasticity of which I had already noticed when studying the drawings of Leonardo. However much I puzzled over the structure of the whole, it was more than I could deal with; my attention always reverted to the study of the skull, which seemed to me to constitute the utmost achievement, as it were, of which this chalky element was capable; it was as if it had been persuaded to make just in this part a special effort to render a decisive service by providing a most solid protection for the most daring feature of all, for something which, though itself narrowly confined, had a field of activity which was boundless. The fascination which this particular structure had for me reached such a pitch finally, that I procured a skull in order to spend many hours of the night with it; and, as always happens with me and things, it was not only the moments of deliberate attention which made this ambiguous object really mine: I owe my familiarity with it, beyond doubt, in part to that passing glance with which we involuntarily examine and perceive our daily environment, when there exists any relationship at all between it and us. It was a passing glance of this kind which I suddenly checked in its course, making it exact and attentive. By candlelight—which is often so peculiarly alive and challenging—the coronal suture had become strikingly visible, and I knew at once what it reminded me of: one of those unforgotten grooves, which had been scratched in a little wax cylinder by the point of a bristle!

And now I do not know: is it due to a rhythmic peculiarity of my imagination that ever since, often after the lapse of years, I repeatedly feel the impulse to make that spontaneously perceived similarity the starting point for a whole series of unheard-of experiments? I frankly confess that I have always treated this desire, whenever it made itself felt, with the most unrelenting mistrust—if proof be needed, let it be found in the fact that only now, after more than a decade and a half, have I resolved to make a cautious statement concerning it. Furthermore, there is nothing I can cite in favor of my idea beyond its obstinate recurrence, a recurrence which has taken me by surprise in all sorts of places, divorced from any connection with what I might be doing.

What is it that repeatedly presents itself to my mind? It is this:

The coronal suture of the skull (this would first have to be investigated) has—let us assume—a certain similarity to the close wavy line which the needle of a phonograph engraves on the receiving, rotating cylinder of the apparatus. What if one changed the needle and directed it on its return journey along a tracing which was not derived from the graphic translation of sound but existed of itself naturally—well, to put it plainly, along the coronal suture, for example. What would happen? A sound would necessarily result, a series of sounds, music. . . .

Feelings—which? Incredulity, timidity, fear, awe—which of all feelings here possible prevents me from suggesting a name for the primal sound which would then make its appearance in the world? . . .

Leaving that aside for the moment: what variety of lines, then, occurring anywhere, could one not put under the needle and try out? Is there any contour that one could not, in a sense, complete in this way and then experience it, as it makes itself felt, thus transformed, in another field of sense?

At one period, when I began to interest myself in Arabic poems, which seem to owe their existence to the simultaneous and equal contributions from all five senses, it struck me for the first time that the modern European poet makes use of these five contributors singly and in very varying degree, only one of them—sight overladen with the world—seeming to dominate him constantly; how slight, by contrast, is the contribution he receives from inattentive hearing, not to speak of the indifference of other senses, which are active only on the periphery of consciousness and with many interruptions within the limited sphere of their practical activity. And yet the perfect poem can only materialize on condition that this world, acted upon by all five levers simultaneously, is seen, under a definite aspect, on the supernatural plane, which is, in fact, the plane of the poem.

A lady, to whom this was mentioned in conversation, exclaimed that this wonderful and simultaneous capacity and achievement of all the senses was surely nothing but the presence of mind and grace of love—incidentally she thereby bore her own witness to the sublime reality of the poem. But the lover is in such splendid danger just because he must depend on the coordination of his senses, for he knows that they must meet in that unique and risky center in which, renouncing all extension, they come together and have no permanence.

As I write this, I have before me the diagram which I have always used as a ready help whenever ideas of this kind have demanded attention. If the world's whole field of experience, including those spheres which are beyond our knowledge, be represented in a complete circle, it will be immediately evident that when the black sectors, denoting that which we are incapable of experiencing, are measured against the lesser, light sections, correspond to that which is illuminated by the senses, the former are very much greater.

Now the position of the lover is this: that he feels himself unexpectedly placed in the center of the circle, that is to say, at the point where the known and the incomprehensible, coming forcibly together at one single point, become complete and simply a possession, losing thereby, it is true, all individual character. This position would not serve for the poet, for individual variety must be constantly present for him; he is compelled to use the sense sectors to their full extent, as it must also be in his aim to extend each of them as far as possible, so that his lively delight, girded for the attempt, may be able to pass through the five gardens in one leap.

As the lover's danger consists in the nonspatial character of his standpoint, so the poet's lies in his awareness of the abysses which divide the one order of sense experience from the other: in truth they are sufficiently wide and engulfing to sweep away from before us the greater part of the world—who knows how many worlds?

The question arises here as to whether the extent of these sectors on the plane assumed by us can be enlarged to any vital degree by the work of research. The achievements of the microscope, of the telescope, and of so many devices which increase the range of the senses upward and downward: do they not lie in *another* sphere altogether, since most of the increase thus achieved cannot be interpreted by the senses, cannot be "experienced" in any real sense? It is perhaps not premature to suppose that the artist, who develops the five-fingered hand of his senses (if one may put it so) to ever more active and more spiritual capacity, contributes more decisively than anyone else to an extension of the several sense fields; only the achievement which gives proof of this does not permit of his entering his personal extension of territory in the general map before us, since it is only possible, in the last resort, by a miracle.

But if we are looking for a way by which to establish the connection so urgently needed between the different provinces now so strangely separated from one another, what could be more promising than the experiment suggested earlier in this recollection? If the writer ends by recommending it once again, he may be given a certain amount of credit for withstanding the temptation to give free reign to his fancy in imagining the results of the assumptions which he has suggested.

Soglio. On the day of the Assumption of the Blessed Virgin, 1919.

Rilke dedicated the most impassionate of reports to phonography. Regardless of the fact that he wrote it on the Assumption, "he was a poet and hated the approximate."[47] Therefore the strange precision with which his text enumerates all the parts of an apparatus that Rilke's physics teacher, employed not coincidentally at an imperial military school, constructed around 1890. As if to confirm the fictional Edison of *Tomorrow's Eve*, who had no supply problems whatsoever when designing the phonograph, a combination of cardboard, paper, the bristles of a clothes brush, and candle wax suffice to open a "new and infinitely delicate point in the texture of reality." Oblivious to the knowledge of the physics teacher and the school drill, students hear their own voices. Not their words and answers as programmed feedback by the education system, but the real voice against a backdrop of pure silence or attention.

And yet the "unforgettable" (in the word's double meaning) phonographic sound recording is not at the center of Rilke's profane illumination. In the founding age of media, the author is captivated more by the technological revolutions of reading than of writing. The "markings traced on the cylinder" are physiological traces whose strangeness transcends all human voices.

Certainly, the writer is no brain physiologist. His amateur status at the Ecole des Beaux-Arts enables him to become acquainted with the vicissitudes of the skeletal structure, but not with the facilitations on which Exner or Freud based their new sciences. But when it comes to mounted and exhibited skeletons, Rilke is fascinated by that "utmost achievement" known as the skull, because "it was as if it had been persuaded to make just in this part a special effort to render a decisive service by providing a most solid protection for the most daring feature of all." During his Parisian nights, Rilke reduces the skull sitting in front of him to a cerebral container. Describing it as "this particular structure" with a "boundless field of activity," he merely repeats the physiological insight that for our central nervous system, "our own body is the outside world."[48] One no less than Flechsig, Schreber's famous psychiatrist, had proven that the cerebral cortex contains a "sphere of physical perception" that neurologically reproduces all parts of the body, distorted according to their importance.[49] Rilke's belief in later years that it was the task of poetry to transfer all given data into an "inner world space" is based on such insights. (Even though literary scholars, still believing in the omnipotence of philosophers, choose to relate Rilke's inner world space to Husserl.)[50]

"Primal Sound" leaves no doubt whatsoever about which contemporary developments were most important to literature in 1900. Instead of lapsing into the usual melancholic associations of Shakespeare's Hamlet or Keller's Green Henry at the sight of a human skull in candlelight, the writer sees phonographic grooves.

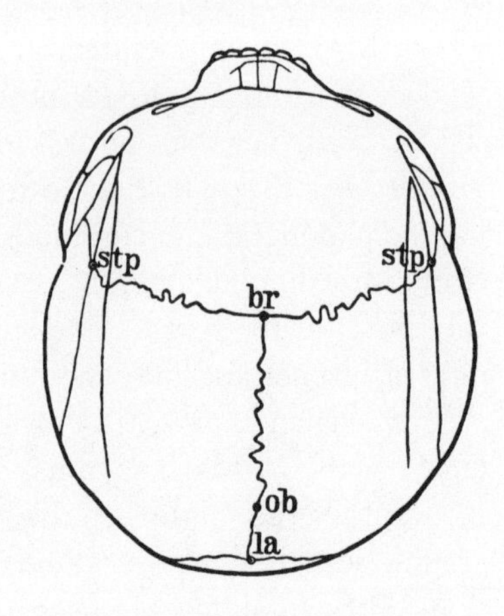

Coronal suture from *stp* to *stp*.

A trace or path or groove appears where the frontal and parietal bones of the "suckling infant"[51]—to use Rilke's anatomically correct term—have grown together. As if the facilitations of Freud and Exner had been projected out of the brain onto its enclosure, the naked eye is now able to read the coronal suture as a writing of the real. A technologically up-to-date author follows in the wake of the brain physiologists, who since the days of Guyau and Hirth have automatically thought of Edison's phonograph when dealing with nerve pathways. Moreover, Rilke draws conclusions more radical than all scientific boldness. Before him, nobody had ever suggested to decode a trace that nobody had encoded and that encoded nothing.

Ever since the invention of the phonograph, there has been writing without a subject. It is no longer necessary to assign an author to every trace, not even God. "Project for a Scientific Psychology" centered on facilitations inscribed by acts of perception, but there is no reason not to set the gramophone needle to random anatomic features. A transgression in the literal sense of the word, which shakes the very words used to phrase it. Acoustics arises from physiology, technology from nature. In Rilke's time, skulls were measured in search of all possible features: intelligence and idiocy, masculinity and femininity, genius and racial characteristics. But their transposition into the acoustic medium remained a challenge that forced dots and question marks onto the writing hand.

What the coronal suture yields upon replay is a primal sound without a name, a music without notation, a sound even more strange than any incantation for the dead for which the skull could have been used. Deprived of its shellac, the duped needle produces sounds that "are not the result of a graphic transposition of a note" but are an absolute transfer, that is, a metaphor. A writer thus celebrates the very opposite of his own medium—the white noise no writing can store. Technological media operate against a background of noise because their data travel along physical channels; as in blurring in the case of film or the sound of the needle in the case of the gramophone, that noise determines their signal-to-noise ratio. According to Arnheim, that is the price they pay for delivering reproductions that are at the same time effects of the reproduced. Noise is emitted by the channels media have to cross.

In 1924, five years after Rilke's "Primal Sound," Rudolph Lothar wrote *The Talking Machine: A Technical-Aesthetic Essay*. Based on the not-very-informed premise that "philosophers and psychologists have hitherto written about the arts" and "neglected" phonography,[52] Lothar drew up a new aesthetic. Its key propositions center exclusively on the relationship between noise and signals.

The talking machine occupies a special position in aesthetics and music. It demands a twofold capacity for illusion, an illusion working in two directions. On the one hand, it demands that we ignore and overlook its mechanical features. As we know, every record comes with interference. As connoisseurs we are not allowed to hear this interference, just as in a theater we are obliged to ignore both the line that sets off the stage and the frame surrounding the scene. We have to forget we are witnessing actors in costumes and makeup who are not really experiencing what they are performing. They are merely playing parts. We, however, pretend to take their appearance for reality. Only if we forget that we are inside a theater can we really enjoy dramatic art. This "as if" is generated by our capacity for illusion. Only when we forget that the voice of the singer is coming from a wooden box, when we no longer hear any interference, when we can suspend it the way we are able to suspend a stage—only then will the talking machine come into its own artistically.

But, on the other hand, the machine demands that we give bodies to the sounds emanating from it. For example, while playing an aria sung by a famous singer we see the stage he stands on, we see him dressed in an appropriate costume. The more it is linked to our memories, the stronger the record's effect will be. Nothing excites memory more strongly than the human voice, maybe because nothing is forgotten as quickly as a voice. Our memory of it, however, does not die—its timbre and character sink into our subconscious where they await their revival. What has been said about the voice naturally also applies to instruments. We see Nikisch conduct the C-minor symphony, we see Kreisler with the violin at his chin, we see trumpets flashing in the sun when listening to military marches.

But the capacity for illusion that enables us to ignore boxes and interference and furnishes tones with a visible background requires musical sensitivity. This is the most important point of phonographic aesthetics: The talking machine can only grant artistic satisfaction to musical people. For only musicians possess the capacity for illusion necessary for every enjoyment of art.[53]

Maybe Rilke, who loved the gong, with its resounding mixture of frequencies, above all other instruments, wasn't a musical person.[54] His aesthetic—"Primal Sound" is Rilke's only text about art and the beautiful in general—subverts the two illusions to which Lothar wants to commit readers or gramophone listeners. From the fact that "every record comes with interferences" he draws opposite conclusions. Replaying the skull's coronary suture yields nothing but noise. And there is no need to add some hallucinated body when listening to signs that are not the result of the graphic translation of a note but rather random anatomical lines. Bodies themselves generate noise. And the impossible real transpires.

Of course, the entertainment industry is all on Lothar's side. But there have been and there still are experiments that pursue Rilke's primal sound with technologically more sophisticated means. In the wake of Mondrian and the Bruitists (who wanted to introduce noise into literature and music), Moholy-Nagy already suggested in 1923 turning "the gramophone from an instrument of reproduction into a productive one, generating acoustic phenomena without any previous acoustic existence by scratching the necessary marks onto the record."[55] An obvious analogy to Rilke's suggestion of eliciting sounds from the skull that were not the result of a prior graphic transformation. A triumph for the concept of frequency: in contrast to the "narrowness" of a "scale" that is "possibly a thousand years old" and to which we therefore no longer must adhere,[56] Moholy-Nagy's etchings allow for unlimited transposition from medium to medium. Any graphisms—including those, not coincidentally, dominating Mondrian's paintings—result in a sound. Which is why the experimenter asks for the "study of graphic signs of the most diverse (simultaneous and isolated) acoustic phenomena" and the "use of projection machines" or "film."[57]

Engineers and the avant-garde think alike. At the same time as Moholy-Nagy's etching, the first plans were made for sound film, one of the first industrially connected media systems. "The invention of Mr. Vogt, Dr. Engel, and Mr. Masolle, the speaking Tri-Ergon-film," was based on a "highly complicated process" of medial transformations that could only be financed with the help of million-dollar investments from

Gramophone record. (Photo: Moholy-Nagy)

the C. Lorenz Company.[58] The inventors say of it, "Acoustic waves emanating from the scene are converted into electricity, electricity is turned into light, light into the silver coloring of the positive and negative, the coloring of the film back into light, which is then converted back into electricity before the seventh and final transformation turns electricity into the mechanical operation of a weak membrane giving off sounds."[59]

Frequencies remain frequencies regardless of their respective carrier medium. The symbolic correlation of sound intervals and planetary orbits, which since [Cicero's] *Dream of Scipio* made up the harmony of the spheres, is replaced by correspondences in the real. In order to synchronize, store, and reproduce acoustic events and image sequences, sound films can let them wander seven times from one carrier to the next. In Moholy-Nagy's own words, his record etchings are capable of generating a "new mechanical harmony": "The individual graphic signs are examined and their proportions are formulated as a law. (Here we may point

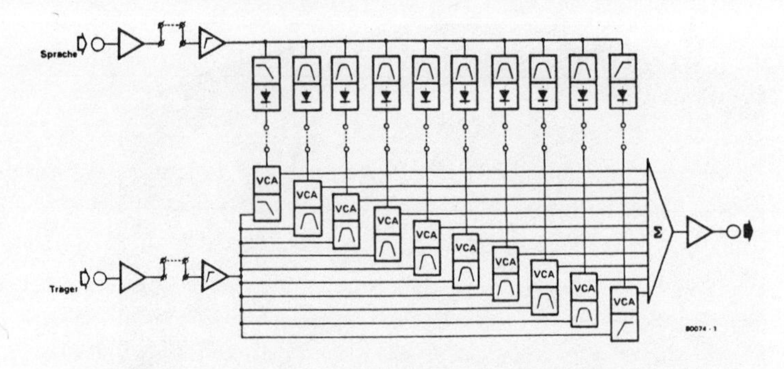

Block schematic of an analog vocoder. The synthesis component is in the lower signal path, the analysis component, in the upper signal path. The latter's low- and high-pass filters limit the input, for example, of "speech," while its band-pass filters break down the audible range into several component frequency channels.Following its coordination as envelope curves, the analysis output—using a switching matrix with arbitrarily chosen correspondences between the signal paths—controls the voltage-controlled amplifiers (VCAs), whose band-pass filters have also broken down the "input" or carrier into several component frequency channels. The sum signal at the exit (of the vocoder) appears as an instrumental sound encoded by a voice (vox).

out a consideration that is at present still utopian: based on strict proportional laws graphic signs can be transposed into music.)"[60]

This idea had lost its utopian character long before it was written down. Fourier's solution of all continuous functions (including musical notes) into sums of pure sine harmonics was achieved before Helmholtz and Edison. Walsh's equally mathematical proof that square wave vibrations may also serve as summands of the Fourier analysis was roughly contemporaneous with Moholy-Nagy's writings. As a result, in 1964 Robert A. Moog, with his electronic talents and the "American vice of modular repetition,"[61] was able to equip all the sound studios and rock bands of this world with synthesizers. A subtractive sound analysis, that is, one controlled by frequency filters, transfers the proportional relationships of graphic depictions (rectangles, saw tooth curves, triangles, trapezoids, and maybe even sine curves) into the music envisioned by Mondrian and Moholy-Nagy.[62]

Rilke's urgent demand to put under the needle and try out a "variety of lines, occurring anywhere," to "complete [it] in this way and then experience it, as it makes itself felt, thus transformed, in another field of sense": it is realized every night in the combination of amplifier and oscillographic display.

But there is more to it. Between 1942 and 1945, while working for Bell laboratories and the British Secret Service, respectively, Shannon and Turing developed the vocoder, a wonder weapon that would make the transatlantic telephone conversations between Churchill and Roosevelt safe from interception by Canaris and the German Abwehr[63] and that, like so many electronic achievements of the Second World War, is now indispensable to popular music. It lives up to its name: it encodes any given data stream *A* with the envelope curves of another sound sequence *B*, for example, the voice of a singer, after a switching matrix has changed the frequency of the envelope curves by way of free permutation. In the case of Laurie Anderson's electronic violin, the one-third octave band between 440 and 550 Hz follows in absolute synchronicity the volume her voice happens to have in the one-third octave band between 1760 and 2200 Hz, while a third one-third octave band of her songs controls a fourth band of her violin, and so on and so forth. Primal sounds do not correspond to anatomical features and sounds do not follow Mondrian's graphics; rather, the paradoxical result is that one and the same controls one and the same: one acoustics controls the other.

To test his vocoder, by the way, Turing first played a record of Winston Churchill's belligerent voice, whose discreet or cut-up sampled values he then mixed with a noise generator using modular addition. Whereupon British officers heard the voice of their prime minister and commander in chief contaminate the speakers as just so much white noise (not to say, primal sound). Appropriately, Turing's vocoder was named after Delilah, who in the Book of Judges tricked another warrior, the Danaite Samson, out of the secret of his strength. Turing's skill as a tinkerer, however, revealed the secret of modern political discourse to be something far worse than weakness: it is "a perfectly even and uninformative hiss,"[64] which offered no regularities and, therefore, nothing intelligible to the ears of British officers or those of German eavesdroppers. And yet, sent through the vocoder a second time, Churchill's original voice emerged from the receiving end.

This is what has become of the "abysses" that, according to Rilke's ingenious formula, "divide the one order of sense experience from the other." In today's media networks, algorithmically formalized data streams can traverse them all. Media facilitate all possible manipulations: acoustic signals control the optical signals of light consoles, electronic language controls the acoustic output of computer music, and, in the case of the vocoder, one set of acoustic signals controls another. Finally, New York disc jockeys turn the esoteric graphisms of Moholy-Nagy into the everyday experience of scratch music.

But Rilke's astute diagnosis only applies to the founding age when the three *ur*-media—phonograph, film, and typewriter—first differentiated acoustics, optics, and writing. Nevertheless, as if anticipating today's media systems, he searched "for a way to establish the connection so urgently needed between the different provinces now so strangely separated from one another." Which is why he fell back on "Arabic poems, which seem to owe their existence to the simultaneous and equal contributions from all five senses," and which let eyes trained in the art of calligraphy enjoy the very materiality of letters. This explains his criticism, historically extremely accurate, of literary epochs such as the Age of Goethe, in which "sight" alone seems to dominate authors and readers because correct reading involves a hallucinatory process that turns words into a real and visible world. This explains as well his proposition for an equally lyrical and scientific coronal suture phonography, which would pay more attention to acoustics than did the "inattentive hearing" of authors from the Age of Goethe.

But before Rilke wrote down his proposal on the Assumption in the alpine solitude of the Bergell, he related it to a woman. Synchronicity of the asynchronic: on the one hand a writer whose "extension" or combination of sensory media goes beyond "the work of research"; on the other a woman who mistakes coronal suture phonography for "love," and love—as involuntary evidence for "the sublime reality of the poem"—for poems. Only as long as the unchallenged and unrivaled medium of the book was able to simulate the storage of all possible data streams did love remain literature and literature love; the ascension of female readers.

But a writer whose school teaches physics instead of philosophy objects. The combination of sensory data streams achieved by love is devoid of "permanence." It cannot be stored by any medium. Moreover, it loses "all individual character." That is, no real can pass through the filter of love. Which is why love does "not serve for the poet, for individual variety must be constantly present for him; he is compelled to use the sense sectors to their full extent," or, simply, to become a media technician among media technicians.

Marinetti's *Technical Manifesto of Futurist Literature* of 1912 proclaimed that crowds of massed molecules and whirling electrons are more exciting than the smiles or tears of a woman (*di una donna*).[65] In other words: literature defects from erotics to stochastics, from red lips to white noise. Marinetti's molecular swarms and whirling electrons are merely instances of the Brownian motion that human eyes can only perceive in the shape of dancing sun particles but that in the real are the noise on all channels. According to Rilke, the "abysses" dividing the orders of sense experience are "sufficiently wide and engulfing to sweep away from before us the greater part of the world—who knows how many worlds?" Which is why love is no longer sufficient for authors who, like Rilke himself, transcribe all the details of sensory perception into an inner world-space known as the brain or literature and, subsequently, phonographically trace the facilitations of this unique container as primal sound itself.

Phonography, notation, and a new eroticism—this is precisely the constellation described by Maurice Renard in a short story of 1907, ten years prior to Rilke's essay. What Rilke saw in the coronal suture Renard's fictitious composer Nerval encounters in a roaring seashell, which, like Rilke's skull, is a physiological substitute for Edison's apparatus. Thirty years later Paul Valéry used almost the same title as that of Renard's story to celebrate shells as architectural works of an artistic nature,[66] but Renard focuses on the central nervous system, on the labyrinth of shells, auricles, and sound. Since machines have taken over the functions of the central nervous system, nobody can say whether the roaring comes from the blood or from the sirens, from the ears or from the sea goddess Amphitrite.

MAURICE RENARD, "DEATH AND THE SHELL" (1907)

> And her shape is of such mysterious nastiness that you brace yourself to listen.
>
> —HENRI DE RÉGNIER, *Contes à soi-même*

Put the shell back where it belongs, Doctor, and do not hold it to your ear for the pleasure of mistaking the roaring of your blood with that of the sea. Put it back. The very man we just buried, our beloved great musician, would still be alive had he not committed the childish mistake of listening to what the mouth of a shell has to say . . . Yes indeed: Nerval, your very own patient . . . You talk of congestion? Maybe. But I am skeptical. Here are my reasons. Keep them to yourself.

Last Wednesday night, on the day before the accident, I dined at Nerval's. His close friends have been meeting there every Wednesday for twenty years. There were five in the beginning. But this time, and for the first time, there were only two of us: a stroke, a contagious flu, and a suicide left Nerval and me facing each other. When you reach sixty such a situation has nothing amusing about it. You keep asking yourself who will be next.

The meal was as gloomy as a funeral feast. The great man remained silent. I did everything possible to cheer him up. Maybe he was mourning other deaths, the secrecy of which made them even more bitter. . . .

Indeed, he was mourning others.

We went to his study. The piano had not been closed; on it there was the first page of a new composition.

"What are you working on, Nerval?"

He raised his finger and spoke like a sad prophet announcing his god:

"*Amphitrite.*"

"*Amphitrite*! At last! For how many years have you been saving her up?"

"Since the Rome prize. I waited and waited. The longer a work is allowed to mature, the better it is, and I wanted to infuse it with the dream and experience of a whole life . . . I believe it is time . . . "

"A symphonic poem, isn't it? . . . Are you satisfied?"

Nerval shook his head.

"No. In a pinch, it might work . . . My thoughts are not distorted beyond recognition . . . "

He interpreted the prelude with great virtuosity: a "Train of Neptune." You will relish it, Doctor; it is a miracle!

"You see," Nerval said to me while striking strange, outrageous, and brutal chords, "up to this fanfare of Tritons it works . . . "

"Marvelous," I answered; "there is . . . "

"But," Nerval continued, "that is all there is to it. The choir that follows . . . a failure. Yes, I can feel my powerlessness to write it . . . It is too beautiful. We no longer know . . . It would have to be composed the way Phidias created his sculptures; it would have to be a Parthenon, as simple as . . . We no longer know . . . Ha!" he suddenly screamed, "to have arrived there, I . . . "

"Listen," I said to him, "you are among the most famous, so . . . "

"So, if this is how I end up, what do others know? But at least their mediocrity is a blessing, which is itself mediocre and satisfied with little. Famous! What is fame when engulfed in sadness! . . . "

"The peaks are always clouded! . . . "

"Enough," Nerval resumed, "a cease-fire for flattery! This is truly a sad hour, so let us, if you wish, dedicate it to real sorrows. We owe it to the departed."

Following these rather mysterious words he took a phonograph from underneath a blanket. I understood.

You can well imagine, Doctor, that this phonograph did not play the "Potpourri from *The Doll*, performed by the Republican Guard under Parès." The very improved, sonorous, and clear machine only had a few cylinders. It merely spoke . . .

Yes, you guessed it: on Wednesday the dead spoke to us . . .

How terrible it is to hear this copper throat and its sounds from beyond the grave! It is more than a photographic, or I had better say cinematographic, something; it is the voice itself, the living voice, still alive among carrion, skeletons, nothingness . . .

The composer was slumped in his chair next to the fireside. He listened with painfully knit brows to the tender things our departed comrades said from the depths of the altar and the grave.

"Well, science does have its advantages, Nerval! As a source of miracles and passions it is approaching art."

"Certainly. The more powerful the telescopes, the larger the number of stars is going to be. Of course science has its good sides. But for us it is still too young. Only our heirs will benefit from it. With the help of each new invention they will be able to observe anew the face of our century and listen to the sounds made by our generation. But who is able to project the Athens of Euripides onto a screen or make heard the voice of Sappho?"

He livened up and played with a large shell he had absentmindedly taken off the chimney mantelpiece.

I appreciated the object that was to revive his spirits, and because I anticipated that the elaboration of the scientific, if not paradoxical, theme would amuse him, I resumed:

"Beware of despair. Nature frequently delights in anticipating science, which in turn often merely imitates it. Take photography, for instance! The world can see the traces of an antediluvian creature in a museum—I believe it is the brontosaurus—and the soil retains the marks of the rain that was falling when the beast walked by. What a prehistoric snapshot!"

Nerval was holding the shell to his ear.

"Beautiful, the roaring of this stethoscope," he said; "it reminds me of the beach where I found it—an island off Salerno . . . it is old and crumbling."

I used the opportunity.

"Dear friend, who knows? The pupils of the dying are said to retain the last image they received . . . What if this ear-shaped snail stored the sounds it heard at some critical moment—the agony of mollusks, maybe? And what if the rosy lips of its shell were to pass it on like a graphophone? All in all, you may be listening to the surf of oceans centuries old . . . "

But Nerval had risen. With a commanding gesture he bid me be quiet. His dizzy eyes opened as if over an abyss. He held the double-horned grotto to his temple as if eavesdropping on the threshold of a mystery. A hypnotic ecstasy rendered him motionless.

After I repeatedly insisted, he reluctantly handed me the shell.

At first I was only able to make out a gurgling of foam, then the hardly audible turmoil of the open sea. I sensed—how I can not say—that the sea was very blue and very ancient. And then, suddenly, women were singing and passing by . . . inhuman women whose hymn was wild and lustful like the scream of a crazed goddess . . . Yes, Doctor, that's how it was: a scream and yet a hymn. These were the insidious songs Circe warned us not to listen to, or only when tied to the mast of a galley with rowers whose ears are filled with wax . . . But was that really enough to protect oneself from the danger? . . .

I continued to listen.

The sea creatures disappeared into the depths of the shell. And yet minute by minute the same maddening scene was repeated, periodically, as if by phonograph, incessantly and never diminished.

Nerval snatched the shell away from me and ran to the piano. For a long time he tried to write down the sexual screaming of the goddesses.

At two in the morning he gave up.

The room was strewn with blackened and torn sheets of music.

"You see, you see," he said to me, "not even when I am dictated to can I transcribe the choir! . . . "

He slumped back into his chair, and despite my efforts, he continued to listen to the poison of this Paean.

At four o'clock he started to tremble. I begged him to lie down. He shook his head and seemed to lean over the invisible maelstrom.

At half past five Nerval fell against the marble chimney—he was dead.

The shell broke into a thousand pieces.

Do you believe that there are poisons for the ear modeled on deadly perfumes or lethal potions? Ever since last Wednesday's acoustic presentation I have not been feeling well. It is my turn to go . . . Poor Nerval . . .

Doctor, you claim he died of congestion . . . and what if he died *because he heard the sirens singing?*

Why are you laughing?

There have been better questions to conclude fantastic tales. But in ways both smooth and comical Renard's fantasy finds its way into technical manuals. In 1902, in the first German monograph on *Care and Usage of Modern Speaking Machines (Phonograph, Graphophone and Gramophone)*, Alfred Parzer-Mühlbacher promises that graphophones—a Columbia brand name also used by Renard—will be able to build "archives and collections" for all possible "memories":

> Cherished loved ones, dear friends, and famous individuals who have long since passed away will years later talk to us again with the same vividness and warmth; the wax cylinders transport us back in time to the happy days of youth—we hear the speech of those who lived countless years before us, whom we never knew, and whose names were only handed down by history.[67]

Renard's narrator clarifies such "practical advice for interested customers" by pointing out that the phonographic recording of dead friends surpasses their "cinematographic" immortalization: instead of black-and-white phantom doubles in the realm of the imaginary, bodies appear by virtue of their voices in a real that once again can only be measured in euphemisms: by carrion or skeletons. It becomes possible to conjure up friends as well as the dead "whose names were only handed down by history." Once technological media guarantee the similarity of the dead to stored data by turning them into the latter's mechanical product, the boundaries of the body, death and lust, leave the most indelible traces. According to Renard, eyes retain final visions as snapshots; according to the scientific-psychological determinations of Benedict and Ribot,[68] they even retain these visions in the shape of time-lapse photography. And if, in strict analogy, the roaring shell only replays its agony, then even the deadest of gods and goddesses achieve acoustic presence. The shell that Renard's fictitious composer listens to was not found on a natural beach; it takes the place of the mouthpieces of a telephone or a loudspeaker capable of bridging temporal distances in order to connect him with an antiquity preceding all discourse. The sound emanating from such a receiver is once again Rilke's primal sound, but as pure sexuality, as *divine clameur sexuelle*. The "rosy lips" and the "double-horned grotto" of its anatomy leave that in as little doubt as the death of the old man to whom they appear.

Thus Renard's short story introduces a long series of literary phantasms that rewrite eroticism itself under the conditions of gramophony and telephony. As a result, apparitions no longer comprise those endearing images of women whom, as Keller put it, the bitter world does not nourish; instead, the temptation of a voice has become a new partial object. In the same letter in which Kafka suggests to his fiancée and her parlograph firm that old-fashioned love letters be replaced by technical relays of telephone and parlograph,[69] he relates a dream:

> Very late, dearest, and yet I shall go to bed without deserving it. Well, I won't sleep anyway, only dream. As I did yesterday, for example, when in my dream I ran toward a bridge or some balustrading, seized two telephone receivers that happened to be lying on the parapet, put them to my ears, and kept asking for nothing but news from "Pontus"; but nothing whatever came out of the telephone except a sad, mighty, wordless song and the roar of the sea. Although well aware that it was impossible for voices to penetrate these sounds, I didn't give in, and didn't go away.[70]

News from "Pontus"—as Gerhard Neumann has shown,[71] in pretechnical days this was news from Ovid's Black Sea exile, the quintessential model for literature as a love letter. Letters of this kind, necessarily received or written in their entirety by women, were replaced by the telephone and its noise, which precedes all discourse and subsequently all whole individuals. In *La voix humaine*, Cocteau's one-act telephone play of 1930, a man and a woman at either end of a telephone line agree to burn their old love letters.[72] The new eroticism is like that of the gramophone, which, as Kafka remarked in the same letter, one "can't understand."[73] "The telephone conversation occupies the middle ground between the rendezvous and the love letter":[74] it drowns out the meaning of words with a physiological presence that no longer allows "human voices" to get through, as well as by superimposing a myriad of simultaneous conversations, which in Kafka's *The Castle*, for instance, reduces the "continual telephoning" to "humming and singing."[75] Likewise, in Renard's short story the superimposition of all the goddesses and sirens that ever existed may have resulted in white noise.

There can be no doubt that Kafka dreamed telephony in all its informational and technological precision: four days prior to his dream he read an essay by Philipp Reis in an 1863 issue of *Die Gartenlaube* on the first telephone experiments.[76] As is clear from the essay's title, "The Music Telegraph," the apparatus was built for the purpose of conveying the human voice. It failed to do so,[77] but like Kafka's imagined telephone mouthpieces it was capable of transmitting music.

Ever since Freud, psychoanalysis has been keeping a list of partial objects that, first, can be separated from the body and, second, excite desires prior to sexual differentiation: breast, mouth, and feces. Lacan added two further partial objects: voice and gaze.[78] This is psychoanalysis in the media age, for only cinema can restore the disembodied gaze, and only the telephone was able to transmit a disembodied voice. Plays like Cocteau's *La voix humaine* follow in their wake.

The only thing that remains unclear is whether media advertise partial objects or partial objects advertise the postal system. The more strategic the function of news channels, the more necessary, at least in interim peace times, the recruitment of users.

In 1980 Dieter Wellershoff published his novella *The Siren*, unfortunately without dedicating it to Renard. A professor from Cologne plans to use his sabbatical to finally complete his long-planned book on communication theory. But he never gets down to writing. An unknown woman who once witnessed Professor Elsheimer's telegenic partial objects on a TV screen starts a series of phone calls that begin like a one-sided suicide hot line and culminate in mutual telephonic masturbation.[79] Written theories of communication stand no chance against the self-advertisement of technological media. Even the most taciturn of European "civil services"[80] recruited for "the profession of telephone operator" and made it "accessible to German women," because from the very beginning its "telephone service" could not "do without" the "clear voices of women."[81]

Therefore, Professor Elsheimer's only means of escaping the spell of the telephonic-sexual mouthpiece is to use one medium to beat another medium. During the last call from the unseen siren he puts on a Bach record and pumps up the volume.[82] And lo and behold, drowned out by Old European notated music the siren magic ceases to exist. Only two technical media communicate between Cologne and Hamburg. "Here," Kafka wrote from Prague to his beloved employee of a phonograph manufacturer, "by the way, is a rather nice idea; a parlograph goes to the telephone in Berlin, while a gramophone does likewise in Prague, and these two carry on a little conversation with each other."[83]

Wellershoff's *The Siren* is an inverted replay of "The Man and the Shell." Renard's fictional composer had not yet acquired the technological skill to employ, of all pieces, the *Art of the Fugue* as a jammer in the

"When telephone and gramophone . . . " Caricature, ca. 1900.

war of the sexes. On the contrary, he wanted to transfer onto musical sheets what was no longer fugue or art: "a goddess's lusty scream," which coincided with the roaring of the sea.

It remained an impossible wish as long as it depended on the five lines of a musical staff, but that changed in the founding age of modern media. In the beginning there was, as always, Wagner, who, by courtesy of ice-cream poisoning in La Spezia, experienced an acoustic fever delirium of "swiftly running water" that suggested to him the *Rhinegold* prelude.[84] Debussy's *Sirènes* for orchestra and female voices followed in

1895, the score of which no longer dictated words, or even syllables or vowels, but sums, as if it were possible to compose the noise of channels or, as Richard Dehmel put it a year later, the "hollow din" of the "telegraph wires."[85] Between 1903 and 1905 Debussy completed the "symphonic poem" which in Renard's tale was named after a Greek sea goddess but which Debussy simply called *La Mer*. Finally, in 1907 Wagner's monotonous, ice cream–induced E-flat chord with all its overtone effects became Nerval's unwritten *Amphitrite*, that "poison for the ear."

Berliner's gramophone is to the history of music what Edison's phonograph is to the history of literature. At the price of being monopolized and mass produced by big industry, records globalized musical noise. Edison's cylinders in turn made the storage of speech a daily enjoyment, even if in each case only a very few copies could be made. As a result, literature's letter-filled papers suffered the same crisis as sheet music.

In 1916, three years before Rilke's "Primal Sound," Salomo Friedlaender delineated the new constellation of eroticism, literature, and phonography. More than any other writer of his time, Friedlaender, better known under the pseudonym Mynona (a palindrome of *anonym*), made stories again out of media history. In 1922 he published the novel *Gray Magic*, which anticipates a technological future in which women are turned into celluloid (and men, incidentally, into typewriters). In 1916 he wrote a short story that conjures up the technological past in the shape of Germany's *ur*-author in order to predict the transformation of literature into sound.

SALOMO FRIEDLAENDER, "GOETHE SPEAKS INTO THE PHONOGRAPH" (1916)

"What a pity," remarked Anna Pomke, a timid middle-class girl, "that the phonograph wasn't already invented in 1800!"

"Why?" asked Professor Abnossah Pschorr. "Dear Pomke, it is a pity that Eve didn't present it to Adam as part of her dowry for their common-law marriage; there is a lot to feel pity for, dear Pomke."

"Oh, Professor, I would have loved to listen to Goethe's voice! He is said to have had such a beautiful organ, and everything he said was so meaningful. Oh, if only he could have spoken into a phonograph! Oh! Oh!"

Long after Pomke had left, Abnossah, who had a weakness for her squeaky chubbiness, still heard her groans. Professor Pschorr, inventor of the telestylus, immersed himself in his customary inventive thoughts. Was it possible retroactively to trick that Goethe (Abnossah was ridiculously jealous) out of his voice? Whenever Goethe spoke, his voice produced vibrations as harmonious as, for example, the soft voice of your wife, dear Reader. These vibrations encounter obstacles and are reflected, resulting in a to and fro which becomes weaker in the passage of time but which does not actually cease. So the vibrations produced by Goethe are still in existence, and to bring forth Goethe's voice you only need the proper receiver to record them and a microphone to amplify their effects, by now diminished. The difficult part was the construction of the receiver. How could it be adjusted to the specific vibrations of Goethe's voice without having the latter at one's disposal? What a fascinating idea! Abnossah determined that it was necessary to conduct a thorough study of Goethe's throat. He scrutinized busts and portraits, but they provided a very vague impression at best. He was on the verge of giving up when he suddenly remembered that Goethe was still around, if only in the shape of a corpse. He immediately sent a petition to Weimar asking for permission to briefly inspect Goethe's remains for the purpose of certain measurements. The petition was rejected. What now?

Furnished with a small suitcase filled with the most delicate measuring and burglary equipment, Abnossah Pschorr proceeded to dear old Weimar; incidentally, in the first-class waiting room he happened to come across the locally known sister of the globally known brother in graceful conversation with some old Highness of Rudolfstadt. Abnossah heard her say, "Our Fritz always had a military posture, and yet he was gentle; with others he was of truly Christian tenderness—how he would have welcomed this war! And the beautiful, sacred book by Max Scheler!"

Abnossah was so shocked he fell flat on his back. He pulled himself up with difficulty and found lodgings in the "Elephant." In his room he carefully examined the instruments. Then he placed a chair in front of the mirror and tried on nothing less than a surprisingly portrait-like mask of the old Goethe. He tied it to his face and exclaimed:

"Verily, you know I am a genius,

"I may well be Goethe himself!

"Step aside, buffoon! Else I call Schiller and my prince Karl August for help, you oaf, you substitute!"

He rehearsed his speech with a deep sonorous voice.

Late at night he proceeded to the royal tomb. Modern burglars, all of whom I desire as my readers, will smile at those other readers who believe that it is impossible to break into the well-guarded Weimar royal tomb.

Please remember that as a burglar Professor Pschorr is ahead of even the most adept professional burglar. Pschorr is not only a most proficient engineer, he is also a psychophysiologist, a hypnotist, a psychologist, and a psychoanalyst. In general, it is a pity that there are so few educated criminals: if all crimes were successful, they would finally belong to the natural order of things and incur the same punishment as any other natural event. Who takes lightning to task for melting Mr. Meier's safe? Burglars such as Pschorr are superior to lightning because they are not diverted by rods.

In a single moment, Pschorr was able to give rise to horror and then immobilize those frozen in terror by using hypnosis. Imagine yourself guarding the royal tomb at midnight: suddenly the old Goethe appears and casts a spell on you that leaves only your head alive. Pschorr turned the whole guard into heads attached to trunks in suspended animation. He had about two hours before the cramp loosened, and he made good use of them. He descended into the tomb, switched on a flashlight, and soon found Goethe's sarcophagus. After a short while he was acquainted with the corpse. Piety is for those who have no other worries. It should not be held against Pschorr that he subjected Goethe's cadaver to some practical treatment; in addition, he made some wax molds and finally ensured that everything was restored to its previous state. Educated amateur criminals may be more radical than professionals, but the radicalness of their meticulous accomplishments furnishes their crimes with the aesthetic charm of a perfectly solved mathematical equation.

After leaving the tomb Pschorr added further elegance to his precision by deliberately freeing a guard from his spell and scolding him in the aforementioned manner. Then he tore the mask off his face and returned to the "Elephant" in the most leisurely fashion. He was satisfied; he had what he wanted. Early next morning he returned home.

A most active period of work began. As you know, a body can be reconstructed by using its skeleton; or at least Pschorr was able to do so. The exact reproduction of Goethe's air passage down to the vocal cords and lungs no longer posed any insurmountable difficulties. The timbre and strength of the sounds produced by these organs could be determined with utmost precision—you merely had to let a stream of air corresponding to the measurement of Goethe's lungs pass through. After a short while Goethe spoke the way he must have spoken during his lifetime.

But since it was not only a matter of recreating his voice but also of having this voice repeat the words it uttered a hundred years ago, it was necessary to place Goethe's dummy in a room in which those words had frequently been spoken.

Abnossah invited Pomke. She came and laughed at him delightfully.

"Do you want to hear him speak?"

"Whom?"

"That Goethe of yours."

"Of mine? Well, I never! Professor!"

"So you do!"

Abnossah cranked the phonograph and a voice appeared:

"Friends, oh flee the darkened chamber . . . " et cetera.

Pomke was strangely moved.

"Yes," she said hastily," that is exactly how I imagined his organ. It is so enchanting!"

"Well, now," cried Pschorr, "I do not want to deceive you, my dear. Yes, it is Goethe, his voice, his words. But it is not an actual replay of words he actually spoke. What you heard was the repetition of a possibility, not of a reality. I am, however, determined to fulfill your wish in its entirety and therefore propose a joint excursion to Weimar."

The locally known sister of the globally known brother was again sitting in the waiting room whispering to an elderly lady: "There still remains a final work by my late brother, but it will not be published until the year 2000. The world is not yet mature enough. My brother inherited his ancestor's pious reverence. But our world is frivolous and would not see the difference between a satyr and this saint. The little people in Italy saw a saint in him."

Pomke would have keeled over if Pschorr had not caught her. He blushed oddly and she gave him a charming smile. They drove straight to the Goethehaus. Hofrat Professor Böffel did the honors. Pschorr presented his request. Böffel became suspicious. "You have brought along a dummy of Goethe's larynx, a mechanical apparatus? Is that what you are saying?"

"And I request permission to install it in Goethe's study."

"Of course. But for what reason? What do you want? What is this supposed to mean? The newspapers are full of something curious, nobody knows what to make of it. The guards claim to have seen the old Goethe, he even roared at one of them. The others were so dazed by the apparition they were in need of medical attention. The incident was reported to the Archduke himself."

Anna Pomke scrutinized Pschorr. Abnossah, however, was astonished. "But what has this got to do with my request? Granted, it is very strange—maybe some actor allowed himself a joke."

"Ah! You are right, that is an explanation worth exploring. I couldn't help but think . . . But how were you able to imitate Goethe's larynx, since you could not have possibly modeled it after nature?"

"That is what I would have preferred to do, but I was unfortunately not given the permission."

"I assume that it would not have been very helpful anyway."

"Why?"

"To the best of my knowledge Goethe is dead."

"I assure you, the skeleton, in particular the skull, would suffice to assemble a precise model; at least it would suffice for me."

"Your skill is well known, Professor. But what do you need the larynx for, if I may ask?"

"I want to reproduce the timbre of the Goethean organ as deceptively close to nature as possible."

"And you have the model?"

"Here!"

Abnossah snapped open a case. Böffel uttered an odd scream. Pomke smiled proudly.

"But you could not have modeled this larynx on the skeleton?" cried Böffel.

"Almost! It is based on certain life-size and lifelike busts and pictures; I am very skilled in these matters."

"As we all know! But why do you want to set up this model in Goethe's former study?"

"He conceivably articulated certain interesting things there; and because the acoustic vibration of his words, though naturally in an extremely diminished state, are still to be found there—"

"You believe so?"

"It's not a question of belief, it's a fact."

"Yes?"

"Yes!"

"So what do you want to do?"

"I want to suck those vibrations through the larynx."

"Pardon me?"

"What I just told you!"

"What an idea—I apologize, but you can hardly expect me to take this seriously."

"Which is why I have to insist all the more forcefully that you give me the opportunity to convince you of the seriousness of this matter. I am at a loss to understand your resistance; after all, this harmless machine won't cause any damage!"

"I'm sure it won't. I am not at all resisting you, but I am officially obliged to ask you a number of questions. I do hope you won't hold it against me?"

"Heaven forbid!"

In the presence of Anna Pomke, Professor Böffel, and a couple of curious assistants and servants, the following scene unfolded in Goethe's study:

Pschorr placed his model on a tripod, ensuring that the mouth occupied the same position as Goethe's had when he was sitting. Then Pschorr pulled a kind of rubber air cushion out of his pocket and closed the nose and mouth of the model with one of its ends. He unfolded the cushion and spread it like a blanket over a small table he had pulled up to the tripod. On this blanket (as it were) he placed a most enchanting miniature phonograph complete with microphone that he had removed from his case. He now carefully wrapped the blanket around the phonograph, leaving a second opening facing the mouth in the shape of an end into which he screwed a pair of bellows. These, he explained, were not to blow air into but to suck it out of the mouth.

When I, as it were, let the nasopharyngeal cavity exhale as it does during speech, Pschorr lectured, this specifically Goethean larynx functions like a sieve that only lets through the acoustic vibrations of Goethe's voice, if there are any; and there are bound to be. The machine is equipped with an amplifier should they be weak.

The buzz of the recording phonograph could be heard inside the rubber cushion. And then an inescapable feeling of horror upon hearing an indistinct, barely audible whispering. "Oh, my God!" Pomke said, holding her delicate ear against the rubber skin. She started. A rasping murmur came from the inside: "As I have said, my dear Eckermann, this Newton was blind with his seeing eyes. How often, my friend, do we catch sight of this when faced with something that appears to be so obvious! Therefore it is in particular the eye and its perceptions which demand the fullest attention of our critical faculties. Without these we cannot arrive at any sensible conclusion. Yet the world mocks judgment, it mocks reason. What it, in truth, desires is uncritical sensation. Many a time have I painfully experienced this, yet I have not grown tired of contradicting the world and, in my own way, setting my words against Newton's."

Pomke heard this with jubilant horror. She trembled and said: "Divine! Divine! Professor, I owe to you the most beautiful moment of my life."

"Did you hear something?"

"Certainly. Quiet, but very distinct!"

Pschorr nodded contentedly. He worked the bellows for a little while and then said, "That should be enough for now."

He put all the instruments back into his case with the exception of the phonograph. All those present were eager and excited. Böffel asked, "Professor, do you honestly believe that you have actually captured words once spoken by Goethe? Real echoes from Goethe's own mouth?"

"I do not only believe so, I am certain of it. I will now replay the phonograph with the microphone and predict that you will have to agree with me."

The familiar hissing, hemming, and squeezing. Then the sound of a remarkable voice that electrified everybody, including Abnossah. They listened to the words quoted above. Then it continued: "Oh ho! So, he, Newton, saw it! Did he indeed? The continuous color spectrum? I, dear friend, I shall reiterate that he was deceived: that he was witness to an optical illusion and accepted it uncritically, glad to resume his counting and measuring and splitting of hairs. To hell with his monism, his continuity; it is precisely the contrast of colors that makes them appear in the first place! Eckermann! Eckermann! Hold your horses! White—neither does it yield any color nor do other colors add up to white. Rather, in order to obtain gray, white must be mechanically combined with black, and it has to be chemically united with gray to produce the varied gray of the other colors. You will never obtain white by neutralizing colors. It merely serves to restore the original contrast of black and white: and of course white is the only one that can be seen in all its brightness. But I, dear friend, I see darkness just as clearly, and if Newton only hit upon white, I, most esteemed comrade, also hit upon black. I should think that a former archer like yourself would greatly appreciate such a feat! That is the way it is, and so be it! From me our distant grandchildren and great-grandchildren populating this absurd world will learn to laugh at Newton!"

Böffel had sat down while everybody was cheering. The servants trampled with delight, like students in the fiery lectures of that upright and demonic graybeard, the smashingly revolutionary, lordly Reucken. But Abnossah sternly said, "Gentlemen! You are interrupting Goethe! He isn't finished yet!"

Silence resumed and the voice continued: "No, Sir, no and again no! Of course you could have if you had so desired! It is the will, the will of these Newtonians, that is pernicious; and a faulty will is a corruptive faculty, an active inability that I abhor even though I catch sight of it everywhere and should be accustomed to it. You may consider it harmless, but the will is the true contriver of all things great and small; it is not the divine power but the will, the divine will, that thwarts man and proves his inadequacy. If you were able to desire in a godlike way, dear friend, the ability would be necessary and not just easy to come by, and a lot of what now dare not show its face for fear of meeting hostility or ridicule would become everyday experience.

"Consider young Schopenhauer, a lad of supreme promise, full of the most magnificent desires, but afflicted by the rot of abundance, by his own insatiability. In the theory of colors he was blinded by the sun to the extent that he did not accept the night as another sun but rather deemed it null and void; likewise, he was captivated by the luster of life in its wholeness, in contrast to which human life struck him as worthless. Behold, Sir, that the purest, most divine will is in danger of failure if it is bent on persisting at all cost; if it is not prepared to wisely and gracefully take into account the exterior conditions as well as the limitations of its own means! Indeed: the will is indeed a magician! Is there anything it cannot do? But the human will is not a will, it is a bad will. Ha! Haha! Hee! Hee!"

Goethe laughed mysteriously and continued in a whisper:

"Very well then, my dearest friend, I shall entrust, indeed reveal, something to you. You will judge it a fairy tale, but to me it has attained the utmost clarity. Your own will can vanquish fate; it can make fate its servant provided—and now listen closely—it does not presume that the tremendous and divinely tense creative intent and exertion within should also be clearly manifest without, especially in a most intense display of muscular strain. Behold earth as it is turned and driven! What mundane industry! What ceaseless motion! But mark my words, Eckermann! It is no more than mundane diligence, nothing but a fatally mechanical driving—while the vibrating, magical will of the sun rests within itself and by virtue of this supreme self-sufficiency gives rise to the electromagnetism that humbles the whole army of planets, moons, and comets into servile submission at its feet. O friend, to understand, to experience and be, in the most serenely spiritual sense of the word, that sublime culprit!—Enough, let us leave it at that. I was accustomed to discipline myself whenever I heard others, and sometimes even Schiller, rhapsodize freely, out of love for such a divine activity, in the face of which one should be silent, because all discourse would not only be useless and superfluous but indeed harmful and obstructive by creating a ridiculously profane understanding, if not the most decisive misunderstanding. Remember this, my friend, and keep it in your heart without attempting to unravel the mystery! Trust that in time it will unravel itself, and this evening go to the theater with Little Wolf, who is eager to go, and do not treat Kotzebue too harshly even though he disgusts us!"

"Oh, God," Pomke said, while the others eagerly congratulated

Abnossah, "oh, God! If only I could listen forever! How much Eckermann withheld from us!"

After a long while a snoring emanated from the machine, then nothing! "Gentlemen!" Abnossah said, "as you can hear, Goethe is obviously asleep. It makes little sense to wait around; there is nothing to expect for a couple of hours, if not for an entire day. Staying around is useless. As you no doubt realize, the apparatus adheres closely to real time. In the most fortunate case we might hear something should Eckermann have returned to Goethe following the performance. I, for one, do not have the time to wait around for that to happen."

"How is it," the slightly skeptical Böffel asked, "that, of all speeches, we were able to listen to this one?"

"Pure chance," Pschorr responded. "The conditions, in particular the makeup of the machine and its positioning, happened to correspond to these and no other sound vibrations. I only took into account the fact that Goethe was sitting and the location of his chair."

"Oh, please, please! Abnossah!" (Pomke, almost maenadic, was as if in a trance; for the first time she called him by his first name.) "Try it somewhere else! I can't hear enough of it—and even if it is only snoring!"

Abnossah put away the machine and locked the suitcase. He had become very pale: "My dear Anna—Madame," he corrected himself, "—another time." (Jealousy of the old Goethe was eating him up inside.)

"How about Schiller's skull?" Böffel asked. "It would decide the dispute over whether it is the real one."

"Indeed," Abnossah responded, "for if we heard Schiller, the Swabian, say in a broad Hessian accent, 'How about a glass of wine?' it wouldn't be Schiller's skull.—I am wondering if the invention couldn't be refined. Maybe I could manufacture a generic larynx that could be adjusted like an opera glass in order to be aligned with all kinds of possible vibrations. We could listen to antiquity and the Middle Ages and determine the correct pronunciation of old idioms. And respected fellow citizens who say indecent things out loud could be handed over to the police."

Abnossah offered Pomke his arm and they returned to the station. They cautiously entered the waiting room, but the locally known one had already left. "What if she let me have the larynx of her famous brother? But she won't do it; she'll claim that the people aren't mature enough and that the literati lack the reverence of the people, and that nothing can be done. Beloved! Beloved! For (oh!) that! That is! That is what you are!"

But Pomke wasn't listening. She appeared to be dreaming.

"How he stresses the *r*s!" she whispered apprehensively.

Abnossah angrily blew his nose; Anna started and asked him distractedly: "You were saying, dear Pschorr?! I am neglecting the master for his work! But the world subsides when I hear Goethe's own voice!"

They boarded the coach for their return journey. Pomke said nothing; Abnossah was brooding silently. After they had passed Halle, he threw the little suitcase with Goethe's larynx out of the window in front of an approaching train. "What have you done?" Pomke shrieked.

"Loved," Pschorr sighed, "and soon I will have lived—and destroyed my victorious rival, Goethe's larynx."

Pomke blushed furiously; laughing, she threw herself vigorously into Abnossah's tightly embracing arms. At that moment the conductor entered and requested the tickets.

"God! Nossah!" murmured Pomke. "You have to get me a new larynx of Goethe, you have to—or else—"

"No or else! Après les noces, my dove!"

Prof. Dr. Abnossah Pschorr
Anna Pschorr, née Pomke
Just married
Currently at the "Elephant" in Weimar

This wedding announcement is truly a happy ending: it puts an end to Classic-Romantic poetry. In 1916 even "timid middle-class girls" like Anna Pomke come under the influence of professors like Pschorr, who as one of the "most proficient" engineers of his day obviously teaches at the new technological institutes so vigorously promoted by Emperor Wilhelm II. Marriage to an engineer vanquishes the middle-class girl's infatuation with Goethe, which lyceums had been systematically drilling into them for over a century.[86] What disappears is nothing less than *The Determination of Women for Higher Intellectual Development*. Under this title, a certain Amalie Holst demanded in 1802 the establishment of girls' schools responsible for turning women into mothers and readers of poets.[87] Without the Anna Pomkes there would have been no German Classicism, and none of its principally male authors would have risen to fame.

Consequently, Pomke can only think of the old century when confronted with the technological innovations of the new one. As if to prove that the Soul or Woman of Classicism and Romanticism was an effect of automata, she laments the unstored disappearance of Goethe's voice with

the very same sigh, "oh" (*ach*), uttered by the talking robot Olympia in Hoffmann's *Sandman*, a sigh that, though it is the only word it can speak, suffices to underscore its soul. In Hegel's words, a female sigh, or a "disappearance of being in the act of being," loves a male poetic capability, or a "disappearance of being in the act of being." And as if to prove that the voice is a partial object, Pomke praises Goethe's voice as "a beautiful organ." Which not coincidentally makes the "psychiatrist" and "psychoanalyst" Professor Pschorr "jealous," for all the power Classical authors had over their female readers rested in the erection of that organ.

Not that middle-class girls were able to hear their master's voice. There were no phonographs "around 1800," and therefore none of the canine obedience for a real that became the trademark of Berliner's gramophone company in 1902. Unlike that of Nipper, the dog that started sniffing at the bell-mouth of the phonograph upon hearing its dead master's voice, and whose vocal-physiological loyalty was captured in oil by the painter Francis Barraud, the brother of the deceased, the loyalty of female Classic-Romantic readers was restricted to the imaginary—to their so-called imagination. They were forced to hallucinate Goethe's voice between the silent lines of his writing. It was not a coincidence that Friedrich Schlegel wrote to a woman and lover that "one seems to hear what one is merely reading." In order for Schlegel wholly to become an author himself, women had to become readers and "appreciate the sacredness of words more than in the past."[88]

"To the extent that graphism"—that is, in the shape of alphabetic writing—"is flattened onto the voice" (while in tribal cultures "it was inscribed flush with the body"), "body representation subordinates itself to word representation." But this "flattening induces a fictitious voice from on high that no longer expresses itself except in the linear flux,"[89] because at least since Gutenberg it has announced the decrees of national bureaucracies.

Thus Anna Pomke's loving sigh confirms the theory of media and writing of the *Anti-Oedipus*.

Once the beautiful and fictitious, monstrous and unique organ of the poet-bureaucrat Goethe, which commanded an entire literary epoch, rose as an acoustic hallucination from the lines of his poems, things proceeded as desired. In 1819, Hoffmann's fairy tale *Little Zaches* noted what "extravagant poets . . . ask for": "First of all, they want the young lady to get into a state of somnambulistic rapture over everything they utter, to sigh deeply, roll her eyes, and occasionally to faint a trifle, or even to go blind for a moment at the peak of the most feminine femininity. Then the aforesaid young lady must sing the poet's songs to the melody that streams forth from her heart"[90] and, finally, in the *Anti-Oedipus*, reveal the secret of its media technology: that it is a fictitious elevated phallus born from the alphabet.

For timid middle-class girls, however, everything depended on literally going "blind" when faced with the materiality of printed letters; otherwise, they could not have provided them with a melody in the imaginary (or at the piano) from their hearts. In doing so, they surrendered unconditionally to the desires of Classic-Romantic poets. "Oh," Anna Pomke sighs from the bottom of her heart, "if only he could have spoken into a phonograph! Oh! Oh!"

A sigh that will hardly reach the ears of engineers. Pschorr can only discern a "groan" in her "oh," mere vocal physiology instead of a heart. Around 1900, love's wholeness disintegrates into the partial objects of particular drives identified by Freud. Phonographs do not only store—like Kempelen's vowel machine or Hoffmann's Olympia—the one signified, or trademark, of the soul. They are good for any kind of noise, from Edison's hearing-impaired screaming to Goethe's fine organ. With the demise of writing's storage monopoly comes to an end a love that was not only one of literature's many possible subjects but also its very own media technology: since 1800 perfectly alphabetized female readers have been able to endow letters with a beloved voice. But tracing primal sounds has, as Rilke put it, nothing to do with "the presence of mind and grace of love."

As a modern engineer who wants to spread his knowledge using everyday language, Professor Pschorr minces no words: "Whenever Goethe spoke, his voice produced vibrations as harmonious as, for example, the soft voice of your wife, dear Reader." However, the fact that what Goethe had to say was "meaningful" enough to fill the 144 volumes of the Großherzogin-Sophien edition is irrelevant. Once again notions of frequency are victorious over works, heartfelt melodies, and signifieds. As if commenting on Pschorr, Rudolph Lothar writes at the outset of his *The Talking Machine: A Technical-Aesthetic Essay*:

> Everything flows, Heraclitus says, and in light of our modern worldview we may add: everything flows in waves. Whatever happens in the world, whatever we call life or history, whatever occurs as a natural phenomenon—everything transpires in the shape of waves.
>
> Rhythm is the most supreme and sacred law of the universe; the wave phenomenon is the primal and universal phenomenon.
>
> Light, magnetism, electricity, temperature, and finally sound are nothing but wave motions, undulations, or vibrations. . . .
>
> The unit of measurement for all wave motions is the meter, the unit of time is the second. Frequencies are the vibrations registered within a meter per second. The frequencies of light, electricity, and magnetism are taken to be identical; with approximately 700 trillion vibrations per second, their speed of propagation is 300 million meters per second.
>
> Sound vibrations exhibit significantly lower frequencies than those described above. The speed of propagation for sound is 332 meters per second. The deepest sound audible to human ears hovers around 8 vibrations, the highest around 40,000.[91]

The new appreciation of waves, those very un-Goethean "primal and universal phenomena," can even result in a poetry that once more stresses the wavelike nature of all that occurs, as in the sonnet "Radio Wave," which the factory carpenter Karl August Düppengiesser of Stolberg submitted to Radio Cologne in 1928:

> Wave, be aware of your many shapes,
> and, all-embracing, weave
> at the world's wheel, entrusted from above,
> the new and wider spirit of the human race.[92]

But engineers like Pschorr are ahead of "other people," even radio wave poets: their "spirits hail"—to quote the engineer-poet Max Eyth—"not from the world that was but from the one that will be." It is more efficient to use waves "to make things that were never made before"[93] than to write sonnets about their many shapes. Pschorr makes use of laws of nature that, unlike the *Panta rei* of Heraclitus or of Goethe's "Permanence in Transition," are valid regardless of the reputation of so-called personalities, because they are based on measurements. The law of waves does not exclude the author of "Permanence in Transition." And because the frequency spectrum and transmission speed of sound are so low, they are easy to measure. (To posthumously film Goethe would require technologies capable of recording in the terahertz range.)

With mathematical precision Pschorr recognizes the frequency of human voices to be a negative exponential function whose value, even after centuries, cannot be zero. In the phonographic realm of the dead, spirits are always present—as sound signal amplitudes "in an extremely diminished state." "Speech has become, as it were, immortal," *Scientific American* pronounced immediately after Edison's invention under the headline, "A Wonderful Invention—Speech Capable of Infinite Repetitions from Automatic Records."[94]

But although he invented a relatively sensitive powder microphone (as opposed to Hughes's carbon microphone), Edison was not able to access the dead. Because it was only equipped with a mechanical amplifier, his phonograph could do no more than record the last gasps of the dying—by using resonance in the recording bell-mouth. The low voltage output of his microphone was increased somewhat by a relayed inductive circuit, but it never approached the recording needle of the phonograph. Goethe's bass frequencies, vibrating in infinity between 100 and 400 hertz in his Weimar abode, remained unmeasurable. A catastrophic signal-to-noise ratio would have rendered all recordings worthless and, at best, provided primal sounds instead of Goethean diction.

Pschorr's optimism, therefore, rests on more advanced technologies. "A microphone to amplify" the "by now diminished" effects of Goethe's voice depends upon the necessary but suppressed premise that infinite amplification factors could be applied. This became possible with Lieben's work of 1906 and De Forest's of 1907. Lieben's controlled hot-cathode tube, in which the amplitude fluctuations of a speech signal influence the cathode current, and De Forest's audion detector, which added a third electrode to the circuit, stood at the beginning of all radio technology.[95] The electrification of the gramophone is due to them as well. Pschorr's miraculous microphone could only have worked with the help of tube-type technology. Short stories of 1916 require the most up-to-date technologies.

Pschorr has other problems. His concerns revolve around filtering, not amplification. Isolated from the word salad produced by visitors to the Goethehaus from Schiller to Kafka, his beloved is supposed to receive only her master's voice. Pschorr's solution is as simple as it is Rilkean: he, too, links media technology and physiology, that is, a phonograph and a skull. As the first precursor of the revolutionary media poets Brecht and Enzensberger, Pschorr assumes that transmitter and receiver are in principle reversible: just as "every transistor radio is, by the nature of its construction, at the same time a potential transmitter,"[96] and, conversely, any microphone a potential miniature speaker, even Goethe's larynx can be operated in normal and inverse fashion. Since speaking is no more than the physiological filtering of breath or noise, and the entry and exit of band-pass filters are interchangeable, the larynx will admit only those frequency mixtures which once escaped from it.

The one thing left for Professor Pschorr to do to implement this selectivity technologically is to grasp the difference between arts and media. His early idea of fashioning a model of Goethe's larynx based on "pictures and busts" is doomed to failure, simply because art, be it painting or sculpture, only conveys "very vague impressions" of bodies.

Malte Laurids Brigge, the hero of Rilke's contemporaneous novel, is asked by his father's doctors to leave the room while they (in accordance with the master of the hunt's last request) perform a "perforation of the heart" on the corpse. But Brigge stays and watches the operation. His reason: "No, no, nothing in the world can one imagine beforehand, not the least thing. Everything is made up of so many unique particulars that cannot be foreseen. In imagination one passes them over and does not notice that they are lacking, hasty as one is. But the realities are slow and indescribably detailed."[97]

From imagination to data processing, from the arts to the particulars of information technology and physiology—that is the historic shift of 1900 which Abnossah Pschorr must comprehend as well. He finds himself, not unlike Brigge at the deathbed of his father and Rilke at the Parisian Ecole des Beaux-Arts, in the company of corpses. His profane illumination, after all, is that "Goethe was still around, if only in the shape of a corpse." Once more, the real replaces the symbolic—those allegedly "life-size and lifelike busts and pictures" that only a Goethehaus director such as Hofrat Böffel could mistake for anatomical exhibits.

The reconstructed respiratory system of a corpse as a band-pass filter, a microphone- and tube-type-enhanced phonograph as a storage medium—Pschorr is ready to go to work. He has engineered a crucial link between physiology and technology, the principal connection that served as the basis for Rilke's "Primal Sound" and all media conceptions at the turn of the century. Only today's ubiquitous digitization can afford to do without such "radicalness," which in Pschorr's case consisted in short-circuiting "cadavers" and machines. Once the stochastics of the real allow for encipherment, that is to say, for algorithms, Turing's laconic statement that there would be "little point in trying to make a 'thinking machine' more human by dressing it up in artificial flesh"[98] is validated.

In the founding days of media technology, however, everything centered on links between flesh and machine. In order to implement technologically (and thus render superfluous) the functions of the central nervous system, it first had to be reconstructed. Rilke's and Pschorr's projects are far removed from fiction.

To begin with, Scott's membrane phonautograph of 1857 was in all its parts a reconstructed ear. The membrane was derived from the eardrum and the stylus with the attached bristle from the ossicle.[99]

Second, "in 1839 the 'great Rhenish physiologist' and conversation partner of Goethe, Johannes Müller, had removed the larynx from various corpses—the acquisition of which tended to be rather adventurous affairs—in order to study *in concreto* how specific vowel sounds were produced. When Müller blew into a larynx, it sounded 'like a fairground whistle with a rubber membrane.' Thus the real answered from dismembered bodies."[100] And thus, with his adventurous acquisition of parts of Goethe's corpse from the sanctuary of the royal tomb, Pschorr perfected experiments undertaken by Goethe's own conversation partner.

Third (and to remain close to Goethe and Pschorr), on September 6, 1839, the Frankfurt birthplace of Germany's primal author witnessed a bold experiment. Philipp Reis had just finished his second lecture on telephone experiments when "Dr. Vogler, the savior of the Goethehaus and founder of the Freie Deutsche Hochschulstift, presented the telephone to Emperor Joseph of Austria and King Maximilian of Bavaria, who were both in Frankfurt attending the royal council."[101] As if the historic shift from literature to media technology had to be localized.

But as Reis himself wrote, his telephone produced "the vibrations of curves that were identical to those of a sound or a mixture of sounds," since "our ear can only perceive what can be represented by similar curves; and this, in turn, is sufficient to make us conscious of any sound or mixture of sounds." However, in spite of all theoretical lucidity, Reis "had not been able to reproduce a human voice with sufficient clarity."[102] Which is why, fourthly and finally, Alexander Graham Bell had to intervene.

A telephone ready for serial production and capable of transmitting not just Reis's musical telegraphy or Kafka's sound of the sea but speeches "in a clarity satisfactory to most everybody" did not exist until 1876.

Two years earlier, the technician Bell, son of a phonetician, had consulted a physiologist and otologist. Clarence John Blake, MD, acquired two middle ears from the Massachusetts Eye and Ear Infirmary. And once Bell realized that "such a thin and delicate membrane" as the eardrum "could move bones that were, relatively to it, very massive indeed," the technological breakthrough was achieved. "At once the conception of a membrane speaking telephone became complete in my mind; for I saw that a similar instrument to that used as a transmitter could also be employed as a receiver."[103]

It is precisely this interchangeability which decades later was to strike Pschorr, Brecht, Enzensberger, *e tutti quanti*. Which is why Bell and Blake did not hesitate to undertake the last step: in the course of a single experimental procedure they coupled technology with physiology, steel with flesh, a phonautograph with body parts. Wherever phones are ringing, a ghost resides in the receiver.

And there is no reason to spare the most illustrious organ in German literature. Pschorr simply reverses the experiment of Blake and Bell a second time: the larynx as the transmitting organ replaces the ear, the receiving organ. And while Pschorr turns the handle, Goethe's reconstructed corpse voices Goethe's verses. As if the "darkened chamber" from which all "friends" are to flee were a grave known as the book.

So far, so good. Anatomical and technical reconstructions of language do not belong to fiction as long as they remain within Pschorr's exactly delineated boundaries: as the "repetition of a possibility, not of a reality." Immediately prior to Pschorr's reconstruction, Ferdinand de Saussure had based a new linguistics on the difference between *langue* and *parole*, language and speech, the possible combinations from a repository of signs and factual utterances.[104] Once it was clear how many phonemes and what distinctive qualities made up Goethe's dialect, any conceivable sentence (and not only the "Tame Xenium" chosen by Pschorr) could be generated. That is all there is to the concept of *langue*.

Once Saussure's *Cours de linguistique générale* turned into a general algorithm of speech analysis and production, microprocessors could extract the phonemic repository of speakers from their speeches without having to fear, as did the media-technological heroes of yore, the blood and poison of corpses. A Turing machine no longer needs artificial flesh. The analog signal is simply digitized, processed through a recursive digital filter, and its autocorrelation coefficients calculated and electronically stored. An analysis that continues Pschorr's band-pass filtering with more advanced means. A second step may involve all kinds of linguistic syntheses—once again the "repetition of a possibility" that computing logic has extracted from language. Instead of lungs and vocal chords we have two digital oscillators, a noise generator for unvoiced consonants and a controlled frequency generator for vowels or voiced sounds. Just as in human speech, a binary decision determines which of the two oscillators connects with the recursive filter. In turn, the autocorrelation coefficients derived from the speech analyses are by way of linear prediction directed towards the filters, an electronic simulation of the oral and nasopharyngeal cavity with all its echoes and running times. Now we only need a simple low-pass filter to translate the signal flow back into analog signals[105]—and we are all as "strangely moved" or "deceived" by the arriving phoneme sequences as Anna Pomke.

But Pschorr wants more. In order to fulfill the desire of timid middle-class girls in its "entirety," he attempts an "actual replay of words actually spoken by Goethe." As if, half a century before Foucault, it were a matter of discourse analysis. As is known, *The Archaeology of Knowledge* is based on the Saussurian notion of language as "a finite body of rules that authorizes an infinite number of performances." "The field of discursive events, on the other hand, is a grouping that is always finite and limited at any moment to the linguistic sequences that have been formulated."[106] Statements, then, "necessarily obey" a "materiality" that "defines possibilities of reinscription and transcription,"[107] as in Pschorr's real repetition.

But how discourse repetition exactly is to be achieved remains (at least in Pschorr's case) a professional secret. For once, Hofrat Böffel's skeptical inquiry, why "of all speeches we were able to listen to this one," is justified. After all, the air is full of sound waves caused by decades of Goethean speechifying. Citing Pschorr, another of Friedlaender's heroes claims that "all the waves of all bygone events are still oscillating in space."[108] Pschorr's phonograph is confronted with a parallel data input that it would first have to convert into a serial arrangement, lest the sum of all Goethean discourses appear as so much white noise on the cylinder.

Stochastic signal analyses such as linear prediction or autocorrelation measurement may enable a technologically enhanced future to assign a time axis even to past events, provided that signal processors have been programmed with certain parameters concerning the language, vocabulary, conversation topics, and so forth, of the object under investigation. The chip production of not–von Neumann machines has begun. But no machine in 1916 could have "adhered so closely" to real time as to have captured Goethe's words in the exact sequence in which they were spoken in the course of one particular evening.

Which merely serves to show that all this electronic discourse proves the obvious: Friedlaender fabricated Goethe's phonographed speech. Mynona, the most nameless of authors, outdoes the most illustrious author by putting new words into his mouth. According to Goethe, literature was a "fragment of fragments," because "the least of what had happened and of what had been spoken was written down," and "of what had been written down, only the smallest fraction was preserved." According to Friedlaender, literature in the media age is potentially everything. His hero could supplement all the conversations Eckermann allegedly "withheld from us."

Especially a chapter from the *Theory of Colors* that (in spite of a common contempt for Newton) has more to do with Friedlaender than with Goethe. Friedlaender borrowed the *Übermensch* notion that "one's own will," united with the "magical sun-will," can "overpower fate" from his teacher Dr. Marcus, who in turn borrowed it from Kant. "We are at the dawn of the magic of reason; it will make a machine of nature itself,"[109] proclaims Dr. Sucram, the hero of Friedlaender's cinema novel and whose name is a palindrome of Marcus, while turning Goethe's theory of color into *Gray Magic*, that is to say, the world into film.

At the same time that technology (to quote Sucram's antagonist, the film producer Morvitius) finally "moves from magic to machine,"[110] philosophy becomes delirious. Machines are supposed to turn back into magic. Pschorr and Sucram are inspired by a technified version of Kant's pure forms of intuition. "All that happens falls into accidental, unintentional receivers. It is stored, photographed, and phonographed by nature itself." United with the spatial and temporal forms of intuition, "these accidental receivers only need to be turned into intentional ones in order to visualize—especially cinematographically, Morvitius—the entire past."[111]

Loyally and deliriously, Friedlaender's philosophy follows in the wake of media technology. On May 19, 1900, Otto Wiener delivered his highly appropriate inaugural lecture on "the extension of our senses" by instruments. As with Friedlaender, his point of departure was the recognition that "in principle it would not be difficult to take stock of our entire knowledge by using self-recording machines and other automatic devices, thus creating a physical museum of automata." This museum would even be able to inform extraterrestrial intelligence of "the level of our knowledge." In conclusion, however, Wiener declared that the "Kantian notions of the a priori nature of the perception of time and space are unnecessary."[112] Media render Man, "that sublime culprit in the most serenely spiritual sense" of his philosophy, superfluous.

Which is why Friedlaender has Goethe's philosophical journey commence with "hissing, hemming, and squeezing," only to end in "snoring." It may not be as random and mathematical as the "perfectly even and uninformative hiss" into which Turing's vocoder turned the radio speech of his commander in chief, but Goethe's "actually recorded" voice, too, belongs to the real. The fictional elevated phallus shrivels up. And once Pschorr has train wheels "defeat his victorious rival, Goethe's larynx," the engineer has finally beaten the author.

"The new phonograph," Edison told the staff of *Scientific American* in 1887, "is to be used for taking dictation, for taking testimony in court, for reporting speeches, for the reproduction of vocal music, for teaching languages," as well as "for correspondence, for civil and military orders" and for "the distribution of the songs of great singers, sermons and speeches, the words of great men and women."[113] Which is why since 1887 those great men and women have been able to do without body snatchers like Pschorr.

To secure the worldwide distribution of these possibilities, Edison sent representatives into all the countries of the Old World. In England, the "willing victims" who "immortalized their voices in wax" included Prime Minister Gladstone, an Edison admirer of long standing, and the poets Tennyson and Browning. In Germany, Edison recruited Bismarck and Brahms, who by recording one of his Hungarian rhapsodies removed it from the whimsy of future conductors.[114] The young emperor Wilhelm II, however, did more than merely provide his voice. He inquired about all the machine's technical details, had it disassembled in his presence, then pushed aside Edison's representative and took it on himself to conduct the assembly and presentation in the presence of an astonished court.[115] The military command—to freely paraphrase Edison—entered the age of technology.

And it was only after the heroic action of their emperor—who for reasons obviously related to naval strategy had studied radio telephony,[116] founded the Telefunken company, and in what almost amounted to military prophecy prompted the construction of the AVUS as the first highway[117]—that Germany's writers paid attention to the alphabetless trace. In 1897, the foreign office legation council and Wilhelmine state poet Ernst von Wildenbruch may have been the first to record a cylinder.

Wildenbruch wrote a poem expressly for the occasion, "For the Phonographic Recording of His Voice." The history of its transmission says it all: it is not collected in the *Collected Works*. Professor Walter Bruch, who as chief engineer of AEG-Telefunken and inventor of the PAL television system had access to the archives of historical recordings, had to transcribe Wildenbruch's verses from the roll. They are quoted here in a format that will horrify poets, compositors, and literary scholars.

> Das Antlitz des Menschen läßt sich gestalten, sein Auge im Bilde fest sich halten, die Stimme nur, die im Hauch entsteht, die körperlose vergeht und verweht.
> Das Antlitz kann schmeichelnd das Auge betrügen, der Klang der Stimme kann nicht betrügen, darum erscheint mir der Phonograph als der Seele wahrhafter Photograph,
> Der das Verborgne zutage bringt und das Vergangne zu reden zwingt. Vernehmt denn aus dem Klang von diesem Spruch die Seele von Ernst von Wildenbruch.

> We may model the human visage, and hold the eye fast in an image, but the bodiless voice, borne by air, must fade away and disappear.
> The fawning face can deceive the eye, the sound of the voice can never lie; thus it seems to me the phonograph is the soul's true photograph,
> Which brings to light what is suppressed and makes the past speak at our behest. So listen to the sound of what I declare, and Ernst von Wildenbruch's soul will be laid bare.[118]

Even the copious writer Wildenbruch did not always rhyme so poorly. His phonographic verses sound as if they had been improvised in front of the bell-mouth without the benefit of any written draft. For the first time since time immemorial, when minstrels combined their formulaic or memorized words into entire epics, bards were in demand again. Which is why Wildenbruch was bereft of written language.

Poetry, the last philosopher and first media theorist Nietzsche wrote, is, like literature, in general simply a mnemotechnology. In 1882, *The Gay Science* remarked under the heading "On the Origin of Poetry":

> In those ancient times in which poetry came into existence, the aim was utility, and actually a very great utility. When one lets rhythm permeate speech—the rhythmic force that reorders all the atoms of the sentence, bids one choose one's words with care, and gives one's thoughts a new colour, making them darker, stranger, and more remote—the utility in question was *superstitious*. Rhythm was meant to impress the gods more deeply with a human petition, for it was noticed that men remember a verse much better than ordinary speech. It was also believed

that a rhythmic tick-tock was audible over greater distances; a rhythmical prayer was supposed to get closer to the ears of the gods.[119]

At the origin of poetry, with its beats, rhythms (and, in modern European languages, rhymes), were technological problems and a solution that came about under oral conditions. Unrecognized by all philosophical aesthetics, the storage capacity of memory was to be increased and the signal-to-noise ratio of channels improved. (Humans are so forgetful and gods so hard of hearing.) The fact that verses could be written down hardly changed this necessity. Texts stored by the medium of the book were still supposed to find their way back to the ears and hearts of their recipients in order to attain (not unlike the way Freud or Anna Pomke had envisioned it) the indestructibility of a desire.

These necessities are obliterated by the possibility of technological sound storage. It suddenly becomes superfluous to employ a rhythmical tick-tock (as in Greece) or rhyme (as in Europe) to endow words with a duration beyond their evanescence. Edison's talking machine stores the most disordered sentence atoms and its cylinders transport them over the greatest distances. The poet Charles Cros may have immortalized the invention of his phonograph, precisely because he was never able to build it, in lyrical rhymes under the proud title "Inscription"—Wildenbruch, that plain consumer, is in a different position. "For the Phonographic Recording of His Voice" no longer requires any poetic means. Rather than dying and fading away, his voice reaches one of today's engineers. Technology triumphs over mnemotechnology. And the death bell tolls for poetry, which for so long had been the love of so many.

Under these circumstances writers are left with few options. They can, like Mallarmé or Stefan George, exorcise the imaginary voices from between the lines and inaugurate a cult of and for letter fetishists, in which case poetry becomes a form of typographically optimized blackness on exorbitantly expensive white paper: *un coup de dés* or a throw of the dice.[120] Or for marketing reasons they can move from imaginary voices, such as those Anna Pomke had hallucinated in Goethe's verses, to real ones, in which case a poetry of nameless songwriters appears, or reappears, on records. Illiterates in particular are their prime consumers, because what under oral conditions required at least some kind of mnemotechnology is now fully automatized. "The more complicated the technology, the simpler," that is, the more forgetful, "we can live."[121] Records turn and turn until phonographic inscriptions inscribe themselves into brain physiology. We all know hits and rock songs by heart precisely because there is no reason to memorize them anymore.

To provide a demographically exact account of *White-Collar Workers*, including their nocturnal activities, Siegfried Kracauer becomes acquainted with a typist, "for whom it is characteristic that she cannot hear a piece of music in a dance hall or a suburban café without chirping along its text. But it is not as if she knows all the hits; rather, the hits know her, they catch up with her, killing her softly."[122]

Only two years or steps separate this sociology "from the newest Germany" from fictional heroes such as those in Irmgard Keun's *Rayon Girl* of 1932, who (obviously under the influence of Kracauer) turn into poets (and in Berlin into prostitutes) when listening to the gramophone or the radio. For it is not the typewriter, in front of which the rayon employee Doris spends her days, that turns an entertainment consumer into a producer. Only when she and her current lover hear "music from the radio" and listen to "Vienna, My One and Only" does she "feel like a poet" who "can also rhyme, . . . if only up to a point." [123] And if "a gramophone next door" should be playing in the moonlight, "something wonderful takes hold of her": listening to a hit, Doris first of all has the feeling "of making a poem" and then decides to write an autobiography or even a novel.

> I think it is good when I describe everything, because I am an uncommon person. I am not thinking of a diary—that would be ridiculous for an up-to-date girl of eighteen. But I want to write like a movie, because that's the way my life is and it will soon be more so. . . . And when I read it later, it will be like a movie—I will see myself in images.[124]

Entertainment novels (including Keun's) describe their own medial conditions of production with great precision. The medium of the gramophone has as its effect a type of poetry that is nothing but the inside of its outside. Skipping all textuality it jumps straight into the medium of film.

> My heart is a gramophone, playing excitedly with a sharp needle in my breast. . . . From the movies comes music, records that are passing on human voices. And all are singing . . . [125]

Novels that flow from hits in order to end in movies are part of the "literature of nonreaders" reviewed in 1926 by, of all journals, *Die literarische Welt*:

> This, the literature of nonreaders, is the most widely read literature in the world. Its history has not yet been written. Nor do I feel quite up to the task myself. I would simply like to make reference to one of its branches: poetry. For the literature of nonreaders, like "our" own, has a special category for poetry.

Every couple of weeks there is a survey: "Who is the most beloved poet of the year?" Every time, the question is answered incorrectly. The ones we know are not even considered. Neither Rilke nor Cäsar Flaischlen, not Goethe, and not Gottfried Benn. Rather, Fritz Grünbaum ("When You Can't, Let Me Do It!"), Schanzer and Welisch ("If You See My Aunt"), Beda ("Yes, We Have No Bananas"), Dr. Robert Katscher ("Madonna, You Are More Beautiful than the Sunshine)"—and who else? A lot more—before Flaischlen, Rilke and Benn come up.

"The 222 Newest Hits"—that is the most popular poetry anthology of all. The contents are revised and expanded every two months. And the whole thing costs just ten cents. Here there is only one genuine type of poem: the love poem. Girls, women, females—other topics are not favored.[126]

Even if all the names on both sides of the debate have long since changed, this remains a very exact appraisal. With the invention of technical sound storage, the effects that poetry had on its audience migrate to the new lyrics of hit parades and charts. Their texts would rather be anonymous than deprived of royalties, their recipients illiterate rather than deprived of love. At the same time, however, media technology's precise differentiation brings about a modern poetry that can do without all supplementary sensualities ranging from song to love because—according to a remark of Oscar Wilde's as ironic as it is appropriate—it is not read.[127] And this remains the case even when Rilke plans poetic coronal suture phonography or Benn writes poems that consciously set themselves apart from the entertainment industry. For Benn's poems can merely note but not verify that records and movies *are* part of a present that outpaces our cultural critics. Otherwise, his poems would be as successful, anonymous, and forgotten as the hits they sing about:

A popular hit is more 1950
than five hundred pages of cultural crisis.
At the movies, to which you can take along hat and coat,
there is more firewater than in the cothurnus
and without the annoying intermission.[128]

Lowbrow and highbrow culture, professional technology and professional poetry: the founding age of modern media left us with those two options. Wildenbruch's third way was eliminated. "So listen to the sound of what I declare, and Ernst von Wildenbruch's soul will be laid bare," the imperial state poet rhymed, as if one could simultaneously speak into technological machines and claim an immortal name. From sound back to poem, from poem back to soul—that is the impossible desire to reduce the real (the physiology of a voice) to the symbolic, and the symbolic (an articulated speech) to the imaginary. The wheel of media technology cannot be turned back to retrieve the soul, the imaginary of all Classic-Romantic poetry. What effectively remains of Wildenbruch in "For the Phonographic Recording of His Voice" is nothing but noise, posthumous already during his lifetime. Record grooves dig the grave of the author. Wildenbruch pulls out all the stops of the imaginary and the symbolic, of his immortal soul and his aristocratic name, so as not to have to speak of his speaking body. "By virtue of our bodies," Paul Zumthor's theory of oral poetry states, "we are time and place: the voice, itself an emanation of our physicality, does not cease to proclaim it."[129] Upon replaying the old cylinder of 1897, it is a corpse that speaks.

Between or before lowbrow and highbrow culture, between hit records and experimental poetry, there is only one third party: science. When Wildenbruch spoke into the bell-mouth, the phonograph stored indices rather than poems. And these indices speak precisely to the extent that their sender cannot manipulate them. The poet performing "The Phonographic Recording of His Voice" seemed at least to have been aware of this: because "the sound of the voice can never lie," its technological storage reveals the "hidden" and makes the "past"—the corpse of a Wildenbruch or a Goethe—speak.

Edison saw his phonograph "pressed into the detective service and used as an unimpeachable witness"[130] in court. With technological media, a knowledge assumes power that is no longer satisfied with the individual universals of its subjects, their self-images and self-representations—these imaginary formations—but instead registers distinguishing particulars. As Carlo Ginzburg has shown in "Clues and Scientific Method," this new knowledge rules Morelli, Freud, and Sherlock Holmes, that is, aesthetics, psychoanalysis, and criminology. However, Ginzburg fails to see that the shift in technologies of power simply follows the switch from writing to media. Books had been able to store and convey the imaginary corporeal self-images entertained by individuals. But unconsciously treacherous signs like fingerprints, pitch, and foot tracks fall into the purview of media without which they could neither be stored nor evaluated. Francis Galton's dactyloscope and Edison's phonograph are contemporaneous allies.

Wildenbruch appears to have suspected as much, or else his verses would not refer to the phonograph as the soul's own true photograph. His paranoia is justified. A phonographically recorded state poet no longer enters a pantheon of immortal writers but rather one of the countless

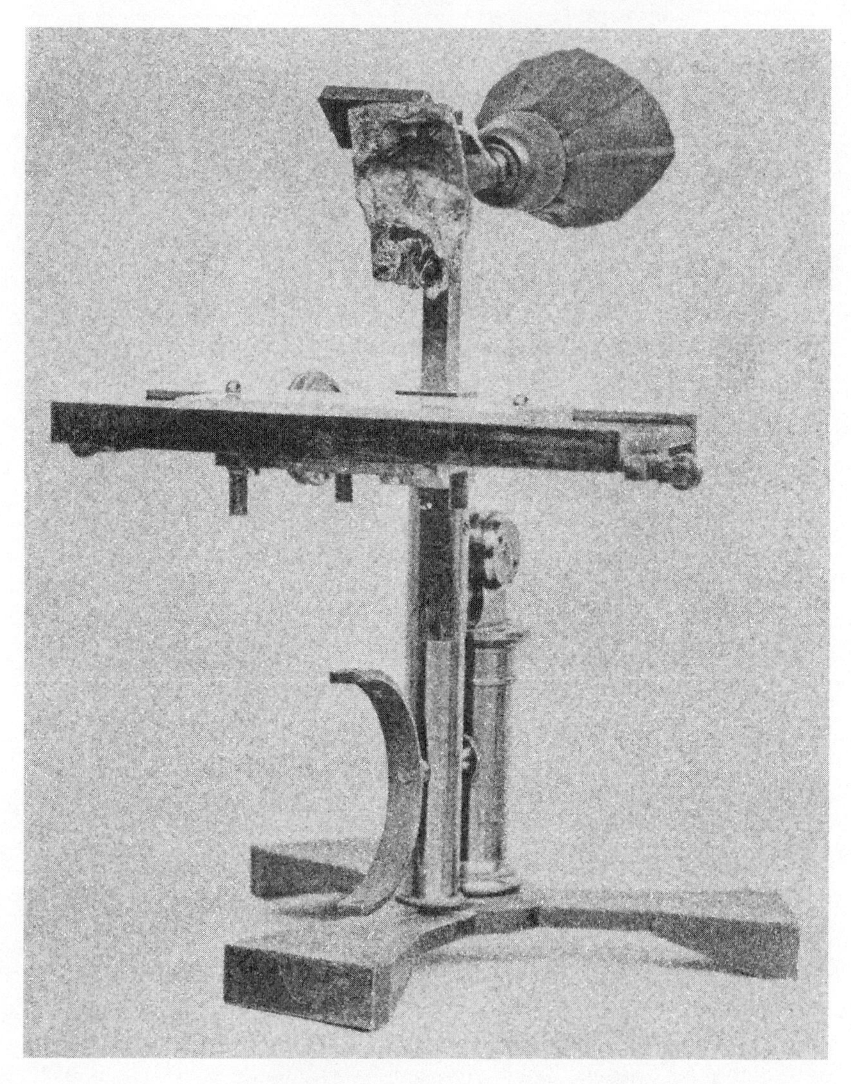

Protoype of receiver (Bell & Clarke, 1874).

evidence-gathering agencies that since 1880 have been controlling our so-called social behavior, that is, all the data and signs that are by necessity beyond our control. The good old days in which a self-controlled and "flattering" face could "fool" eyes equally bereft of media are over. Rather, all the sciences of trace detection confirm Freud's statement that "no mortal can keep a secret" because "betrayal oozes out of him at every pore."[131] And because (we may add) since 1880, there has been a storage medium for each kind of betrayal. Otherwise there would be no unconscious.

In 1908, the psychologist William Stern publishes a "Summary of Deposition Psychology." This new science is designed to cleanse the oral depositions of court protocols, medical reports, personal files, and school reports from all guile and deceit on the part of the speakers. Old European, that is to say, literary, means of power are not immune from deception. Whether for criminals or for the insane, the traditional "stylized depositions often produce a false impression of the examination and obscure the psychological significance of individual statements." As each answer "is, from the point of view of experimental psychology, a reaction to the operative stimulus in the question,"[132] experimenters and investigators provoke countermeasures in their subjects as long as they use the bureaucratic medium of writing. An argument made by the stimulus-response psychologist Stern that, sixty years later, is reiterated by interaction psychologists like Watzlawick (despite all criticism of the stimulus-response scheme).[133] Which is why examiners of 1908 recommend "the use of the phonograph as an ideal method"[134] and those of 1969 recommend tape decks.[135]

In 1905, the Viennese psychiatrist Erwin Stransky, quietly anticipating his colleague Stern, published a study, *On Speech Disturbances*. In order to contribute to the knowledge of such disturbances among the "mentally ill and mentally healthy," German psychiatry for the first time availed itself of the ideal method of phonography. Stransky had his subjects "look and speak directly into the black tube" for one minute (the recording time for one roll) after "all extraneous sense stimuli," that is, all the psychological problems of deposition, had been eliminated.[136] Whatever they said was completely irrelevant. The "aim" of the whole experiment "consists in shutting out all general concepts."[137] To test "concepts like 'speaking at odds,' 'hodgepodge,' 'thinking out loud,' 'hallucination,' etc.,"[138] the subjects had to abandon their so-called thinking. In Stransky's phonographic experiment, "language," in its "relative autonomy from the psyche,"[139] takes the place of general concepts or signifieds, as if intending to prepare or facilitate a key concept of modern literature.

Media technology could not proceed in a more exact fashion. Thanks to the phonograph, science is for the first time in possession of a machine that records noises regardless of so-called meaning. Written protocols were always unintentional selections of meaning. The phonograph, however, draws out those speech disturbances that concern psychiatry. Stransky's fine statement that "the formation of general concepts" could be inhibited "for pathological or experimental reasons"[140] is a euphemism. The "or" should be replaced by an equal sign. All the more so because the splendidly consistent Stransky places in front of the machine not only psychiatric patients but also, to collect comparative data, his own colleagues, the doctors. For the latter, the ensuing hodgepodge was related to experimental reasons, needless to say, while the patients had their pathological reasons. But the fact that psychiatrists, too, immediately produce a whole lot of nonsense when speaking into a phonograph, thereby relinquishing the professional status that distinguishes them from madmen, fully demonstrates the machine's power. Mechanization relieves people of their memories and permits a linguistic hodgepodge hitherto stifled by the monopoly of writing. The rules governing rhyme and meter that Wildenbruch employs to arrange his words when speaking into the phonograph; the general concepts that Stransky's colleagues use to arrange theirs during the first test runs—Edison's invention renders them all historically obsolete. The epoch of nonsense, our epoch, can begin.

This nonsense is always already the unconscious. Everything that speakers, because they are speaking, cannot also think flows into recording devices whose storage capacity is only surpassed by their indifference. "The point could be made"—a certain Walter Baade remarked in 1913 in "On the Recording of Self-Observations by Dictaphone"—

> that such an exertion is unnecessary, because it is not a matter of recording *all* remarks but only the important ones—this, however, fails to realize first of all that utterances of great importance are often made by subjects in moments when they themselves believe only to have made a casual remark and the examiner is altogether unprepared for an important comment, and secondly that even when both parties are aware that at least some part of a remark is "important," the decision what should and should not be recorded by the protocol is frequently very difficult and, subsequently, has a disturbing effect. For the most part, these two aforementioned reasons make the uninterrupted, indiscriminate recording of all utterances appear as an ideal.[141]

Presumably the first to follow this ideal is a fictional psychiatrist of 1897. Bram Stoker's *Dracula*, that perennially misjudged heroic epic of the final victory of technological media over the blood-sucking despots of old Europe,[142] features a certain Dr. Seward, who is baffled by the nonsensical discourse produced by his schizophrenic patient Renfield. The latter keeps screaming that the master is approaching, but Dr. Seward has no way of knowing that this refers to Dracula's arrival in England.

However, in the wake of a profane illumination anticipating Dr. Stransky, Dr. Seward resorts to media technology. He purchases one of the recently mass-produced phonographs, not to record the patient (as Stransky did) but rather his own associations triggered by the latter's speeches. The grooves store, to quote Seward's succinct and precise description, an "unconscious cerebration" that divines the subconscious of the schizophrenic but cannot advance all the way to the psychiatrist's ego. It is only (as Baade put it) the uninterrupted, indiscriminate recording of all utterances or associations that will allow Dr. Seward's unconscious cerebration "to give the wall to [its] conscious brother."[143] And only the typed transcription of all cylinders, recommended as early as 1890 by Dr. Blodgett,[144] by a certain Mina Harker will reveal to him and all the others hunting Dracula that the Count himself was behind Renfield's schizophrenic nonsense.

Since 1897, the year of *Dracula*'s publication, this procedure has no longer belonged to the realm of fiction. A science has emerged that turns it and all its particulars into a method: psychoanalysis.

As is known, Freud's "talking cure" is based on a segmentation of speech. On the one hand, patients lying on the couch speak—at least that's what they believe—according to classic discourse rules: A Kantian ego has to be able to accompany all my representations and provide for correct words and sentences—sentences that, unfortunately, say nothing about the patient. On the other hand, many minor symptoms emerge in the flow of speech—interruptions and paralalia, nonsensical words and puns—in which (to paraphrase Stransky) for pathological or experimental reasons the formation of general concepts has not occurred and a subconscious appears. Subsequently, the attentive doctor need only separate nonsense from sense like wheat from chaff (and not the other way around). He feeds the parapraxes back to the patient, thus triggering new associations and parapraxes, which once again are fed back, and so on until an ego in control of speech has been dethroned and the unspeakable truth can be heard.

Around 1900, only media theoreticians play as revolutionary a part as the physician Freud. Experimenting with telephones and phonographs, Hermann Gutzmann, a lecturer in speech disorders in Berlin, discovers that the prompting of nonsense words to his patients produces nothing but parapraxes. Precisely because both machines—due to transmission economy or technical imperfections—limit the frequency band of language on either end, what subjects "understand" can differ from what they "hear." Gutzmann speaks nonsense syllables like "bage" or "zoses" into the mouthpiece, and the ear at the other end receives "lady" or "process."[145] A simple question brings to light an unconscious. And the research in "On Hearing and Understanding" is able to "answer the question what such experiments may mean for experimental psychology":

> First of all, it is evident that using fake words stimulates the combinatory powers to such a degree that even against his will the listener is forced to replace the nonsense syllables he has heard with those words closest in his mind, in the pertinent constellation of ideas; that is, he is forced to hear the latter in the former. This can be seen very clearly in the protocol of Subject 1, a fickle eighteen-year-old who is deeply in love; he is attracted to everything feminine, and the many girls' names and an additional "lady" make his constellation of ideas easily recognizable. This also applies to the fake French words of the two "well-educated young ladies." If we wanted to conduct phonographic tests aimed at discovering certain suspected trains of thought, we would only need to use syllables sounding like the corresponding words as stimuli in order to arrive at the positive or negative result.[146]

Freud turns Gutzmann's simple suggestion into his explicit goal and imaginary constellations into the subconscious. In other words, he himself takes the place of phonographic tests. And for good reason: the psychoanalyst in his chair would also be faced with the problem of repressing or filtering the communication of an alien subconscious with his own subconscious had he not from the very beginning turned his ears into a technical apparatus. Unlike Gutzmann's subjects, Freud's patients fall from sense into nonsense, yet their doctor is not allowed to use his understanding to turn it back into sense. For that reason, Freud's "Recommendations to Physicians Practising Psycho-Analysis" simply amount to telephony:

> Just as the patient must relate everything that his self-observation can detect, and keep back all the logical and affective objections that seek to induce him to make a selection from among them, so the doctor must put himself in a position to make use of everything he is told for the purposes of interpretation . . . without substituting a censorship of his own for the selection that the patient has foregone. To put it in a formula, he must turn his own unconscious like a receptive organ towards the transmitting unconscious of the patient. He must adjust himself to the patient as a telephone receiver is adjusted to the transmitting microphone. Just as the receiver converts back into sound waves the electric oscillations in the telephone line which were set up by sound waves, so the doctor's unconscious is able, from the derivatives of the unconsciousness which are communicated to him, to reconstruct that unconscious, which has determined the patient's free associations.[147]

The fictional Dr. Seward had been obliged to first record his unconscious associations, which traced those of another unconscious, before he was able to arrive at a conscious interpretation upon replaying them. In exactly the same way the historical Dr. Freud turns into a telephone receiver. Following the nationalization of the Vienna telephone exchange in 1895, he not only had a telephone installed in his study[148] but also described the work that went on in that study in terms of telephony. As if "psychic apparatus," Freud's fine neologism or supplement for the antiquated soul, were to be taken literally, the unconscious coincides with electric oscillations. Only an apparatus like the telephone can transmit its frequencies, because each encoding in the bureaucratic medium of writing would be subject to the filtering and censoring effects of a consciousness. Under media conditions, however, "selection and refusal," to quote Rilke, are no longer permissible.[149] Which is why the conscientious deposition psychologist Freud abstains from note-taking during his sessions; instead—and much like Dr. Seward listening to his cylinders—he produces them later.[150]

The question remains, however, how the telephone receiver Freud can retain the communication from another unconscious. The phonograph owners Drs. Seward, Stransky, and Gutzmann are not faced with this problem, since they are in possession of a storage medium. Producing psychoanalytic case studies, that is, putting into writing what patients said, requires that one record whatever the two censors on and behind the couch want to render unsaid: parapraxes, puns, slips, signifier jokes. Only technological media can record the nonsense that (with the one exception of Freud) technological media alone were able to draw out into the open. Freud's telephone analogy elides this point. Nonetheless, his principle that consciousness and memory are mutually exclusive[151] formulates this very media logic. For that reason, it is consistent to define psychoanalytic case studies, in spite of their written format, as media technologies. Freud introduces his "Fragment of an Analysis of a Case of Hysteria" with the audacious avowal that his written "record" of hysterical speeches has a "high degree of trustworthiness," though it is "not absolutely—phonographically—exact."[152]

Evidently, psychoanalysis competes with technological sound recording. Its enemy or image is the phonograph, not film as Benjamin concluded from global parallels.[153] Neither as a word nor as a subject does film occur in Freud's writings. Rather, psychoanalytic texts are haunted by the absolute faithfulness of phonography. Thus, Freud's method of detecting unconscious signifiers in oral discourse and then interpreting these signifiers as letters of a grand rebus or syllable puzzle[154] appears as the final attempt to establish writing under media conditions. Whereas women, children, and madmen simply stop reading assigned novels and desert to the movies as a "couch of the poor,"[155] psychoanalysis once again teaches them letters that, however, are signifiers devoid of all meaning and phantasms. As a science it performs what Mallarmé or George inaugurate as modern literature.

In Berliner's own words, his Gramophone holds on to "the sound of letters";[156] conversely, Freud's psychoanalysis holds on to the letters of sound. While the entertainment industry transmits speech flows, the factual data input of every talking cure, and Freud's teacher Brücke, the ancestor of German speech physiology, analyzes them as such, Freud writes down their signifiers. His justification: unlike any street urchin, he "could not imitate"[157] all the stuttering, clicking of the tongue, gasping, and groaning[158] of his female hysterics. Which is why psychoanalysis is "not absolutely—phonographically—exact"; and why "reality will always remain 'unknowable.'"[159]

A global success that falls short of the absolute or real has only one precondition: patients, who, thanks to the telephonic and equidistant receptivity of Freud's' unconscious, may indulge in any kind of babble as long as they stick to the everyday medium of orality, are themselves not allowed to make use of storage technologies, lest they incur the wrath of psychoanalysis, the discrete textual recording of contractually arranged indiscretions.[160]

Concerning "The Handling of Dream-Interpretation in Psycho-Analysis," its inventor notes that it would be a mistake to let patients write down their own dreams. "For even if the text of a dream is in this way laboriously rescued from oblivion, it is easy enough to convince oneself that nothing thereby has been achieved for the patient. Associations will not come to the text, and the result is the same as if the dream had not been preserved."[161] The storage medium of writing fails once it is utilized by the patient and not by the analyst. Turning speech flows into syllable puzzles or "letters," which "do not occur in nature,"[162] remains the monopoly of the scientist seated in his chair. Precisely because a dream *text* already amounts to half an interpretation, it can no longer draw ideas or speech flows out of the sick unconscious. As a result of this drainage, writing assumes the transitoriness of orality; it is consumed by oblivion. And thus psychoanalysis establishes with self-recursive elegance the renown and status of its own text. In 1932, Freud's writings receive the Goethe prize.

Transcription of the phonogram of a schizophrenic, 1899.

"Should We Let Patients Write Down Their Own Dreams?" Karl Abraham asks in an essay of 1913 that appears to confirm Freud's authoritarian words with examples from the couch practice. "Against the doctor's orders," one of Abraham's patients "put writing materials next to his bed" and, following a "a very extensive, eventful, and highly charged dream," brought "two quarto pages full of notes" to the session. But to his own shame and to the delight of Abraham, he realizes "that the notes are almost completely illegible."[163] The psychoanalyst's love of nonsensical speeches has no written or cryptographic equivalent. As is well known, only printed works of literature, not illegible commonplaces, solicit interpretations.

But in spite of its title and its veneration of Freud, Abraham's essay does not limit itself to the old medium of writing. What brought the essay to writing or to shock was something far more modern and "ingenious": a phonograph in the hands of a patient.

Observation. 2d patient, who in response to his question was advised by me not to write down dreams, produces a whole series of dreams in the following nights. Upon awakening—in the middle of the night—he ingeniously tries to save from oblivion the dreams he considers important. He owns an apparatus for recording dictations and proceeds to speak the dreams into the bell-mouth. Characteristically, he forgets that for the last couple of days the machine has been malfunctioning. As a result the dictation is difficult to understand. Patient is forced to fill in a lot from memory. The dictation had to be complemented by the dreamer's memory! The dream analysis proceeded without notable resistance, thus we can assume that in this particular case the dream would have been retained even without any recording.

The patient, however, was not convinced by this experience and instead repeated the experiment one more time. Following a dream-filled night, the machine, which in the meantime had been repaired, delivered a clearly audible dictation. But according to the patient its content was so confused that he had difficulties enforcing some kind of order. As the succeeding nights furnished a bounty of dreams which centered on the same complexes and could be reproduced without artificial aid, this case, too, proved the uselessness of immediate recording.[164]

In terms of deposition psychology, a patient who no longer writes down but phonographically records his dreams is on the same level as his psychoanalyst. No writing material or filter interposes itself between the unconscious and its storage, no consciousness making the "selections" disdained by Freud creates order. Reason enough to bring along the repaired machine to the session and set it up next to the couch. Then the patient would be free to go for a walk while his phonograph—to paraphrase Kafka—could exchange dream-related information with the telephone receiver called a doctor. But no, preprogrammed by the analyst's instructions, Abraham's patient for once reverses the judgment that deposition psychology had passed on phonography, its ideal method: audible to the ear and to the unconscious, but confused and useless when it comes to content and level of consciousness. Thus is the historic opportunity missed to test, during Freud's lifetime and without artificial aid, what distinguishes absolute—phonographic—faithfulness from medical reproduction.

The test did not take place until 1969, when Edison's awkward machine was replaced by mass-produced magnetophones. Jean-Paul Sartre received (and published) an anonymous tape with an enclosed letter that suggests that the recording be entitled "Psychoanalytic Dialogue."[165] A., a 33-year-old patient in a lunatic asylum, smuggled a tape recorder into his last session and recorded everything: associations, interpretations, and ultimately the terror of the doctor upon discovering the machine:

Dr. X. Help! Murder! Helllp! Helllp!

A. Shut up and sit down.

Dr. X. Hellllllp! (*screams again*)

A. You're afraid I'm going to cut off your weenie?

Dr. X. Hellllllllp! (*That's the most beautiful scream of them all.*)

A. That's a funny recording![166]

Indeed. For the first time a machine in the patient's hands has replaced case studies, that is, essays from the doctors' hands. A "large part" of the conversation may be lost "due to the noise of the recording,"[167] but in the end are recorded all those data that Freud, orally or on paper, was unable to imitate. Subject to neither selection nor refusal, a speech flow—that of the psychoanalyst himself—is perpetuated as pure voice physiology.

As a result of which—according to the editor, Sartre—"the analyst now becomes an object" and "the encounter of man with man is thwarted once again." (From an existentialist perspective, psychoanalysis was itself already a form of alienation.)[168]

Writers faced with media and philosophers faced with technology are blind. As if so-called face-to-face communication could do without rules or interfaces, storage or channels, Man once again has to see to it that information systems are ignored. What Sartre calls the second alienation is simply the demolition of a monopoly. In the patient's hands, the tape recorder advances on a notation technique that could never be "absolutely—phonographically—exact" and therefore once more reenacts Old Europe under technologically advanced conditions: on the one hand, patients, who unlike bygone illiterates can read and write but are not allowed to; on the other, highly professional writers, who guard and monopolize their archives as if universal literacy or even media technology were some pie in the sky. According to Foucault, "the political credit of psychoanalysis" rested on the fact that it set "the system of law, the symbolic order, and sovereignty" against the unrestricted "extension and intensification of micro-powers"—powers not even Foucault revealed as media technologies.[169] This law, however, from Freud's "Mystic Writing-Pad" to Lacan's "Insistence of the Letter in the Unconscious," is writing about writing, alphabetized monopoly squared. Only psychoanalysts (they say) can write what does not cease not to write itself.

But the beat must go on. Technology and industry do not tolerate any delay simply because a couple of writers or psychoanalysts stick to white paper. From Edison's primitive phonograph cylinders all the way to popular music, the true poetry of the present, everything has gone like clockwork. Berliner's gramophone record of 1887, which no longer allowed consumers to make their own recordings but which since 1893 has allowed producers infinite reproductions of a single metal matrix, became the "prerequisite of the record mass market,"[170] with a return that exceeded the 100 million dollar mark before the advent of radio.[171] The mass-produced sound storage medium only needed mass-produced communication and recording media to gain global ascendancy. Far removed from old notions of sovereignty, all the powers of this and only of this century strive to reduce the "population's leadership vacuum"[172] (to quote a German media expert of 1939) to zero.

Broadcasting of weightless material came about for the purpose of the mass transmission of records: in 1921 in the United States, in 1922 in Great Britain, and in 1923 in the German Reich. "The uniting of radio with phonograph that constitutes the average radio program yields a very special pattern quite superior in power to the combination of radio and telegraph press that yields our news and weather programs."[173] Whereas Morse signs are much too discrete and binary to be a symbolic code for radio waves, the continuous low frequencies of records are ideal for the amplitude and frequency modulations known as broadcasting.

In 1903 a principal switch for transmitting such records was developed by Professor Slaby of the Berlin Technical University, whose *Voyages of Discovery into the Electric Ocean* delighted "His Imperial Majesty's dinner table at tranquil Hubertusstock."[174] The same Imperial Majesty put Slaby's assistant Count von Arco in charge of Telefunken GmbH. Building on Valdemar Poulsen's procedure, the two Berliners were able to produce a high frequency whose wireless oscillations "were no longer in the range of audibility but delighted the electrician as much as the thrice-accented C of a famous tenor would a music lover."[175] On this radio carrier frequency, "Caruso's singing, though emanating from the bell-mouth of a gramophone, could be transmitted in all its purity to our ears through the roaring metropolis";[176] that is, all the way from Sakrow to Potsdam.[177] Slaby's choice of tenors was not coincidental: on March 18, 1902, Caruso had revamped his immortality—from the hearsay of future opera audiences to gramophony.

Slaby and Arco, however, were conducting their research in the service of the emperor and his navy. But soon civilians, too, came to enjoy electrically transmitted records. A recording of Handel's *Messiah* is said to have been part of the first actual radio broadcast, hosted by Reginald A. Fessenden of the University of Pennsylvania on Christmas Eve, 1906.[178] Long before the St. Petersburg revolutionaries, Brant Rock, Massachusetts, had started its broadcast with "CQ, CQ—to all, to all"—but only wireless operators on ships[179] were able to receive the call and the Christmas record.

A world war, the first of its kind, had to break out to facilitate the switch from Poulsen's arc transmission to Lieben or De Forest's tube-type technology and the mass production of Fessenden's experimental procedure. It was not only in Germany, where the signal corps created in 1911 went to war with 550 officers and 5,800 men but returned with 4,831 officers and 185,000 men,[180] that the development of amplifier tubes was given the highest priority.[181] Fighter planes and submarines, the two new weapons systems, required wireless communications, just as military command required vacuum tube technology for the control of high and low frequencies. Tanks, however, which were equally in need of communications, kept losing their antennas in the barbed wire of the trenches and for the time being had to make do with carrier pigeons.[182]

But the exponentially growing radio troops were also in need of entertainment, because apart from machine-gun skirmishes and drumfire offensives, trench warfare is nothing but sensory deprivation—or *Combat as Inner Experience*, as Jünger so succinctly put it.[183] After three years in the wasteland between Flanders and the Ardennes, the military staffs—

De Forest's audion.

the British ones in Flanders[184] and a German one in Rethel in the Ardennes—took pity on their troops. Though trench crews had no radios, they were in possession of "army radio equipment." Beginning in May 1917, Dr. Hans Bredow, an AEG engineer before the war and afterward the first undersecretary for the national German radio network, was able to "use a primitive tube transmitter to broadcast a radio program consisting of records and the reading of newspaper articles. The project, however, was canceled when a superior command post got wind of it and prohibited the 'abuse of army equipment' for any future broadcast of music or words!"[185]

But that's the way it goes. The entertainment industry is, in any conceivable sense of the word, an abuse of army equipment. When Karlheinz Stockhausen was mixing his first electronic composition, *Kontakte*, in the Cologne studio of the Westdeutscher Rundfunk between February 1958 and fall 1959, the pulse generator, indicating amplifier, band-pass filter, as well as the sine and square wave oscillators were made up of discarded U.S. Army equipment: an abuse that produced a distinctive sound. A decade later, when the Cologne studio had at its disposal professionally developed audio electronic equipment and the record industry demanded that *Kontakte* attain hi-fi stereo quality, Stockhausen attempted in vain to reproduce the sound: as an echo of a world war it could not do without the abuse of military equipment.

And what is true microcosmically is also true macrocosmically. In November 1918, the 190,000 radio operators of the imperial German army were demobilized but kept their equipment. Supported or supervised by the executives of the USPD (Independent Socialist Party), the inspectorate of the technical division of the signal corps (Itenacht) founded a Central Broadcasting Bureau (ZFL), which on November 25 was granted a broadcasting license by the executive committee of the workers and soldiers council.[186] A "radio specter" that could have nipped the Weimar Republic in the technological bud triggered the immediate "counterattack" by Dr. Bredow.[187] For the simple purpose of avoiding the anarchistic abuse of military radio equipment, Germany received its entertainment radio network. Records that hitherto had been used to liven up military communication in the trenches of the Ardennes now came into their own. Otherwise people themselves, rather than the government and the media industry, could have made politics. In December 1923, two months after the first Berlin broadcast, Postal Minister Dr. Höfle, a member of the centrist party, listed (in order of increasing importance) the three tasks of the "Entertainment Broadcasting Network":

1. Wireless music, lectures etc. are to provide the general public with quality entertainment and education.
2. It is to be a new and important source of national revenue.
3. The new installations are to provide a convenient means for the nation and the states to convey whenever necessary official information to the public at large; the latter may be of importance with regard to state security.

In the interest of state security it is necessary to ensure that only those citizens own and operate equipment who have secured an official license to operate radio stations, and that, in addition, owners of radio equipment only record that which is intended for them.[188]

But what is intended for consumers is determined not only by state security but also by technology. "Even at the risk of losing to radio all they have earned with their records,"[189] the record industry had to submit to the standards of the new medium. *Struggle in the Ether* was the fitting name of Arnolt Bronnen's novel dealing with the establishment of the radio networks and the music industry—a novel that cunningly puts the desires of postal ministers into the mouths of the people and in particular into that of a Berlin typist: "'Records, gramophones, money,' she smiled, lost in a dream, 'if one could sit here without records, gramophones, money but still hear music . . .'"[190]

In order to fulfill these wishes, the major arms and communications technology corporations had to get rid of the old shellac craft. Pioneering tinkerers like Edison and Berliner left the stage. The vacuum-tube amplifier proceeded from high to low frequencies, from radio to records. In 1924, Bell Labs developed electromagnetic cutting amplifiers for recording and an electromagnetic pickup for replaying and thereby delivered sound recordings from the mechanical scratching of Edison's needle. In the same year, Siemens presented the recording studios of the media conglomerates with equally electric ribbon microphones, as a result of which grooves were finally able to store frequencies ranging from 100 bass hertz to 5 kilohertz overtones, thus rising to the level of medium-wave transmitters.

Edison's prototype had for good reasons preferred human voices to orchestras. Only with electrical sound processing are records ready for Höfle's "wireless music." "At last," the *Sunday Times* wrote, mistaking frequency bandwidth for sensuality, "an orchestra really sounds like an orchestra; we get from these records what we rarely had before—the physical delight of passionate music in the concert room or opera house. We do not merely hear the melodies going this, that, or the other way in a sort of limbo of tonal abstraction; they come to us with the sensuous excitement of actuality."[191]

And actuality itself can be produced once composers are up to date. For the third movement of *Pini di Roma*, Respighi wrote or rather demanded the recorded voice of a nightingale played against the backdrop of composed-out string arpeggios. Villiers de l'Isle-Adam's fictional Edison had already surrounded his woman of the future with metallic birds of paradise, who "by using the Microphone" make "an immense volume of sound" with their songs.[192] But only Bell Labs nightingales were capable of outplaying entire symphonic orchestras. Thus, Arturo Toscanini was able to premiere Respighi's sound poem as a media link combining an orchestral score with phonographic kilohertz sensuality.[193]

And the band played on. In the same year, 1924, U.S. researchers hit upon the idea of applying to sound processing the technique of producing intermediate frequencies. Thanks to frequency reduction, bat voices outside of the range of human audibility were caught on record. At least that is what was reported by newspapers in Prague; the same Prague in which a story was written immediately afterward entitled "Josefine the Singer, or The Mouse Folk." "Is Josefine's art singing at all?" Kafka's mice ask.

> Is it not perhaps just a piping? And piping is something we all know about, it is the real accomplishment of our people, or rather no mere accomplishment, but a characteristic expression of our life. We all pipe, but of course no one dreams of making out that our piping is an art, we pipe without noticing it, and there are even many among us who are quite unaware that piping is one of our characteristics.[194]

"The universe of sound," Cocteau's radio theory concludes, "has been enriched by that of ultrasound, which is still unknown. . . . We shall know that fish shout, that the sea is full of noises and that the void is peopled with realistic ghosts in whose eyes we are the same."[195]

In order to locate Cocteau's submarine ghosts, a world war, the second one, had to break out. Today realism is in any event strategic. An unparalleled surge of innovations that from 1939 on filled land, sea, and air with noise finally provided us (beyond Bell Labs) with records whose frequency range approached both limits of the audibility range; that is, with high fidelity. In 1940, four years before consumers were also able to purchase "FFRR" (full frequency range recording) records and seven years before Ansermet's hi-fi *Petrouchka* helped drive up annual record production to four hundred million, the Decca Record Company succeeded in capturing the ghostly noises on shellac. Quietly anticipating "Yellow Submarine" and the sound quality of the Beatles,

> the RAF Coastal Command had approached the English-owned Decca Record Company with a secret and difficult assignment. Coastal Command wanted a training record to illustrate differences between the sounds of German and British submarines. Such aural distinctions were extremely delicate, and to reproduce them accurately on a record called for a decided enlargement of the phonograph's capabilities. Intensive work under the supervision of Decca's chief engineer, Arthur Haddy, led to new recording techniques and the kind of record Coastal Command desired.[196]

But the enemy was not left standing behind. German record companies participated in the Battle of the Bulge. To avoid Allied suspicions when the chief of Army Communications ordered a sudden radio silence for all areas of troop concentration south of the Cologne-Aachen line on November, 12, 1944, the enemy had to be fed simulated attack preparations at other parts of the front. The Army High Command's propaganda division developed special recordings for army loudspeakers, "which, among other things, simulated: tank noises, marching troops, departing and arriving trucks, the unloading of equipment, etc."[197]

The whole spectrum of sound from infra- to ultrasound is, as was the case with Kafka's mice, not art but an expression of life. It finally allows modern detection to locate submarines wherever they may be, or tank brigades where they are not. The great musicologist Hornbostel had already spent the First World War at the front: sound location devices with huge bell-mouths and superhuman audibility ranges were supposed to enable ears to detect enemy artillery positions even at a distance of 30 kilometers. Ever since, human ears have no longer been a whim of nature but a weapon, as well as (with the usual commercial delay) a source of money. Long before the headphone adventures of rock'n'roll or original radio plays, Heinkel and Messerschmitt pilots entered the new age of soundspace. The Battle of Britain, Göring's futile attempt to bomb the island into submission in preparation for Operation Sea Lion, began with a trick for guiding weapon systems: radio beams allowed Luftwaffe bombers to reach their destinations without having to depend on daylight or the absence of fog. Radio beams emitted from the coast facing Britain, for example from Amsterdam and Cherbourg, formed the sides of an ethereal triangle the apex of which was located precisely above the targeted city. The right transmitter beamed a continuous series of Morse dashes into the pilot's right headphone, while the left transmitter beamed an equally continuous series of Morse dots—always exactly in between the dashes—into the left headphone. As a result, any deviation from the assigned course resulted in the most beautiful ping-pong stereophony (of the type that appeared on the first pop records but has since been discarded). And once the Heinkels were exactly above London or Coventry, then and only then did the two signal streams emanating from either side of the headphone, dashes from the right and dots from the left, merge into one continuous note, which the perception apparatus could not but locate within the very center of the brain. A hypnotic command that had the pilot—or rather, the center of his brain—dispose of his payload. Historically, he had become the first consumer of a headphone stereophony that today controls us all—from the circling of helicopters or Hendrix's *Electric Ladyland* all the way to the simulated pseudo-monophony, in the

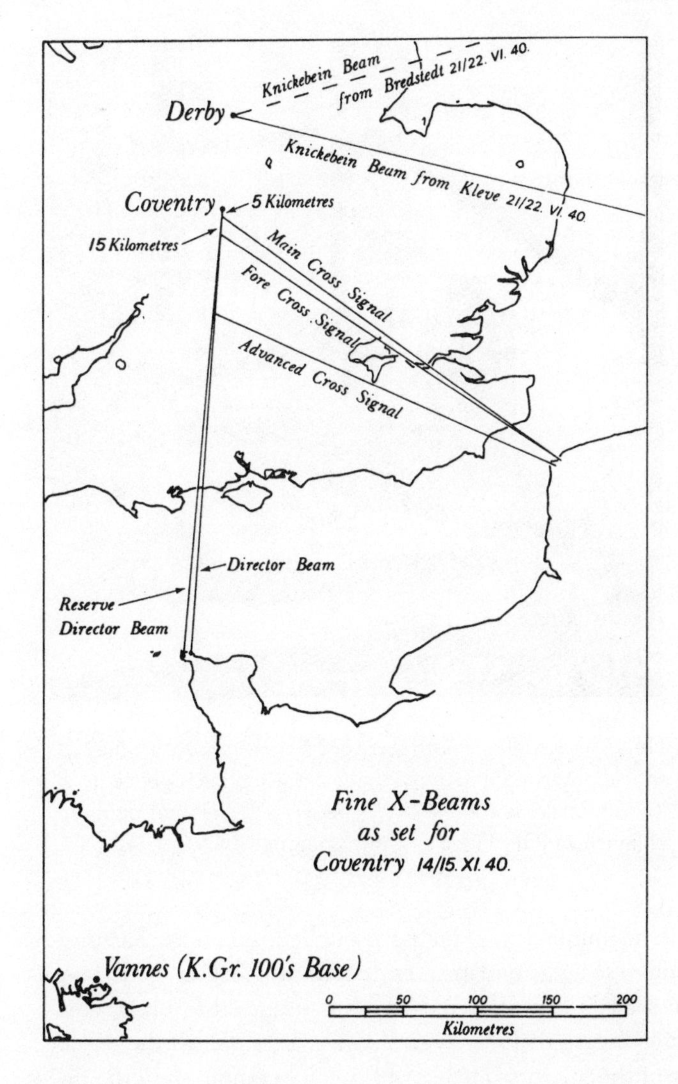

midst of the soundspace of Pink Floyd's *Wish You Were Here*, that once more wishes for the acoustics of targeted bombing.[198]

The difficulty British intelligence had in countering stereophonic remote control is explained by its chief technical officer, Professor Reginald Jones. Because the Luftwaffe's radio beam transmitters operated in frequency ranges even beyond VHF, which in 1940 the Secret Service was incapable of receiving and of which it had no conception, help could only

Hughes microphone with recorded fly. The same fly whose footstep was amplified by Hughes's carbon microphone in 1878 to make it audible circles between the left and right channels in Pink Floyd's "Ummagumma."

come from a profane illumination. An incident occurred on the Farnborough airfield while testing a loudspeaker system attached to a fuselage, which, just as in today's Pentagon project, was designed to blast rebellious natives in northwestern India with divine voices. When the officer standing in front of the microphone heard his voice coming from the distant loudspeaker two seconds later, he laughed about this acoustic delay. His laughter, in turn, was returned as another echo until the feedback affected all the participants and Farnborough resounded with a noise similar to that heard when rock musicians lean their guitars against the speakers. A "system that laughed by itself," Jones called it. But instead of laughing along, he chose to understand: Feedback, the principle of all oscillators, can also generate centimetric wave frequencies, something the experts refused to believe.[199] Jones ordered the construction of synchronized receivers, which, in turn, located the Luftwaffe's radio beam transmitters and their targets. The Battle of Britain was won. (Even if the warlord Churchill, not wanting to reveal to the enemy that his secrets had been revealed, disallowed the evacuation of Coventry, which had already been identified as a target city.)

Survivors and those born later, however, are allowed to inhabit stereophonic environments that have popularized and commercialized the trigonometry of air battles. Ever since EMI introduced stereo records in 1957,[200] people caught between speakers or headphones have been as controllable as bomber pilots. The submarine location duties of aspiring air force officers or the bombing target locations of Heinkel pilots turn into hypnosis, which in Stoker's 1897 *Dracula* still had to be used to solve, without the help of radio technology, a very strategic submarine detection problem.[201] But in 1966, following two world wars and surges in innovation, hypnosis and recording technology finally coincide: engine noises, hissing steam, and a brass band move across the walls from left to right and back while a British voice sings of the literal chain that linked Liverpool's submarine crews to postwar rock groups.

> In the town where I was born
> lived a man who sailed to sea
> and he told us of his life
> in the land of submarines.
>
> . . .
>
> So we sailed up to the sun
> till we found the sea of green
> and we lived beneath the waves
> in our yellow submarine.
>
> And our friends are all aboard
> many more of them live next door
> and the band begins to play
>
> "We all live in a yellow submarine . . . "[202]

The Beatles simply transported everybody to that impossible space that once concealed Count Dracula in his black coffin in the black belly of his ship, floating in the Black Sea until he was located, and subsequently destroyed, by hypnotic sound detection. Hi-fi stereophony can simulate any acoustic space, from the real space inside a submarine to the psychedelic space inside the brain itself. And should locating that space either fail or be a ruse designed to fool the consumer, it is only because the supervising sound engineer has proceeded as shrewdly as the disinformation campaign prior to the Battle of the Bulge.

Once again, these deceptions were programmed by the admirable Villiers de l'Isle-Adam. By design or accident, his Edison places "his hand on the central control panel of the laboratory," whereupon the telephonic voice of his agent in New York "seemed to come from all the corners of the room at once." A dozen speakers scattered across the laboratory—obviously modeled on the first soundspace experiments conducted between the Paris Opera and the Palace of Industry in 1881—make it possible.[203]

With the help of stereo recordings and stereo, VHF acoustic deceptions can invade operas completely. When, in 1959, John Culshaw produced Solti's beautifully overmodulated *Rhinegold*, the homelessness of spirits was implemented. Of course the other gods and goddesses, male and female singers, were each assigned their own space between the stereo channels. But Wagner's great technician Alberich, upon tearing the newly completed Tarnhelm out of his brother Mime's hands and demonstrating in hands-on fashion the advantages of invisibility, appears to be coming, like Edison's telegrapher, from all corners at once. "Thus, in scene III, Alberich puts on the Tarnhelm, disappears, and then thrashes the unfortunate Mime. Most stage productions make Alberich sing through a megaphone at this point, the effect of which is often less dominating than that of Alberich in reality. Instead of this, we have tried to convey, for thirty-two bars, the terrifying, inescapable presence of Alberich: left, right, or centre there is no escape for Mime."[204]

Culshaw's stereo magic simply puts into practice what the great media technician Wagner had in mind for his dramatic doppelgänger. "Everywhere now he lies in wait," sings Alberich, lost in acoustic space, making those he keeps "under guard" "subject to him forever. "[205] In other words, Wagner invented the radio play, as Nietzsche immediately realized: "His art always carries him in two directions, out of a world of auditory drama into a mysteriously kindred world of visual drama, and vice versa."[206] *The Ring of the Nibelung*, that zero series of all word wars, could just as well be called *Struggle in the Ether*.

To broadcast the ethereal struggle, radio merely had to take over the innovations of the world wars and, in a move that reversed the one following the First World War, adapt itself to the standard of records. Because amplitude modulation did not leave enough frequency range, the old AM radio would have been unable to transmit hi-fi songs or stereo radio plays.

The spectacular growth of FM is attributable to its technical superiority to AM, and relative cheapness as an investment medium. In the late fifties, it was found that the great range of FM channels could not only sustain a higher fidelity for single transmissions, but could in fact also be used to broadcast separate signals simultaneously in a process called "multiplexing." This discovery made possible stereo musical broadcast. Stereo broadcast was particularly attractive to those audiences discriminating and wealthy enough to prefer high fidelity music. . . . As the rock audience grew in size and sophistication, it came to demand the same sound quality which it could get from records at home (reflected in the tremendous increase in the middle and late sixties in the stereo component market), but could not get from AM radio.[207]

Frequency modulation and signal multiplexing, the two components of VHF, are of course not a U.S. commercial discovery of the 1950s. Without "his ingenious technical decision" in favor of signal multiplexing, General Fellgiebel, chief of Army Communications, would not have been able to control the invasion of Russia, that is, "the most immense task ever faced by any signal corps in the world."[208] Without Colonel Gimmler of Army Ordnance and his refutation of the delusion "that very high frequencies (between 10m and 1m) propagate in a straight line and are therefore of no use in the battle field,"[209] Colonel General Guderian, the strategist of the tank blitzkrieg, would have been forced to resort to World-War-I-era carrier pigeons. Instead, his armored wedges, "from the tanks in the most forward position back to divisional, corps, and army command," were, unlike his enemies, equipped with VHF.[210] "The engine is the soul of the tank," Guderian used to say, "and radio," General Nehring added, "its number one." Then as now VHF radio reduces the leadership vacuum to zero.

On September 11, 1944, American tank vanguards liberated the city of Luxembourg and its radio station. Radio Luxembourg returned to its prewar status as the largest commercial broadcaster and advertiser of records on a continent of postal, telegraphic, and radio state monopolies.[211] But four years as an army station had left its traces: traces of a new way of storing traces.

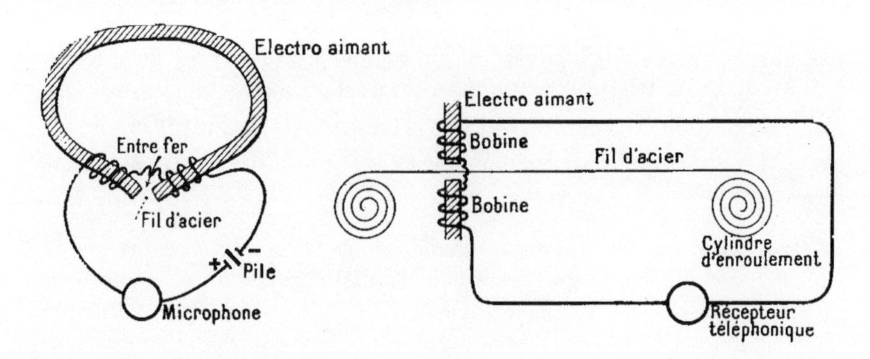

Basic diagram of Poulsen's telegraphone.

By the early 1940s, German technicians had made some startling advances. Radio monitors who listened to the German broadcasting stations day after day for British and United States intelligence soon realized that many of the programs they were hearing could not possibly derive from live studio broadcasts. Yet there were a fidelity and a continuity of sound, plus an absence of surface scratching, in the German transmissions that ordinary transcription records could never have yielded. The mystery was solved . . . when the Allies captured Radio Luxembourg . . . and discovered among the station's equipment a new Magnetophone of extraordinary capabilities.[212]

It was not until 1940 that technicians at BASF and AEG had by chance hit upon the technique of radio frequency premagnetizing, thus turning Valdemar Poulsen's experimental telegraphone of 1898 into an operational audiotape with a 10 kilohertz frequency bandwidth. Up until then, the record-radio media link had operated as a one-way street. Transmitters and gramophone users replayed what Berliner's master disc had once and for all recorded, even if radio stations—in a late vindication of Edison—made use of special phonographs developed for the specific purpose of program storage.[213] But under combat conditions those wax cylinders, which, since 1930, were allowed to record parliamentary sessions strictly for "archival purposes," were useless.[214] A propaganda ministry that turned radio into "the cultural SS of the Third Reich"[215] needed a recording and storage medium as modern and mobile as Guderian's tank divisions.

Major General von Wedel, chief of Army Propaganda, recounts:

We were also essentially dependent on developments of the propaganda ministry with regard to radio equipment for war correspondents. That also applied to the appropriate vehicles. When it came to tank divisions, the Luftwaffe, or parts of the navy, the opportunities for original combat recordings were hampered by the fact that we could not obtain the stable and horizontal supports necessary for producing discs. At first, we were forced to make do with belated dispatches.

A significant change occurred after the Magnetophone was invented and thoroughly designed for the purpose of war reports. Original combat reports from the air, the moving armored vehicle, or the submarine, etc., now became impressive firsthand accounts.[216]

As Ludendorff had pointed out, it is a truth of *Total War* that "the mass usage of technological equipment can be tested much better in wartime than would ever be possible in peace."[217] The motorized and mobilized audiotape finally delivered radio from disc storage; "Yellow Submarine," or "war as acoustic experience," became playable.

But reaching beyond the acoustic experiences of the so-called general public, the magnetic tape also revolutionized secret transmissions. According to Pynchon, "operators swear they can tell the individual sending-hands."[218] As a consequence, the Abwehr [German Counterintelligence Service], as part of the German Army High Command, had the "handwriting" of every single agent recorded at the Wohldorf radio station close to Hamburg before they went abroad on their secret missions. Only magnetic tapes guaranteed to Canaris and his men that it "was really their agent sitting at the other end and not an enemy operator."[219]

Inspired by this success, the Abwehr switched from defense to offense. Because the enemy was not yet in possession of magnetic tapes, the Abwehr was in a position to transmit its famous *Funkspiele* (radio games), which in spite of their name resulted not in the entertainment of millions in front of speakers but in the death of 50 British agents. The Abwehr managed to capture and turn around agents who had parachuted into the Netherlands. As if nothing had happened, they were forced to continue their transmissions in their own handwriting. The transmission of German *Funkspiel* messages to London (or, in one parallel case, to Moscow) lured additional agents into the Abwehr trap. Normally, intelligence agencies arrange emergency signals with their agents for such situations, "such as using an old code, making absurd mistakes, or inserting or omitting certain letters of punctuation."[220] Each Morse message of the converted agents was taped, analyzed, and, if need be, manipulated before it was transmitted. This procedure continued uninterrupted for years in the hardly civilian ether.

The world-war audiotape inaugurated the musical-acoustic present. Beyond storage and transmission, gramophone and radio, it created empires of simulation. In England, Turing himself considered using a captured German Magnetophone as the storage mechanism for his projected large computer. Like the paper strip of the universal discrete machine, tapes can execute any possible manipulation of data because they are equipped with recording, reading, and erasing heads, as well as with forward and reverse motion.[221] Which is why early, cheap PCs work with attached tape decks.

In a far more practical vein, captured magnetic tapes aroused sleepy U.S. electric and music giants who had, naturally, taken on duties other than commercial ones between 1942 and 1945.[222] Inserted into the signal path, audiotapes modernized sound production; by replacing gramophones they modernized sound distribution. Tape decks made music consumers mobile, indeed automobile, as did the radio producers in the Magnetophone-equipped German lead tanks of old. Thus, the "American mass market" was "opened up" by "the car playback system."[223] To minimize the leadership vacuum and exploit the possibilities of stereophony, the only things missing were new VHF stations with rock'n'roll and traffic reports on the transmitting end and car radios with FM and decoders on the receiving end. Six-cylinder engines whisper, but the stereo equipment roars. Engine and radio are (to paraphrase Guderian and Nehring) also the soul of our tourist divisions, which under so-called postwar conditions rehearse or simulate the blitzkrieg.

The central command, however, has moved from general staffs to engineers.[224] Sound reproduction revolutionized by magnetic tape has rendered orders unnecessary. Storing, erasing, sampling, fast-forwarding, rewinding, editing—inserting tapes into the signal path leading from the microphone to the master disc made manipulation itself possible. Ever since the combat reports of Nazi radio, even live broadcasts have not been live. The delay that in the case of tapes is due to separate head monitoring (and that is now more elegantly achieved by digital shift registers)[225] suffices for so-called broadcast obscenity policing lines. It appears that listeners, once they have been called by a disc jockey and are on the air, are prone to exhibit an unquenchable desire for obscenities. Today everybody can and (according to Andy Warhol) wants to become famous, if only for two minutes of airtime. In the blind time to which media, as opposed to artists, are subject, chance is principally unpredictable. But the 6.4 seconds of dead time the broadcast obscenity policing line inserts between telephone call and actual broadcast make censorship (if not art) possible in the data flow of the real.

That is precisely the function of audiotapes in sound processing. Editing and interception control make the unmanipulable as manipulable as symbolic chains had been in the arts. With projects and recourses, the time of recurrence organizes pure random sequences; Berliner's primitive recording technology turns into a *Magical Mystery Tour*. In 1954, Abbey Road Studios, which not coincidentally produced the Beatles' sound, first used stereo audiotapes; by 1970 eight-track machines had become the standard; today discos utilize 32 or 64 tracks, each of which can be manipulated on its own and in unison.[226] "Welcome to the machine," Pink Floyd sang, by which they meant, "tape for its own ends—a form of collage using sound."[227] In the *Funkspiele* of the Abwehr, Morse hands could be corrected; in today's studios, stars do not even have to be able to sing anymore. When the voices of Waters and Gilmour were unable to hit the high notes in "Welcome to the Machine," they simply resorted to time axis manipulation: they dropped the tape down half a semitone while recording and then dropped the line in on the track.[228]

But neither is tape technology always an end in itself, nor does editing always amount to correction or beautification. If media are anthropological a prioris, then humans cannot have invented language; rather, they must have evolved as its pets, victims, or subjects. And the only weapon to fight that may well be tape salad. Sense turns into nonsense, government propaganda into the white noise of Turing's vocoder, impossible fillers like *is/or/the* are edited out:[229] precisely the ingredients of William Burroughs's tape cut-up technique.

"Playback from Eden to Watergate" begins (like all books) with the word, and in the beginning that word was with God. But not only in the shape of speech, which animals, too, have at their command, but also as writing, the storage and transmission of which made culture possible in the first place. "Now a wise old rat may know a lot about traps and poison but he cannot write 'Death Traps in Your Warehouse' for the *Reader's Digest*."[230] Such warnings, or "tactics," are restricted to humans—with the one exception that they were not capable of warning of the warning system of writing, which subsequently turned into a deadly trap. Because apes never mastered writing the "written word" mastered them: a "killer virus" that "made the spoken word possible. The word has not yet been recognized as a virus because it has achieved a state of stable symbiosis with the host," which now seems to be "breaking down."[231] Reconstructing the apes' inner throat, which was not designed for speech, the virus created humans, especially white males, who were stricken with the most malignant infection: they mistook the host itself for its linguistic parasite. Most apes died from sexual frenzy or because the virus caused "death through strangulation and vertebral fracture."[232] But with two or three survivors the word was able to launch a new beginning.

Let us start with three tape recorders in the Garden of Eden. Tape recorder one is Adam. Tape recorder two is Eve. Tape recorder three is God, who deteriorated after Hiroshima into the Ugly American. Or, to return to our primeval scene: tape recorder one is the male ape in a helpless sexual frenzy as the virus strangles him. Tape recorder two is the cooing female ape who straddles him. Tape recorder three is DEATH.[233]

What began as a media war has to end as a media war so as to close the feedback loop linking Nixon's Watergate tapes to the Garden of Eden. "Basically, there is only one game and that game is war."[234] World war weapons like the Magnetophone have been put to commercial use in the shape of tape recorders, as a result of which ex-writers like Burroughs can take action. The classic rift between the production and reception of books is replaced by a single military interception.[235]

We now have three tape recorders. So we will make a simple word virus. Let us suppose that our target is a rival politician. On tape recorder one we will record speeches and conversations, carefully editing in stammers, mispronunciations, inept phrases—the worst number one we can assemble. Now, on tape recorder two we will make a love tape by bugging his bedroom. We can potentiate this tape by splicing it with a sexual object that is inadmissible or inaccessible or both, say, the Senator's teenage daughter. On tape recorder three we will record hateful, disapproving voices. We'll splice the three recordings in together at very short intervals and play them back to the Senator and his constituents. This cutting and playback can be very complex, involving speech scramblers and batteries of tape recorders, but the *basic principle is simply splicing sex tape and disapproval tapes together.*[236]

As simple as any abuse of army equipment. One just has to know what Shannon's and Turing's scrambler or the German Magnetophone can be used for.[237] If "control," or, as engineers say, negative feedback, is the key to power in this century,[238] then fighting that power requires positive feedback. Create endless feedback loops until VHF or stereo, tape deck or scrambler, the whole array of world war army equipment produces wild oscillations of the Farnborough type. Play to the powers that be their own melody.

Which is exactly what Burroughs does after having described "a number of weapons and tactics in the war game":[239] he joins Laurie Anderson in producing records. Which is exactly what rock music does in the first place: it maximizes all electro-acoustic possibilities, occupies recording studios and FM transmitters, and uses tape montages to subvert the writing-induced separation into composers and writers, arrangers and interpreters. When Chaplin, Mary Pickford, D. W. Griffith, and others founded United Artists following the First World War, a movie executive announced that "the lunatics have taken charge of the asylum." The same thing happened when Lennon, Hendrix, Barrett and others started recording their *Gesamtkunstwerke* by making full use of the media innovations of the Second World War.[240]

Funkspiel, VHF tank radio, vocoders, Magnetophones, submarine location technologies, air war radio beams, etc., have released an abuse of army equipment that adapts ears and reaction speeds to World War $n+1$. Radio, the first abuse, lead from World War I to II, rock music, the next one, from II to III. Following a very practical piece of advice from Burroughs's *Electronic Revolution*,[241] Laurie Anderson's voice, distorted as usual on *Big Science* by a vocoder, simulates the voice of a 747 pilot who uses the plane's speaker system to suddenly interrupt the ongoing entertainment program and inform passengers of an imminent crash landing or some other calamity. Mass interception media like rock music amount to mobilization, which makes them the exact opposite of Benjamin's distraction.[242] In 1936, only the unique "*Reichsautozug Deutschland*, a motorcade consisting of eighty vehicles," was able to "broadcast party congresses and mass rallies without any local help by setting up speaker systems on a giant scale, erecting stands, and so on":[243] today, the same is achieved night after night by the trucks and kilowatt systems of any rock group. Filled to the brim with electronics or army equipment, they carry us away to *Electric Ladyland*. The theme of love, that production secret of the literature for nonreaders, has run its course. Rock songs sing of the very media power which sustains them.

Lennon and McCartney's stereo submarine is not the only postwar lyric in the literal sense of the word. *The Final Cut*, Pink Floyd's last record, was written by Roger Waters (born 1944) for Eric Fletcher Waters (1913–1944), that is, for a victim of a world war. It begins, even before the first sound, with tape cut-ups of news broadcasts (on the Falklands, NATO fleet transporters, nuclear power stations), which all simply serve to point out that "postwar," both the word and the thing itself, is a "dream," a distortion made to mollify consumer ears. "Post War Dream" is followed by "The Hero's Return." The cut-up returns to its origins: when army communication equipment, the precursor of the mass medium radio, cuts up the symbolic and the real, orders and corpses. A commemoration that is the flip side of postwar, love and Muzak.

Sweetheart, sweetheart, are you fast asleep, good
'cos that's the only time I can really talk to you
and there is something that I've locked away
a memory that is too painful
to withstand the light of day.

When we came back from the war
the banners and flags hung on everyone's door
we danced and we sang in the street
and the church bells rang.
But burning in my heart
a memory smoulders on
of the gunner's dying words
on the intercom.[244]

Interception, chopping, feedback, and amplification of war reports: "Sympathy for the Devil" means nothing else. Legend has it that the Rolling Stones used cut-up techniques to produce the lyrics for *Beggars Banquet*. They cut out newspaper headlines, pasted them to the studio wall, and shot at them. Every hit was a line. Anticipating modern statistics, the precondition of cut-up and signal processing in general, Novalis remarked: "The individual facts are random events—the combination of random events—their concurrence is itself not subject to chance, but to laws—a result of the most profound systematic wisdom."[245]

Thus, the random distribution of newspaper headlines results in the law of information technology and a martial history of rock music. The devil, whose voice is immortalized by "Sympathy for the Devil," was there when the revolutionaries of St. Petersburg killed the czar and, with their radio transmission "CQ—to all," turned army equipment into global AM radio; he was there when television broadcast both Kennedy assassinations, turned "you and me" into murderers, and exorcised all radio magic. But above all, Lucifer screams out that radio specter, ghost army, or tank general which VHF and rock music are indebted to:

I rode a tank
held a gen'rals's rank
when the blitzkrieg raged
and the bodies stank.[246]

The blitzkrieg, as is well known, raged from 1939 to 1941, when Guderian rode his lead tank. The bodies stank longer.

From "War Heroes" to *Electric Ladyland*: a mnemotechnology of rock music. Nietzsche's gods had yet to receive the sacrifice of language;

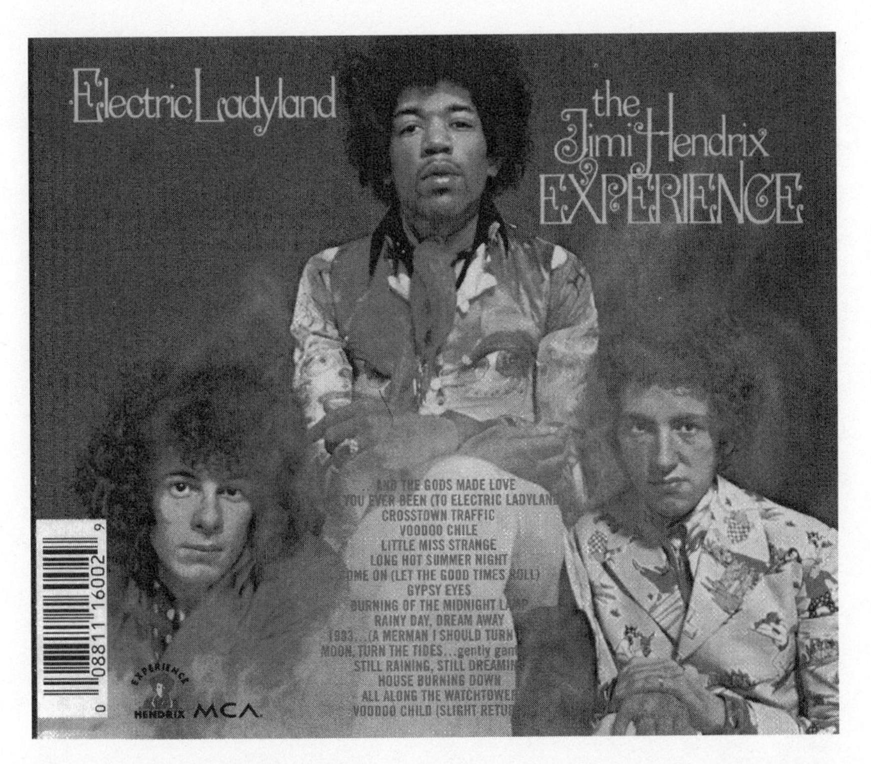

The Jimi Hendrix Experience, *Electric Ladyland*, 1968. (Courtesy of Authentic Hendrix, LLC, and MCA Records, Inc.)

AND THE GODS MADE LOVE

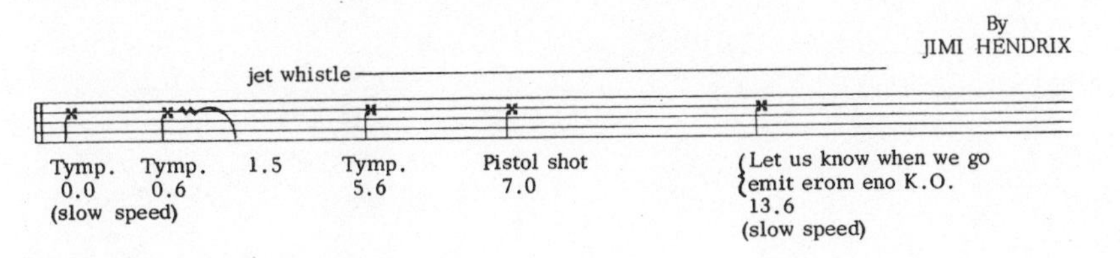

21.0 backward & forward tapes of speech

29.8 Harmonics run up and down at high speed

cut-up techniques have done away with that virus. Before Hendrix, the paratrooper of the 101st Airborne, cuts his machine-gun-like guitar to the title song, tape technology operates for its own sake: tympana, jet engines, pistol shots. Writing can write nothing of that. The *Songbook* for *Electric Ladyland* notes the tape's forward and backward motion as well as its changing speed and the test points of a blind but manipulable time.[247] The title on the cover—that which does not cease not to write itself.

FILM

Media cross one another in time, which is no longer history. The recording of acoustic data was accomplished with sound tricks, montage, and cuts; it is with film tricks, montage, and cuts that the recording of optical processes began. Since its inception, cinema has been the manipulation of optic nerves and their time. This is proved, among other ways, by the now-prohibited trick of repeatedly splicing individual frames of a Coca-Cola ad into feature films: because its flashlike appearance for 40 milliseconds reaches the eyes but not consciousness, the audience develops an inexplicable yet irresistible thirst. A cut has undercut its conscious registration. The same is true of film. Beginning with Eastman in 1887, when celluloid superseded Daguerre's photographic glass plates and provided the material basis for feature films, such manipulations became feasible. Cinema, in contrast to sound recording, began with reels, cuts, and splices.

It is said that the Lumière brothers documented simply and incessantly what their lens could record and what the type of projection they developed could reproduce. Legend has it, however, that Georges Méliès, the great film pioneer, ran out of celluloid while shooting a street scene. He left the tripod and camera in position and loaded a new reel, but in the meantime so-called life naturally went on. Viewing the fully spliced film, its director was consequently surprised by the magical appearance and disappearance of figures against a fixed background. Méliès, who as former director of the Théâtre Robert Houdin had already projected many a magical trick onto the technological screen,[1] had accidentally also stumbled upon the stop trick. Hence in May 1896, "before the eyes of an astonished and dumbfounded audience," he presented "*L'Escamotement d'une dame*, the disappearance of a woman from the picture."[2] Technological media (following Villiers and his Edison) liquidate that "great

Jean Cocteau, *Le Sang d'un poète*, 1930.

Lady, Nature," as it had been described, but never viewed, by the nineteenth century. Woman's sacrifice.

And castration. For what film's first stop tricks did to women only repeated what the experimental precursors of cinema did to men. Since 1878 Edward Muggeridge (who changed his name to Eadweard Muybridge to commemorate old Saxon kings)[3] had been experimenting with twelve special cameras on behalf of the California railroad tycoon and university founder, Leland Stanford. The location was Palo Alto, which later saw the invention of the vacuum tube, and the assignment was the recording of movements whose speed exceeded the perception of any painter's eye. Racehorses and sprinters dashed past the individually and sequentially positioned cameras, whose shutters were triggered successively by an electromagnetic device supplied by the San Francisco Telegraph Supply Company—1 millisecond for every 40 milliseconds.[4]

With such snapshots (literally speaking) Muybridge's handsome volumes on *Animal Locomotion* were meant to instruct ignorant painters in what motion looks like in real-time analysis. For his serial photographs testified to the imaginary element in human perception, as in the positions of horses' legs on canvas or on English watercolor paper. To speak of cinema as Muybridge's historical goal would, however, be inaccurate, since celluloid was not yet available. The technological medium was meant to modernize a venerable art form, as indeed happened when impressionists like Degas copied photographs in their paintings. Hence Stanford University's fencers, discus throwers, and wrestlers posed as future models for painters, that is, nude—at least as long as they turned their backs to one of the twelve cameras. In all the milliseconds of frontal shots, however, Muybridge reached one last time for the painter's brush in order to practice (long before Méliès) the disappearance of the male anatomy with retouched gymnastic shorts.

Had they been copied onto celluloid and rolled onto a reel, Muybridge's glass plates could have anticipated Edison's kinetoscope, the peephole precursor to the Lumières' cinematic projection. The astonished visitors to the 1893 World's Fair in Chicago would then have been witness to the first trick film: the jumpy appearance and disappearance of moral remains, which in the age of cinema approximate the condition of pure image-flickering.

The trick film therefore has no datable origin. The medium's possibilities for cutting and splicing assail its own historiography. Hugo Münsterberg, the private lecturer at the University of Freiburg whom William James called to the Harvard Psychological Laboratory, clearly recognized this in 1916 in the first history of cinema written by a professor:

> It is arbitrary to say where the development of the moving pictures began and it is impossible to foresee where it will lead. What invention marked the beginning? Was it the first device to introduce movement into the pictures on a screen? Or did the development begin with the first photographing of various phases of moving objects? Or did it start with the first presentation of successive pictures at such a speed that the impression of movement resulted? Or was the birthday of the new art when the experimenters for the first time succeeded in projecting such rapidly passing pictures on a wall?[5]

Münsterberg's questions remain unanswered because the making of films is in principle nothing but cutting and splicing: the chopping up of continuous motion, or history, before the lens. "Discourse," Foucault wrote when he introduced such caesuras into historical methodology itself, "is snatched from the law of development and established in a discontinuous atemporality: . . . several eternities succeeding one another, a play of fixed images disappearing in turn, do not constitute either movement, time, or history."[6] As if contemporary theories, such as discourse analysis, were defined by the technological a priori of their media.

Methodological dreams flourish in this complication or implication. Theory itself since Freud, Benjamin, and Adorno has attempted to pseudo-metamorphose into film.[7] It is also possible, however, to understand technological a prioris in a technological sense. The fact that cuts stood at the beginning of visual data processing but entered acoustic data processing only at the end can then be seen as a fundamental difference in terms of our sensory registration. That difference inaugurated the distinction between the imaginary and the real.

The phonograph permitted for the first time the recording of vibrations that human ears could not count, human eyes could not see, and writing hands could not catch up with. Edison's simple metal needle, however, could keep up—simply because every sound, even the most complex or polyphonous, one played simultaneously by a hundred musicians, formed a single amplitude on the time axis. Put in the plain language of general sign theory, acoustics is one-dimensional data processing in the lower frequency range.[8]

The continuous undulations recorded by the gramophone and the audiotape as signatures of the real, or raw material, were thus passed on in an equally continuous way by sound engineers. Cutting and splicing would have produced nothing but crackling noises, namely, square-curve jumps. Avoiding them presupposes great skill on the part of recording engineers, if not the computer algorithms of digital signal processing. Therefore, when pioneers of the radio play such as Breslau's Walter Bischoff were looking for genuinely "radio-specific" (*funkisch*) means of expression, they studied the parallel medium of silent films and considered only the fade-out, not the cut, as a possible model: "The man working the amplifier," as Bischoff argued in *Dramaturgy of the Radio Play*, "is in charge of a function similar to that of the camera man. He fades in and out, as we say in the absence of a radio-specific terminology. By slowly turning down the condenser at the amplifier, he lets the scene, the finished sequence of events, fade into the background, just as he can, by gradually turning the condenser up, give increasing form and shape to the next acoustic sequence."[9] By following such continuity, which is diametrically opposed to the film cut, things worked well for thirty years. But ever since VHF radio began transmitting stereophonically, that is, two amplitudes per unit of time, fade-outs have been "more difficult to execute": "the mise-en-scène, invisible yet localizable, cannot be dismantled and replaced by a new one in front of the listener as easily as in the case of a monophonic play."[10] Once tethered, such are the constraints produced by the real.

For one thing, optical data flows are two-dimensional; for another, they consist of high frequencies. Not two but thousands of units of light per unit of time must be transmitted in order to present the eye with a two- or even three-dimensional image. That requires an exponential magnification of processing capacities. And since light waves are electromagnetic frequencies in the terahertz range, that is, a trillion times faster than concert pitch (A), they outpace not only human writing hands but even (unbelievably) today's electronics.

Two reasons why film is not directly linked to the real. Instead of recording physical waves, generally speaking it only stores their chemical effects on its negatives. Optical signal processing in real time remains a thing of the future. And even if, following Rudolph Lothar's rather timely metaphysics of the heart, everything from sound to light is a wave (or hertz),[11] optical waves still don't have a storage or computing medium—not, at any rate, until fiber technologies running at the speed of light have put today's semiconductors out of business.

A medium that is unable to trace the amplitudes of its input data is permitted a priori to perform cuts. Otherwise, there would be no data. Since Muybridge's experimental arrangement, all film sequences have been scans, excerpts, selections. And every cinematic aesthetic has developed from the 24-frame-per-second shot, which was later standardized. Stop trick and montage, slow motion and time lapse only translate technology into the desires of the audience. As phantasms of our deluded eyes, cuts reproduce the continuities and regularities of motion. Phonography and feature film correspond to one another as do the real and the imaginary.

But this imaginary realm had to be conquered. The path of invention, from Muybridge's first serial photographs to Edison's kinetoscope and the Lumière brothers, does not merely presuppose the existence of celluloid. In the age of organic life stories (as poetry) and organic world histories (as philosophy), even in the age of mathematical continuity, caesuras first had to be postulated. Aside from the material precondition, the spliceable celluloid, there was a scientific one: the system of possible deceptions of the eye had to be converted from a type of knowledge specific to illusionists and magicians (such as Houdini) to one shared by physiologists and engineers. Just as the phonograph (Villiers de l'Isle-Adam notwithstanding) became possible only after acoustics had been made an object of scientific investigation, so "cinematography would never have been invented" had not "researchers been occupied with the consequences of the stroboscopic effect and afterimages."[12]

Afterimages, which are much more common and familiar than the stroboscopic effect, were already present in Goethe's *Theory of Colors*—but only, as in *Wilhelm Meister's Years of Apprenticeship*, to illustrate the effects of Classic-Romantic literature on souls: a woman hovers in front of the inner eye of the hero or the readers as the optical model of perfect alphabetization, even though her beauty simply cannot be recorded in words. Wilhelm Meister observes to himself and his like-minded readers, "If you close your eyes, she will present herself to you; if you open them, she will hover before all objects like the manifestation which a dazzling image leaves behind in the eye. Was not the quickly passing figure of the Amazon ever present in your imagination?"[13] For Novalis, imagination was the miraculous sense that could replace for readers all of their senses.

At least as long as Goethe and his *Theory of Colors* were alive. For it was Fechner who first examined the afterimage effect with experimental rigor. Experimenter and subject in one, he stared into the sun—with the result that he went blind in 1839 for three years and had to resign from his physics chair at the University of Leipzig. The historical step from psychology to psychophysics (Fechner's beautiful neologism) was as consequential as the emergence of modern media from the physiological handicaps of its researchers was literal.

No wonder, then, that the aesthetics of the afterimage effect is also due to a half-blind person. Nietzsche, the philosopher with −14 diopters,[14] produced a film theory before its time under the pretext of describing both *The Birth of Tragedy* in ancient Greece and its German rebirth in the mass spectacles of Wagner.[15] In Nietzsche, the theater performances that were produced in the shadeless midday sun of an Attic setting were transformed into the hallucinations of inebriated or visionary spectators, whose optic nerves quite unconsciously processed white-and-black film negatives into black-and-white film positives: "After an energetic attempt to focus on the sun, we have, by way of remedy almost, dark spots before our eyes when we turn away. Conversely, the luminous images of the Sophoclean heroes—those Apollonian masks—are the necessary productions of a deep look into the horror of nature; luminous spots, as it were, designed to cure an eye hurt by the ghastly night."[16]

Prior to Fechner's historical self-experiment, blinding was not a matter of desire. An eye hurt by the ghastly night that requires for its remedy inverted afterimage effects is no longer directed toward the stage of the Attic amphitheater but onto the black surface of soon-to-come movie screens, as the Lumière brothers will develop them in defiance of their name. Nietzsche's ghastly night is the first attempt to christen sensory deprivation as the background to and other of all technological media.[17] That the flow of data takes place at all is the elementary fact of Nietzsche's aesthetic, which renders interpretations, reflections, and valuations of individual beauty (and hence everything Apollonian) secondary. If "the world" can be "justified to all eternity . . . only as an aesthetic product,"[18] it is simply because "luminous images" obliterate a remorseless blackness.

The Nietzsche movie called *Oedipus* is technological enough to predate the innovation of the Lumières by a quarter century. According to *The Birth of Tragedy*, a tragic hero, as inebriated spectators visually hallucinate him, is "at bottom no more than a luminous shape projected onto a dark wall, that is to say, *appearance* through and through."[19] It is precisely this dark wall, which allows actors to turn into the imaginary, or film stars, in the first place, that has been opening theater performances since 1876, the year of the inauguration of the theater in Bayreuth, whose prophecy *The Birth of Tragedy* undertook. Wagner did what no dramaturg before him had dared to do (simply because certain spectators insisted on the feudal privilege of being as visible as the actors themselves): during opening night, he began *The Ring of the Nibelung* in total darkness, before gradually turning on the (as yet novel) gaslights. Not even the presence of an emperor, Wilhelm I, prevented Wagner from reducing his audience to an invisible mass sociology and the bodies of actors (such as the Rhine maidens) to visual hallucinations or afterimages against the background of darkness.[20] The cut separating theater arts and media technologies could not be delineated more precisely. Which is why all movie theaters, at the beginning of their screenings, reproduce Wagner's cosmic sunrise emerging from primordial darkness. A 1913 movie theater in Mannheim, as we know from the first sociology of cinema, used the slogan, "Come in, our movie theater is the darkest in the whole city!"[21]

Already in 1891, four years prior to the projection screen of the Lumière brothers, Bayreuth was technologically up to date. Not for nothing did Wagner joke that he would have to complete his invention of an invisible orchestra by inventing invisible actors.[22] Hence his son-in-law, the subsequently notorious Chamberlain, planned the performance of symphonies by Liszt that would have become pure feature films with equally pure film music: accompanied by the sound of an orchestra sunk in Wagnerian fashion, and situated in a "nightclad room," a camera obscura was supposed to project moving pictures against a "background" until all spectators fell into "ecstasies."[23] Such enchantments were unthinkable with old-fashioned viewing: eyes did not mix up statues or paintings, or the bodies of actors, for that matter—those basic stage props of the established arts—with their own retinal processes. Thanks only to Chamberlain's plans and their global dissemination by Hollywood, the physiological theory of perception becomes applied perceptual practice: moviegoers, following Edgar Morin's brilliant formulation, "respond to the projection screen like a retina inverted to the outside that is remotely connected to the brain."[24] And each image leaves an afterimage.

In order to implement the stroboscopic effect, the second theoretical condition of cinema, with the same precision, one needs only to illuminate moving objects with one of the light sources that have become omnipresent and omnipotent since the 1890s. As is widely known, back then Westinghouse won out over Edison, alternating current over direct current, as a public utility. The glow of light alternates fifty times per second in European lightbulbs, sixty times in American ones: the uncomplicated, and hence imperceptible, rhythm of our evenings and of an antenna called the body.

The stroboscopic illumination transforms the continuous flow of movement into interferences, or moirés, as can be seen in the wheeling spokes of every Western. This second and imaginary continuity evolved from discontinuity, a discovery that was first made by physiologists during the founding age of modern media. We owe a large part of the theory of alternating current to Faraday, as well as to the study *On a Peculiar Class of Optical Deceptions* (1831).[25] Coupled with the afterimage effect, Faraday's stroboscopic effect became the necessary and sufficient condition for the illusions of cinema. One only had to automatize the cutting mechanism, cover the film reel with a wing disk between moments of exposure and with a Maltese cross during moments of projection, and the eye saw seamless motion rather than 24 single and still shots. One perforated rotating disk during the recording and projection of pictures made possible the film trick preceding all film tricks.

Chopping or cutting in the real, fusion or flow in the imaginary—the entire research history of cinema revolves only around this paradox. The problem of undermining the threshold of audience perception through Faraday's "deceptions" reflected the inverse problem of undermining the threshold of perception of psychophysics itself to avoid disappointment or reality. Because real motion (above and beyond optical illusions) was to become recordable, the prehistory of the cinema began exactly as that of the gramophone. Étienne-Jules Marey, professor of natural history at the Collège de France in Paris, and later (following his successful film experiments) president of the French photographic society,[26] earned his initial fame with a sphygmograph copied from the work of German physiologists that was capable of recording pulse rates onto soot-covered glass plates as curves.[27] In the same way, Weber and Scott had mechanically stored sounds (musical intervals themselves) that were not acoustic illusions.

Beginning with heart muscle contractions, Marey investigated move-

Marey's chronophotographic gun.

ment in general. His chronographic experiments on humans, animals, birds—published as *La machine animale* (1873), a title that does justice to La Mettrie—inspired Governor Stanford of California to give Muybridge his assignment. The professional photographer only had to replace Marey's mechanized form of trace detection with a more appropriate, or professional, optical one—and where eyes had always seen only poetic wing-flaps could begin the analysis of the flight of birds, the precondition for all future aircraft constructions. It was no coincidence that pioneers of photography such as Nadar opted against the *montgolfières* of 1783 and in favor of literal airships: for flying machines heavier than air.[28] "Cinema Isn't I See, It's I Fly,"[29] says Virilio's *War and Cinema*, in view of the historically perfect collusion of world wars, reconnaissance squadrons, and cinematography.

In the meantime, the first photographs from *Animal Locomotion* had hardly appeared when Marey began work on improving Muybridge's improvement of his own work. The time was ripe for engineers to work together, for innovations of innovations. Marey also stored motion optically, but he reduced the number of cameras from the twelve of his predecessor to one and constructed—first with fixed photo glass plates, and, from 1888 on, with modern celluloid[30]—the first serial-shot camera. Instead of indulging in what Pynchon called "the American vice of modular repetition,"[31] he realized that for moving objects, a single, movable apparatus was enough. Its name—the chronophotographic gun—spoke nothing but the real truth.

It was in 1861, whilst traveling on a paddle-steamer and watching its wheel, that the future Colonel Gatling hit upon the idea of a cylindrical, crank-driven machine gun. In 1874 the Frenchman Jules Janssen took inspiration from the multi-chambered Colt (patented in 1832) to invent an astronomical revolving unit that could take a series of photographs [when attached to a telescope]. On the basis of this idea, Étienne-Jules Marey then perfected his chronophotographic rifle, which allowed its user to aim at and photograph an object moving through space.[32]

The history of the movie camera thus coincides with the history of automatic weapons. The transport of pictures only repeats the transport of bullets. In order to focus on and fix objects moving through space, such as people, there are two procedures: to shoot and to film. In the principle of cinema resides mechanized death as it was invented in the nineteenth century: the death no longer of one's immediate opponent but of serial nonhumans. Colt's revolver aimed at hordes of Indians, Gatling's or Maxim's machine-gun (at least that is what they had originally been designed to do) at aboriginal peoples.[33]

With the chronophotographic gun, mechanized death was perfected: its transmission coincided with its storage. What the machine gun annihilated the camera made immortal. During the war in Vietnam, U.S. Marine Corps divisions were willing to engage in action and death only when TV crews from ABC, CBS, and NBC were on location. Film is an immea-

André Malraux, *Espoir.*

surable expansion of the realms of the dead, during and even before bullets hit their targets. A single machine-gun (according to Jünger's observation on *Der Arbeiter*) finishes off the fraternity-based heroism of entire Langemarck regiments of 1914;[34] a single camera does the same with the dying scenes thereafter.

It was then only a matter of combining the procedures of shooting and filming to take Marey's brand name literally. The chronophotographic gun became reality in the cinema of artificial, that is, lethal, bird flights. Reconnaissance pilots of the First World War such as Richard Garros constructed an on-board machine-gun whose barrel was pointed parallel to the axis of the propeller while they filmed its effects.[35] During the Second World War, which according to General von Fritsch was supposed to have been won by superior reconnaissance, "the construction of recording devices within aircraft yielded still better results." Major General von Wedel, chief of Army Propaganda, was "especially delighted that Inspector Tannenberg was successful in having developed a camera unit that could be built into fighter planes, Stukas, and other aircraft and that, synchronized with the weapon, made possible very impressive combat pictures."[36]

As if targeting Inspector Tannenberg and his appropriate name,[37] Pyn-

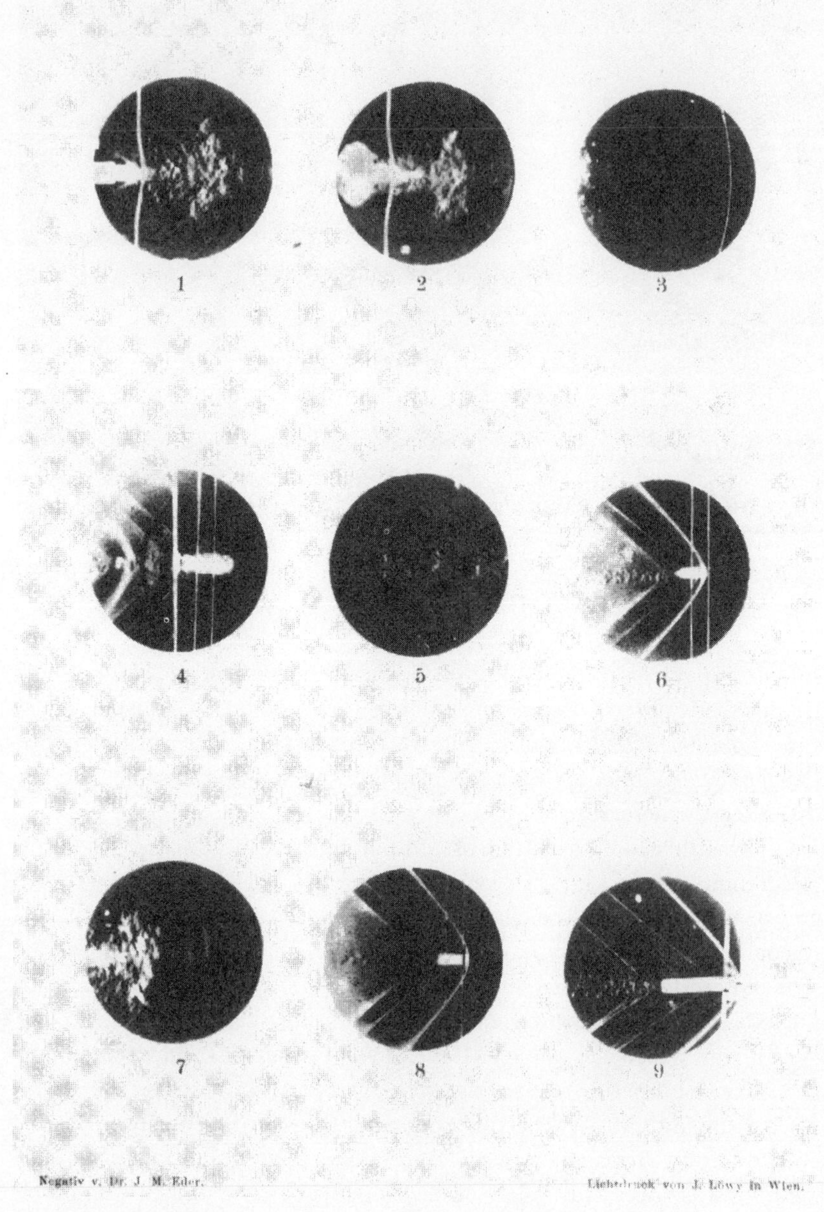

Ernst Mach, freeze-frame photos of bullets.

chon describes in *Gravity's Rainbow* "this strange connection between the German mind and the rapid flashing of successive stills to counterfeit movement, for at least two centuries—since Leibniz, in the process of inventing calculus, used the same approach to break up the trajectories of cannonballs through the air."[38] That is how venerable (in strict accordance with Münsterberg) the prehistory of cinema is. But it makes a difference whether ballistic analysis appears on the paper of a mathematician or on celluloid. Only freeze-frame photographs of flying projectiles, developed in 1885 by one no less than Ernst Mach, made visible all interferences, or moirés, in the medium of the air. Only freeze-frame photographs run automatically and as real-time analysis (since then, TV cameras have reduced the processing time of pictures to near-zero). Which is why Inspector Tannenberg's propaganda weapon still had or has a future: toward the end of the Second World War, when even 8.8 millimeter anti-aircraft guns with their teams of operators were ineffective against the Allied carpet bombings of Germany, the first developments toward our strategic present took place—the search by technicians for weapons systems with automatic target searching.[39] The chronophotograph was made for that.

Built into aircraft, TV cameras or infrared sensors are no longer the owls of Minerva, lagging behind so-called real history like Hegel's nightly philosophy. The kinds of infinitesimal movement they process through integration and differentiation are much more efficient: with servomotors electrically linked to a missile guidance system, they can hone in on the enemy target. Until camera and target, intercept missile and fighter aircraft, explode in a flash of lightning, a blitz.

Today's cruise missiles proceed in the same fashion, for they compare a built-in film of Europe's topography (from Hessia to Belarus, from Sicily to Ukraine) with their actual flight path in order to correct any possible deviations. Marey's chronophotographic gun has reached its target in all its senses. When a camera blows up two weapons systems simultaneously, and more elegantly than kamikaze pilots did, the analysis and synthesis of movement have become one.

At the end of *Gravity's Rainbow*, a V2—the first cruise missile in the history of warfare, developed at the Peenemünde Army Test Site—explodes over the Orpheus movie theater in Los Angeles. In grandiose time axis manipulation, which a fictitious drug by the name of Oneirine grants the whole novel,[40] the launch is correctly dated March 1945, but the rocket does not hit its target until 1970, when the novel was written. That is how interminably world wars go on, not least because of German-American technological transfer. The off-ground detonator of the V2 kicks in, and a ton of Amatol, the rocket's payload, explodes. Shortly thereafter, the image on the screen dissolves, as if the projection bulb were blowing out, but only so that its orphic truth can shine forth. We, "old fans, who've always been at the movies," are finally reached by a film "we have not learned to see"[41] but have been hankering after since Muybridge and Marey: the melding of cinema and war.

Nothing, therefore, prevented the weapons-system movie camera from aiming at humans as well. On the three fronts of war, disease, and criminality—the major lines of combat of every invasion by media—serial photography entered into everyday life in order to bring about new bodies.

As is widely known, during the First World War the barrels of machine-guns moved away from the black, yellow, and red skins against which they had been developed and started aiming at white targets. Movie cameras, however, kept pace and experienced a boom that might have been a misuse of army property (as with AM radio). At any rate, Münsterberg, who had to know about it, since he sought to prevent the outbreak of the German-American war in futile fireside chats with President Wilson up to the very end (and who, for that reason, remains unacknowledged by his colleagues at Harvard to this very day)[42]—Münsterberg wrote in 1916:

> It is claimed that the producers in America disliked these topical pictures because the accidental character of the events makes the production irregular and interferes too much with the steady preparation of the photoplays. Only when the war broke out, the great wave of excitement swept away this apathy. The pictures from the trenches, the marches of the troops, the life of the prisoners, the movements of the leaders, the busy life behind the front, and the action of the big guns absorbed the popular interest in every corner of the world. While the picturesque old-time war reporter has almost disappeared, the moving picture man has inherited all his courage, patience, sensationalism, and spirit of adventure.[43]

And as with the reporters, so with the stars of the new medium. Shortly after the trench war, when the *Soul of the Cinema* was in demand again, Dr. Walter Bloem, S.J., explained what was at the center of the sensationalism critiqued by Münsterberg: "During the war, film actors busily studied the thousands of dead, the results of which we can now admire on the screen."[44]

Since April 1917, the founding days of radio entertainment for army radio operators as well, such studies had been resting on a solid foundation. The chiefs of the new Army High Command, Hindenburg and Ludendorff, were serious about total war, and for that reason (among others) they advanced to the top of Germany's film directors. What evolved in the Grand General Staff was a Bureau of Pictures and Film [BUFA; *Bild- und Filmamt*] "whose founding and mode of operation was kept rather secret." Still, it is known that the bureau's "range of operations" included "supplying the inland and the front with films, setting up field movie theaters, the placement of war reporters, . . . censoring all films to be imported and exported, as well as providing all censoring agencies with instructions from the governing military censorship authorities."[45]

The way Ludendorff justified these changes is more than just memorable; it has made film history. A memo by the general quartermaster led via the chain of command to the founding of the UFA. As a major corporation, UFA was to take over the classified assignments of the Bureau of Pictures and Film in a much more public and efficient way—from the end of the First World War until, as is widely known, the end of the Second:

> Chief of the General Staff of the Army. HQ. 4 July 1917
> M.J. No. 20851P.
> To the
> Imperial War Ministry Berlin
>
> The war has demonstrated the overwhelming power of images and films as a form of reconnaissance and persuasion. Unfortunately, our enemies have exploited their know-how in this area so thoroughly that we have suffered severe damage. Even for the more distant continuation of the war, film will not lose its significance as a political and military means of influence. Precisely for that reason, for a successful conclusion to the war it is absolutely imperative that film have a maximal effect in those areas where German intervention is still possible.
>
> *signed Ludendorff*[46]

Thus, film as a means of reconnaissance and persuasion has been explained, or reconnoitered, in the strictest (that is, military) sense of the term.

The path leads, as with radio, from interception to reception and mass mediality. And Ludendorff donated 900 of his movie theaters at the front to this reception, making it possible to decode Lieutenant Jünger's *Combat as Inner Experience*.

Positional warfare prohibits inner experience in Goethean terms, that is, sensory substitutions between the lines of literature. In both his title and his subject, Jünger announces a very different type of sensuousness: "When red life clashes against the black cliffs of death, what we get are sharp pictures composed of bright colors. . . . There is no time to read one's Werther with teary eyes."[47] For media-technological reasons, poetry comes to an end in the trenches, those "pure brainmills": "This failure even appears to be a matter of writing," says a fellow officer and friend of Jünger whose "intellectual faculties, in the daily rhythm between watch duty and sleep, gradually dwindle toward zero." Which the troop leader and recipient of the Ordre pour le Mérite demonstrates and confirms with his telegram-style answer, "that this war is a chokehold on our literature."[48]

But ghosts, a.k.a. media, cannot die at all. Where one stops, another somewhere begins. Literature dies not in the no-man's-land between the trenches but in that of technological reproducibility. Again and again, Lieutenant Jünger asserts how completely the inner experience of the battle has become a matter of neurophysiology. After the "baptism by fire" of 1914, soldiers had become "so cerebral that the landscape and the events, in retrospect, managed to escape from memory only as dark and dreamlike shadows."[49] Even more clearly, and in terms of radio: "Every brain, from the simplest to the most complicated, vibrated with the waves of the monstrous, which propagated itself over the landscape."[50] The war, even though "it was so palpable, and rested heavily, like lead, on our senses"—as when, for example, "an abandoned group traversed unknown territory under the canopy of night"—was hence and simultaneously "perhaps only a phantasm of our brains."[51]

Brain phantasms, however, "glowing visions"[52] that "burden anxious brains"[53] like the trenches: they exist only as the correlatives to technological media. The soul becomes a neurophysiological apparatus only when the end of literature draws near. Hence, the "screams from the dark" that "touch the soul most immediately, . . . since all languages and poets, by contrast, are only stammerings," combine the "clamor of fighters" with "the automatic play of the barrel-organ."[54] And as with acoustics, so with the optics of war: "Once again, one's individual experience, the individual, . . . was compressed, once again the colorful world rolled like a swift film through the brain."[55]

In the days of the founding age of modern media, the neurologist Benedict described how the dying visualize their past as time-lapse photography. Lieutenant Jünger could do this without pseudomorphosis. After one of his "fourteen"[56] war injuries, he was, for purposes of reconvalescence, relocated to Douchy, a village and communications site in Flanders, "the headquarters of the 73d [light-infantry regiment]."[57] "There was a reading room, a café, and later even a cinema in a large barn skillfully converted."[58]

Only in *Storm of Steel*, his fact-based *Diary of a German Storm-Troop Officer*, does Jünger speak of the BUFA and its work: "supplying the inland and the front with films, setting up field movie theaters," and so on. In *Combat as Inner Experience*, this hymn to the trench worker, he does not so much ignore media technology as translate its effects expressionistically. Writing itself relocates in the projection room of Douchy. That is why and why alone "the blossoms of the world, blinding and benumbing, cities on waters of light, southern coasts where blue waves washed against the shore, women cast in satin, queens of boulevards," and the whole range of feature-film archives of inner experiences, "opened themselves" up to the "wandering brains"[59] of soldiers in the trenches, even in their darkest moments of sensory deprivation.

One year before the outbreak of the war, Kurth Pinthus's *Movie Book* announced: "One has to get used to the thought that kitsch will never be eliminated from the world of humans. After we've been trying for decades to get rid of kitsch in the theater, it resurfaces in cinema. And one is led to believe that the masses have found the kitsch expelled from the stage somewhere else."[60]

In a world war, for example: "All hearts pound with excitement when the armies of soldiers line up for battle with desperately harsh faces; when grenades burst, releasing a shower of smoke; and when the camera relentlessly traverses the battlefield, ingesting the stiff and mutilated bodies of senselessly killed warriors."[61]

A prophecy that Jünger, the mythic war reporter, realizes or recognizes. To recognize combat as an inner experience means (following Ludendorff) understanding that the use of film "in those areas where German intervention is still possible" is "absolutely imperative . . . for a successful conclusion to the war." For although historical prose suggests, as is widely known, that the other side won, Jünger's camera style drives forward German attacks again and again, only to freeze the continuation of history or the movies in a last still. In the final analysis, such a film trick becomes possible simply because in mechanized warfare, machine-gun operators kill without seeing any corpses,[62] and storm troopers—Ludendorff's newly formed precursors to the blitzkrieg[63]—storm without seeing into enemy trenches.

That is why the British, when their attack tears Jünger out of his filmic "castle in the air," appear only "for one second . . . like a vision engraved . . . on my eyes."[64] That is why the novel succeeds in letting its end, its goal and wish fulfillment—namely, the failed Ludendorff offensive of "March 21, 1918"[65]—succeed in the world of hallucination. As a camera shot, and after "an eternity in the trenches,"[66] an attack is nothing short of redemption:

Only rarely does the enemy appear to us . . . in flesh and blood, even though we are separated only by a narrow, torn-up field strip. We've been hunkering down in the trenches for weeks and months, swarms of projectiles showering down upon us, surrounded by thunderstorms. It happens that we almost forget we are fighting against human beings. The enemy manifests itself as the unfolding of a gigantic, impersonal power, as fate that thrusts its fist into the unseen.

When we storm forward and climb out of the trenches, and we see the empty, unknown land in front of us where death goes about its business between flaring columns of smoke, it appears as if a new dimension has opened up to us. Then we suddenly see up close, in camouflage coats and in faces covered with mud like a ghostly apparition, what awaits us in the land of the dead: the enemy. That is an unforgettable moment.

How differently one had envisioned the scene. The blooming edge of a forest, a flowery meadow, and guns banging into the spring. Death as a flurrying back and forth between the two trench lines of twenty-year-olds. Dark blood on green blades of grass, bayonets in the morning light, trumpets and flags, a happy, shimmering dance.[67]

But contemporary technologies of the body have done their duty, in military as well as choreographic terms. When war and cinema coincide, a communications zone becomes the front, the medium of propaganda becomes perception, and the movie theater of Douchy the scheme or schemes for an otherwise invisible enemy. "When our storm signals flash across, [the English] get ready for a wrestling match about bits and pieces of trenches, forests, and the edge of villages. But when we clash in the haze of fire and smoke, then we become one, then we become two parts of one force, fused into one body."[68] Lieutenant Jünger meets his imaginary other, as Lacan will define it in 1936: as a mirror image that might restore the body of the soldier, dismembered fourteen times, back to wholeness.[69] If only were there no war and the other not a doppelgänger. For "all cruelty, all the compilation of the most ingenious brutalities, cannot fill a human being with as much horror as the momentary apparition of his mirror image appearing in front of him, [with] all the fiery marks of prehistory reflected in his distorted face."[70]

Jünger's film breaks off at precisely this image, long before *Gravity's Rainbow* ends in the blackout of a real or filmed rocket hit above Universal Studios of California. For once the enemy was recognized as a doppelgänger, "then, in the last fire, the dark curtain of horror may well have lifted in the brains, but what was behind, lying in wait, the rigid mouth could no longer speak."[71]

Ludendorff and Jünger's falling storm troopers are silent, either because (following a hermeneutic tautology) they are falling, or because (following a media-technological analysis) their a priori is the silent film. Now, however, we have war films with sound that can spell out the puzzle behind the dark curtain of horror. What was lying in wait were first of all facts that Jünger systematically bypassed: the failure of the Ludendorff offensive, the retreat to the Siegfried position, and capitulation. Second, and more horrific still, the film doppelgänger harbored the possibility of fiction. A cinematic war may not even take place at all. Invisible enemies that materialize only for seconds and as ghostly apparitions can hardly be said any longer to be killed: they are protected from death by the false immortality of ghosts.

In *Gravity's Rainbow*, the novel about the Second World War itself, GI von Held asks celebrated film director Gerhardt von Göll (alias Springer, Lubitsch, Pabst, etc.) about the fate of a German rocket technologist who had fallen into the hands of the Red Army:

> "But what if they did *shoot* him?"
>
> "No. They weren't supposed to."
>
> "Springer. This ain't the fucking *movies* now, come on."
>
> "Not yet. Maybe not quite yet. You'd better enjoy it while you can. Someday, when the film is fast enough, the equipment pocket-size and burdenless and selling at people's prices, the lights and booms no longer necessary, *then* . . . then . . . "[72]

Total use of media instead of total literacy: sound film and video cameras as mass entertainment liquidate the real event. In *Storm of Steel* nobody except for the diary keeper survives, in *Gravity's Rainbow* all the people pronounced dead return, even the rocket technician of Peenemünde. Under the influence of the fictitious drug Oneirine, the writing of world-war novels turns into movie fiction.

It is widely known that war—from the sandbox models of the Prussian General Staff to the computer games of the Americans—has become increasingly simulable. "But there, too," as these same general staffs wisely recognized, "the last question remains unanswered, because death and the enemy cannot 'be factored in realistically.'"[73] Friedlaender, media-technologically as always, has drawn from this the daring, inverted conclusion: for death in battle to coincide with cinema would be its own death.

SALOMO FRIEDLAENDER, "FATA MORGANA MACHINE" (CA. 1920)

For many years Professor Pschorr had been preoccupied with one of the most interesting problems of film: his ideal was to achieve the optical reproduction of nature, art, and fantasy through a stereoscopic projection apparatus that would place its three-dimensional constructs into space without the aid of a projection screen. Up to this point, film and other forms of photography had been pursued only in one-eyed fashion. Pschorr used stereoscopic double lenses everywhere and, eventually, indeed achieved three-dimensional constructs that were detached from the surface of the projection screen. When he had come that close to his ideal, he approached the Minister of War to lecture him about it. "But my dear Professor," the Minister smiled, "what has your apparatus got to do with our technology of maneuvers and war?" The Professor looked at him with astonishment and imperceptibly shook his inventive head. It was incredible to him that the Minister did not have the foresight to recognize how important that apparatus was destined to become in times of war and peace. "Dear Minister," he insisted, "would you permit me to take some shots of the maneuver so that you can convince yourself of the advantages of my apparatus?" "I'd rather not," the Minister contemplated, "but you are trustworthy. You know the dangerous article on high treason, of course, and will surely keep the secret." He granted the Professor unlimited access. A couple of weeks after the maneuver, all the generals gathered in open terrain that was in part rolling, mountainous, and wooded, and that contained several large ponds and ravines, slopes, and a couple of villages. "First, dear Minister and honored generals, allow me to tell you that the whole landscape, including our own bodies, appears as nothing but a single, purely optical phantasmagoria. What is purely optical in it I will make disappear by superimposing projections of other things onto it." He variously combined beams of floodlights and switched on a film reel, which began to run. Immediately the terrain transformed: forests became houses, villages became deserts, lakes and ravines became charming meadows; and suddenly one could see bustling military personnel engaged in battle. Of course, as they were stepping or riding into a meadow, they disappeared into a pond or a ravine. Indeed, even the troops themselves were frequently only optical illusions, so that real troops could no longer distinguish them from fake ones, and hence engaged in involuntary deceptions. Artillery lines appeared as pure optical illusions. "Since the possibility exists of combining, precisely and simultaneously, optical with acoustic effects, these visible but untouchable cannons can boom as well, making the illusion perfect," said Pschorr. "By the way, this invention is of course also useful for peaceful purposes. From now on, however, it will be very dangerous to distinguish things that are only visible from touchable ones. But life will become all the more interesting for it." Following this he let a bomber squadron appear on the horizon. Well, the bombs were dropped, but they did their terrible damage only for the eye. Strangely enough, the Minister of War in the end decided against purchasing the apparatus. Full of anger, he claimed that war would become an impossibility that way. When the somewhat overly humanistic Pschorr exalted that effect, the Minister erupted: "You cannot turn to the Minister of War to put a dreadful end to war. That falls under the purview of my colleague, the Minister of Culture." As the Minister of Culture prepared to buy the apparatus, his plans were vetoed by the Minister of Finance. In brief: the state was unwilling to buy. Now the film corporation (the largest film trust) helped itself. Ever since this moment, film has become all-powerful in the world; but only through optical means. It is, quite simply, nature once again, in all its visibility and audibility. When a storm is brewing, for example, it is unclear whether this storm is only optically real or a real one through and through. Abnossah Pschorr has been exercising arbitrary technological power over the fata morgana, so that even the Orient fell into confusion when a recent fata morgana produced by solely technical means—conjuring Berlin and Potsdam for desert nomads—was taken for real. Pschorr rents out every desired landscape to innkeepers. Surrounding Kulick's Hotel zur Wehmut these days is the Vierwaldstätter Lake. Herr v. Ohnehin enjoys his purely optical spouse. Mullack the proletarian resides in a purely optical palace, and billionaires protect their castles through their optical conversion into shacks.

Not too long ago, a doppelgänger factory was established. . . . In the not too distant future, there will be whole cities made of light; entirely different constellations not only in the planetarium, but everywhere in nature as well. Pschorr predicts that we will also be able to have technological control over touch in a similar way: not until then will radio traffic with real bodies set in, which means not just film but life, and which will leave far behind all traffic technologies . . .

The Minister of War's question about what Pschorr's apparatus has to do with the technology of maneuvers and war is the only fiction in Friedlaender's text. Even in its experimental prehistory, that is, even before it became cinema, film conditioned new bodies. But ministers of war were in touch with current developments.

In 1891, Georges Demeny, Marey's assistant and anatomist at the Institute, began work on his *Photography of Speech*. Initially, his purpose in conducting this strange exercise was to advance the breakdown of discourse into separate subroutines. In his experiments, motoric and optic data were to be on an equal footing with the sensory and acoustic data derived from Edison's phonograph. And Marey's silent chronophotograph was the perfect instrument for their storage.

Hence a serial camera with shutter speeds in the milliseconds was aimed at Demeny himself, who adhered to the honor—common during the founding age of modern media—of performing simultaneously as experimenter and subject, priest and victim of the apparatus. A human mouth opened, expectorated the syllables "Vi-ve la Fran-ce!" and closed again, while the camera dissected, enlarged, stored, and immortalized its successive positions, including the "fine play of all facial muscles," in component parts with a frequency of 16 Hz. To contemporaries, "many of these oral movements appeared exaggerated because our eye cannot perceive fleeting movements such as these, but the camera makes them visible by bringing motion to a standstill."[74] But that was precisely the point. Edison was rumored to have been enamored of the enlarged shots of his colleague's mouth.[75]

Based on the data of a freeze-framed patriotism, Demeny (fascinated by physical impairments, as is every media technologist) first revolutionized instruction for deaf and mute people. Patients of the Hôtel de Ville in Paris were asked to synthesize acoustically the mouth positions film had analyzed optically. Then they could—in "*oral* examinations" that proved sensational[76]—scream "Vi-ve la Fran-ce!" without ever hearing a syllable. In the material battles soon to come, when the Joffre divisions stormed and died like flies, self-perception was hardly necessary anymore.

"As early as 1892," Demeny "envisioned all the procedures that have since been in use in so-called cinematographic apparatuses and which are nothing but reversible chronophotographs."[77] One only would have had to follow the principles of the revolver and supplement a rotating photo storage device with a rotating photo projector. But even though Demeny was envious of the Lumières' success, research into slow motion was more important than the illusions of feature films. He remained faithful

Demeny says, “Vi-ve la Fran-ce!”

to the chronophotographic gun and moved from studying single patriotic mouths to masses of patriotic legs. On official assignment from the French army, he filmed the traditional goose step in order to optimize it.[78]

What *physiologists of art* (of all people) announced in 1897 as a new feedback loop between psychophysics, maneuver drills, and the unconscious was realized to the letter. Regarding the "condition we call '*thinking*,'" Georg Hirth wrote:

> That condition as well becomes automatized following frequent repetition; namely, when optical, acoustic, and other stimuli—which effect every closing apperception—recur in roughly periodic intervals and in a known intensity. Recall, for example, the activity of a marksman in a shooting gallery. At the beginning of his service, the man is thoroughly infused with the condition of conscious and prospective attentiveness: gradually, however, he becomes sure and relaxed; after each bullet hits he steps mechanically up front to show the mark. His attention can go for a walk—it returns to business only if the impact is delayed long enough for his automatic-rhythmic feeling to subside. The same is true for the recruit during his exercises. Indeed, the whole debate surrounding the length of active military service revolves around the question: how long does it take to *automatize* the military (moral as well as technical) memory structure of the average twenty-year-old in such a way that the apparatus does not fail in the real-life event and that the attention (attentiveness)—which every man must be equipped with at any time in times of war and peace—is not absorbed by mindless service?[79]

Mechanization Takes Command—Sigfried Giedion could not have come up with a better title for a book that retraces the path from Marey's chronophotographic gun via modern art to military-industrial ergonomics. The automatized weapons of world wars yet to come demanded similarly automatized, average people as "apparatuses" whose motions—in terms of both precision and speed—could only be controlled by filmic slow motion. Since they were introduced during revolutionary civil wars, exclamations such as "Vive la France!" had nurtured the death drive only psychologically and had left the reaction time at the gun to a "thinking" that exists only in quotation marks for physiologists of art and film.

Storm-troop leaders such as Jünger, however, have since Ludendorff been trained to work in time frames below any threshold of perception. The apparition of the enemy appears to them only "for one second," barely perceptible, but measurable. As Jünger notes immediately prior to the Ludendorff offensive, "phosphoric digits are glowing on the watch on my wrist. Watch digits, an unusual word.[80] It is 5:30. We'll begin to storm in one hour."[81] Two common items of today, trench coats (or, literally, "coats for the trenches") and watches with second hands, are the prod-

Giacomo Balla, *Ragazza che corre sul balcone* (study), 1912.

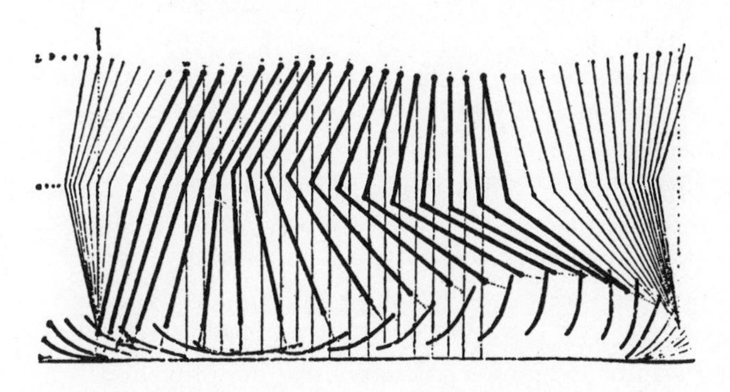

E. J. Marey, "Amplitudes of the Leg While Walking," before 1885.

ucts of the First World War.[82] In the standardized jump of the second hand, film transport imposes its rhythm upon average people. No wonder that storm-troop leader Jünger hallucinated the body of the enemy—that unreality hidden for months in the trenches—in the medium of film. The opponent could only be a film doppelgänger. Demeny, we recall, had standardized the movements of a whole army through chronophotography.

And Professor Pschorr, as always, only had to do his excessive share to transform the "bodies" of soldiers and the entire landscape of their maneuver into "a single, purely optical phantasmagoria," which, moreover, could be combined precisely and simultaneously with acoustic effects.

A fata morgana machine that can now be had around the globe. Without war, simply by paying an admission fee. For mechanization has also taken command over so-called times of leisure and peace. Night after night, every discotheque repeats Demeny's goose-step analyses. The stroboscopic effect at the beginning of film has left physiological labs and now chops up dancers twenty times per second into film images of themselves. The barrage of fire has left the major lines of combat and these days echoes from security systems—including their precise and simultaneous combination with optical effects. Demeny's photography of speech continues as a videoclip, his "Vi-ve la Fran-ce!" as a salad of syllables: "Dance the Mussolini! Dance the Adolf Hitler!"[83]

Deaf, mute, and blind, bodies are brought up to the reaction speed of World War n+1, as if housed in a gigantic simulation chamber. Computerized weapons systems are more demanding than automatized ones. If the joysticks of Atari video games make children illiterate, President Reagan welcomed them for just that reason: as a training ground for future bomber pilots. Every culture has its zones of preparation that fuse lust and power, optically, acoustically, and so on. Our discos are preparing our youths for a retaliatory strike.

War has always already been madness, film's other subject. Body movements, as they are provoked by the stroboscopes of today's discotheques, went by a psychopathological name a century ago: a "large hysterical arc." Wondrous ecstasies, twitchings without end, circus-like contortions of extremities were reason enough to call them up with all the means of hypnosis and auscultation. A lecture hall full of medical students, as yet all male, was allowed to watch the master, Charcot, and his female patients.

> A handwritten note [in the as yet unpublished archives of the Salpêtrière] gives an account of the session of November 25, 1877. The subject exhibits hysterical spasms; Charcot suspends an attack by placing first his hand, then the end of a baton, on the woman's ovaries. He withdraws the baton, and there is a fresh attack, which he accelerates by administering inhalations of amyl nitrate. The afflicted woman then cries out for the sex-baton in words that are devoid of metaphor: "G. is taken away and her delirium continues."[84]

The Salpêtrière makes iconographs of its hysteria.

But this performance was not, or not any longer, the truth about hysteria: what was produced by psychopathic media was not allowed simply to disappear in secret memories or documents. Technological media had to be able to store and reproduce it. Charcot, who transformed the Salpêtrière from a dilapidated insane asylum into a fully equipped research lab shortly after his appointment, ordered his chief technician in 1883 to start filming. Whereupon Albert Londe, later known as the constructor of the Rolleiflex camera,[85] anatomized (strictly following Muybridge and Marey) the "large hysterical arc" with serial cameras. A young physiology assistant from Vienna visiting the Salpêtrière was watching.[86] But Dr. Freud did not make the historical connection between films of hysteria and psychoanalysis. As in the case of phonography, he clung (in the face of other media) to the verbal medium and its new decomposition into letters.

For this purpose, Freud first stills the pictures that the bodies of his female patients produce: he puts them on his couch in the Berggasse. Then a talking cure is deployed against the images seen or hallucinated. Without mentioning the gender difference between male obsessive-neurosis and female hysteria, in *Studies on Hysteria* he observes:

> When memories return in the form of pictures our task is in general easier than when they return as thoughts. Hysterical patients, who are as a rule of a "visual" type, do not make such difficulties for the analyst as those with obsessions.
>
> Once a picture has emerged from the patient's memory, we may hear him say that it becomes fragmentary and obscure in proportion as he proceeds with the description of it. *The patient is, as it were, getting rid of it by turning it into*

> *words.* We go on to examine the memory picture itself in order to discover the direction in which our work is to proceed. "Look at the picture once more. Has it disappeared?" "Most of it, yes, but I still see this detail." "Then this residue must still mean something. Either you will see something new in addition to it, or something will occur to you in connection with it." When this work has been accomplished, the patient's field of vision is once more free and we can conjure up another picture. On other occasions, however, a picture of this kind will remain obstinately before the patient's inward eye, in spite of his having described it; and this is an indication to me that he still has something important to tell me about the topic of the picture. As soon as this has been done the picture vanishes, like a ghost that has been laid.[87]

Naturally, such sequences of images in hysterics or visually oriented people are an inner film: as in the case of psychoanalytical dream theory, a "pathogenic recollection," notwithstanding the patient's "forms of resistance and his pretexts," provokes its optical "reproduction."[88] When Otto Rank subjected *The Student of Prague*, as the second German *auteur* film, to psychoanalytical examination in 1914, he observed that "cinematography . . . in numerous ways reminds us of dream-work." Which, conversely, meant that internal images were modeled, as with hysterics, after the "shadowy, fleeting, but impressive scenes" of film. Consequently, "the technique of psychoanalysis," which "generally aims at uncovering deeply buried and significant psychic material, on occasion proceeding from the manifest surface evidence, . . . need not shy away from even some random and banal subject"—such as "the film-drama"—"if the matter at hand exhibits psychological problems whose sources and implications are not obvious."[89]

But this rather filmic uncovering, the return from the cinema to the soul, from manifest surface or celluloid skin to unconscious latency, from a technological to a psychic apparatus, only replaces images with words. While optical data in film are storable, they are also "shadowy, fleeting": one cannot look them up, as with books (or today's videotapes). This intangibility governs Rank's methodology. "Those whose concern is with literature may be reassured by the fact that the scenarist of this film, *The Student of Prague*, is an author currently in vogue and that he has adhered to prominent patterns, the effectiveness of which has been tested by time."[90] Which is why psychoanalysis (to paraphrase Freud) basically imitates the doppelgänger film by translating it into words. Rank's discussion of the doppelgänger quotes all available sources from 1800 on and turns movies back into literature.[91]

For a talking cure, nothing else is left to do. Still, after attending

Londe's filmings of hysteria, Freud did just the opposite with it. Literally, psychoanalysis means chopping up an internal film, in steps that are as methodical as they are discrete, until all of its images have disappeared. They break to pieces one by one, simply because female patients have to translate their visions into depictions or descriptions. In the end the medium of the psychoanalyst triumphs, because he stills bodily movements and slays the remaining, internal sightings like so many ghosts or Draculas. When Freud "unlocks images," he does so not to store them, as Charcot does, but to decode the puzzles of their signifiers. Thus, the emergence particularly of nonverbal storage technologies around 1900 leads to a differentiation that establishes discourse as a medium among media. Freud the writer is still willing to admit the competition of the phonograph, because gramophony (despite all its differences with the talking cure and its case-study novels) deals with words. The competition of silent film, however, Freud does not even acknowledge. And even if Abraham and Sachs operate as "psychoanalytical collaborators" on a 1926 project that makes *The Mysteries of the Unconscious* into a film, and hence teaches contemporaries "the necessities of modern-day education without pain and job training,"[92] Freud himself flatly denies an offer from Hollywood.

This differentiation of storage media decides the fate of madness. Psychoanalytical discourse, which, following Lacan's thesis, is a consequence and displacement of hysterical discourse, translates the most beautiful pathology into the symbolic. At the same time, the serial photography of psychiatry, understood as the trace detection it is, stores the real along the "great hysterical arc." Londe's still shots of each individual twitch and ecstasy travel (due to a lack of opportunities for projecting films) into the multivolume *Iconography of the Salpêtrière*. There they rest, but only to emigrate henceforth from the real and to return to the imaginary, for which Freud had no use. For although the "great hysterical arc" can no longer be found in the lecture halls of today's medical schools, the countless *Jugendstil* images of women, with their bows and twists, can only derive from this *iconographie photographique*.[93] Works of art of the *Jugendstil* did not simply suffer from the age of their technological reproducibility; in their style, they themselves reproduced measured data and hence practiced the precise application Muybridge had ascribed from the very beginning to his study of *Animal Locomotion*.

Hysteria, however, became as omnipresent as it became fleeting. In the real, it gave rise to archives of trace detection that returned in the imaginary of the paintings of the *Jugendstil*; in the symbolic, it gave rise to a science that returned in the female hysterics of Hofmannsthal's dra-

The *Jugendstil* makes iconographs of its hysteria.

mas.[94] One reproduction chased the other. With the result that madness might not take place under conditions of high technology. It becomes, like war, a simulacrum.

A successor to Londe, Dr. Hans Hennes of the Provinzial-Heil- und Pflegeanstalt Bonn, almost managed to figure out this ruse. His treatise on *Cinematography in the Service of Neurology and Psychiatry* identified only one appropriate medium for the "wealth of *hysterical motoric malfunctions*": filming. In a manner "more visual and complete than the best description" (and presumably photographs as well),[95] technological media reproduced psychopathological ones. But since serial photographs could be projected as films by 1909, Hennes went one step beyond Londe. Not until psychiatry was in a position "to convert a rapid succession of movements into a slow one through cinematographic reproduction" was it possible to see things "whose precise observation is, in real life, hardly or not at all possible."[96] As if cinema had enlarged the madness (of both patients and physicians) through the whole realm of unreality and fiction; as if Hennes had, in vague anticipation of McLuhan, understood the medium as the message. For "in all cases . . . it was typical that distraction from the symptoms of the disease and the suspension of external stimuli were sufficient to reduce, or almost completely eliminate, [hysterical] movements. By contrast, it is enough to draw attention to phenomena, or for the physician to examine the patient, even just step up to him, in order for dysfunctions to appear with greater intensity."[97]

That is how psychiatry—whose attention had lately been running on automatic pilot, that is, filmically—itself discovered Charcot's simple secret, long before Foucault ever did; namely, that every test produces what it allegedly only reproduces. According to Dr. Hennes, who is fearless about contradicting himself and could even describe the doctor's attention as a contraindication after recommending it a moment before, it is quite likely that there would be no madness without filming it:

> How often does it happen to the professor that a patient fails during a lecture, that a manic suddenly changes his mood, a catatonic suddenly fails to perform his stereotyped movements. Although he executed his pathological movements without disturbance on the ward, the changed environment of the lecture hall has the effect of not letting him produce his peculiarities—so that he does not display precisely what the professor wanted him to demonstrate. Other patients show their interesting oddities "maliciously," only when there are no lectures, continuing education courses, and so on. Such occurrences, which are frequently disturbing to the clinical lecturer, are almost completely corrected by the cinematograph. The person doing the filming is in a position to wait calmly for the *best possible moment* to make the recording. Once the filming is done, the pictures are available for reproduction at any moment. Film is always "in the mood." There are no failures.[98]

That means that films are more real than reality and that their so-called reproductions are, in reality, productions. A psychiatry beefed up by media technologies, a psychiatry loaded with scientific presumptions, flips over into an entertainment industry. In view of the "rapid dissemination of this invention and the unmatched popularity it has attained in such a short time,"[99] Hennes advises his profession to create, "through collective participation and collaboration, a *cinematographic archive* analogous to the phonographic one."[100]

Hence it is no wonder that the "great hysterical arc" disappears from nosology or the world shortly after its storage on film. Since there are "no failures" and mad people on film are "always 'in the mood,'" inmates of insane asylums can forgo their performances and withhold their "interesting oddities 'maliciously'" from all storage media. At the same time, psychiatrists no longer have to hunt for their ungrateful human demonstration material. The only thing they have to do is shoot silent films, which as such (through the isolation of movements from the context of all speech) already envelop their stars in an aura of madness. To say nothing of the many possible film tricks that could chop up and reassemble these body movements, until the simulacrum of madness was perfect.

The age of media (not just since Turing's game of imitation) renders indistinguishable what is human and what is machine, who is mad and who is faking it. If cinematographers can "correct in an almost perfect way" disturbing occurrences of non-madness, they might as well film paid actors instead of asylum inmates. Although the historiography of film presumes a line of development from fairground entertainment to expressionist film art, it is closer to the truth to speak of an elegant leap from experimental setups into an entertainment industry. Actors, that is, the doppelgängers of the psychiatrically engineered insane, visited the movie screen.

Certainly, Dr. Robert Wiene's *Cabinet of Dr. Caligari* (1920) seems to see cinema itself as part of the genealogy of the circus. The action as a whole confronts small-town life and vagrants. The titular hero appears as a traveling circus artist accompanied by a somnambulist medium who predicts the future for Caligari's paying customers. But the paths leading from the fairgrounds to Caligari are as tenuous as those leading (according to Siegfried Kracauer's simplified sociological reading) *From Caligari to Hitler*. In film and/or history, mass hysterias are, rather, the effect of massively used media technologies, which in turn have solid scientific foundation in theories of the unconscious. Caligari's wagon moves toward the motorcade of the Third Reich.

That is why Caligari's title of "Dr." remains the vacuous presumption of a charlatan only in Carl Mayer's and Hans Janowitz's draft of the screenplay, a charlatan who misuses his medium Cesare as a remote-controlled murder weapon and who ends up in a straitjacket in an insane asylum once his ruse has been found out. The fairground is conquered by an order whose disruptions not coincidentally have cost the lives of a municipal office worker and a youthful aesthete, two people, moreover, who are interested in books. As if screenplays as well had to defend script as their medium.

Following an idea of the great Fritz Lang,[101] however, the completed film frames the action in a way that represents not only the transvaluation of all values but also their enigmatization. Citizens and mad people exchange their roles. In the framed story, the youthful hero kills Caligari and in the process underscores his bourgeois media love for female readers and books. In the framing story, he turns mad, and driven by his crazed love he stalks another person in the asylum, the alleged lover of the female reader. His private war against Caligari shrinks to the optical hallucination of a paranoid. As if the film attempted to uncover the pathology of a medium that entwined reading and loving but has abdicated its power to film. The madwoman simply does not register loving glances anymore.

Caligari, however (or, at any rate, a face that looks just like his), towers above the insane asylum of the framing story as director and psychiatrist. No murder charges can prevail against his power to make a diagnosis such as paranoia. Apparently, "while the original story exposed the madness inherent in authority," the eventual film "glorified authority" simply because it "convicted its antagonist of madness."[102] But Kracauer's attack against undefined authorities fails to take into account a psychiatry whose effects have produced new beings, not just Carl Mayer's biographical experiences with German military psychiatrists during the war.[103]

It is precisely this indistinguishability between framed and framing story, between insanity and psychiatry, that does justice to film technology. Nothing prevents the asylum director in the narrative frame to act simultaneously as the mad Caligari. It is only that such ascriptions are communicated via the symbolic order of doctoral titles or the stories of patients, which are not part of the silent film. The identity between psychiatrist and murderer remains open-ended because it is offered to the eyes only and is not institutionalized by any word. A never-commented-upon similarity between faces renders all readings indistinguishable.

That is how faithfully Wiene's film follows cinematographically modernized psychiatry. When professorial media technologists of the founding age conduct their experiments, they simultaneously play project director and subject, murderer and victim, psychiatrist and madman, but storage technologies do not want to, and cannot, record this difference. Dr. Jekyll and Mr. Hyde, Stevenson's fictitious doppelgänger pair of

1886, are only the pseudonyms of actual privy councilors. A gramophone records the words of Stransky, the psychiatrist, as a salad of syllables; a chronophotograph records the patriotic grimacing of Demeny. The situation in Wiene's feature film is no different. Filmed psychiatrists go mad of necessity, especially if they, like the director of the asylum, declare an old book explicable in psychiatric as well as media-technological terms.

Somnambulism. A Compendium of the University of Uppsala. Published in the year 1726: Thus reads the *Fraktur*-lettered title of the book that the asylum director studies in order to learn everything about a historical "mystic, Dr. Caligari," and his "somnambulist by the name of Cesare." Likewise, Charcot and his assistants studied dust-covered files on witches and the obsessed as they were transforming mysticism into a psychiatrically proper diagnosis of hysteria.[104] The researchers of hypnosis, Dr. Freud and Dr. Caligari, are thus doppelgängers.[105] The one "found" the Oedipus complex for purposes of diagnosis and therapy initially "in my own case";[106] the other, according to the film's subtitle, "under the domination of a hallucination" reads a sentence in white letters written on the walls of the asylum: "YOU MUST BECOME CALIGARI." Charges that "the director" must, for one, be mad and, second, "be Caligari" are to no avail, because modern experimenters say or do the same thing much more clearly—namely, immorally—than bourgeois heroes do. The similarity between psychiatrists and madmen, an enigma throughout the whole film, originates from research strategies and technologies.

The fact that an asylum director is directed by hallucinated writings to become Caligari in the framed story is simply a film trick. An actor plays both roles. With celluloid and cuts (the weapons of Dr. Wiene) Dr. Caligari or his official doppelgänger emerges victorious.

It is only because of a life-size puppet that simulates Cesare sleeping in somnambulist stiffness that the title hero can provide his medium with protective alibis while executing nightly murders under the influence of hypnotic orders. The puppet deceives the bourgeois hero (as contemporary theories *On the Psychology of the Uncanny* predicted).[107] Prior to the introduction of stuntmen (and much to the dismay of aesthetes), films engaged in the "frequently used practice of replacing the artist with a puppet in particularly dangerous scenes."[108] Thus, Cesare is always already a silent movie medium, and it is for this reason alone that he can be a somnambulistic and murderous medium. The photograph taken with a camera obscura (the cabinet in the title of the film itself) learns to move; the *Iconographie photographique de la Salpêtrière* enters into Albert Londe's filming stage. As the mobilization of his puppet alibi, Cesare walks stiffly and with raised arms; he stumbles, tries to regain his balance, and finally rolls down a slope. Dr. Hennes describes in virtually identical terms the "accident hysteria" of his patient Johann L., who is "61 years old" and a "workhorse": "He walks in straddle-legged and stiff fashion, and often tilts as he turns around; moreover, he patters and walks in small steps; this gait is accompanied by grotesque ancillary movements of his arms, and is, in general, so bizarre that it appears artificially exaggerated." Nothing but indescribables, for which, however, "the cinematographic image presented a very vivid illustration and supplement."[109]

And that—when bizarreness and artificial exaggeration originate in a hypnotic command—is above all when pathology and experiment coincide once again. Cesare operates as the weapon of Caligari the artist. Psychiatrists constructed the first cruise missile systems, reusable systems to boot, long before cyberneticists did. With the serial murders of Cesare (and his numerous descendants in cinema), the seriality of film images enters plot itself. That is why his hypnosis hypnotizes moviegoers. In Wiene's pictures, they fall victim to a trompe l'oeil whose existence Lacan demonstrated through historical periods of painting: the incarnate look of a power that affected pictures long before it created them,[110] or that even produced that look as pictures. Yesterday the accident hysteric Johann L., today Cesare, tomorrow movie fans themselves. With the somnambulism of his medium, Dr. Caligari already programs "the collective hypnosis" into which the "darkness of the theatre and the glow of the screen"[111] transport an audience.

Film doppelgängers film filming itself. They demonstrate what happens to people who are in the line of fire of technological media. A motorized mirror image travels into the data banks of power.

Barbara La Marr, the subtitle heroine of a novel by Arnolt Bronnen with the cynical title *Film and Life* (1927), experienced it herself. She had just finished doing her first screen tests for Hollywood and was sitting next to the director Fitzmaurice in the darkened projection room while film buyers were examining her body.

> Barbara suddenly got frightened. She stopped breathing. She clutched her chest; was her heart still beating; what happened on the screen? Something terrible stared at her, something strange, ugly, unknown; that wasn't she, that couldn't be she who stared at her, looked to the left, to the right, laughed, cried, walked, fell, who was that? The reel rolled, the projectionist switched on the light. Fitzmaurice looked at her.
>
> "Well?" She regained her composure, smiled. "Oh. That is how angels up in

heaven must look down upon us, the way I look in this picture." Fitzmaurice disagreed laughingly: "I would never have thought of you as an angel. But that is not bad at all. In fact, just the opposite. Better than I thought. Much better." But she got up, trembling, it erupted from within her, almost screaming: "totally bad," she screamed, "terrible, ghastly, mean, I am completely untalented, nothing will become of me, nothing, nothing!"[112]

Film transforms life into a form of trace detection, just as literature during Goethe's time transformed truth into an educational discipline. Media, however, are ruthless, while art glosses over. One does not have to be hypnotized, like the mad Cesare, to become strange, ugly, unknown, terrible, ghastly, mean, in brief, "nothing" on the screen. It happens to each and everyone, at least before the plots of feature films (following the logic of phantasms and the real) begin to obscure the undesirable. A protagonist of one of Nabokov's novels goes to the movies with his girlfriend, unexpectedly sees his "doppelgänger" (following his brief engagement as a movie extra months earlier), and feels "not only shame but also a sense of the fleeting evanescence of human life."[113] Bronnen's title *Film and Life* hence repeats the classic line of the stick-up man, "Your money or your life!" Whoever chooses money loses his life anyway; whoever chooses life without money will die shortly thereafter.[114]

The reason is technological: films anatomize the imaginary picture of the body that endows humans (in contrast to animals) with a borrowed I and, for that reason, remains their great love. Precisely because the camera operates as a perfect mirror, it liquidates the fund of stored self-images in La Marr's psychic apparatus. On celluloid all gesticulations appear more ridiculous, on tapes, which bypass the skeletal sound transmission from larynx to ear, voices have no timbre, on ID cards (according to Pynchon, of whom no photo exists) a "vaguely criminal face" appears, "its soul snatched by the government camera as the guillotine shutter fell."[115] And all that not because media are lying but because their trace detection undermines the mirror stage. That is to say: the soul itself, whose technological rechristening is nothing but Lacan's mirror stage. In Bronnen's work, budding starlets must experience that, too.

Film is not for tender souls, Miss, . . . just like art in general. If you insist on showing your soul—which nobody else is interested in, by the way; we are far more interested in your body—you need to have a tough and hard-boiled soul; otherwise it won't work. But I don't think you will achieve any particularly great footage with your little indication of a soul. Let go of your soul without getting bent out of shape. I had to learn it myself, to let go of my inner self. Today I do films; back then I was poet.[116]

The true words of a deserter who has grasped the difference between media and the arts. Even the most poetic of words could not store bodies. The soul, the inner self, the individual: they all were only the effects of an illusion, neutralized through the hallucination of reading and widespread literacy. (*Alphabêtise*, as Lacan put it.)[117] When, in the last romantic comedy, Büchner's King Peter of the kingdom of Popo searched for his son Leonce, who was at large, he put the police of the Archduchy of Hessia once more in an embarrassing situation. They could only go by "the 'wanted poster,' the description, the certificate" of a "person," "subject," "individual," "delinquent," and so on: "walks on two feet, has two arms, also a mouth, a nose, two eyes. Distinguishing features: a highly dangerous individual."[118]

That is how far literature went when it came to storing bodies—to the point of individual generality, but no further. Which is why literary doppelgängers, which began to show up in Goethe's time, appeared principally to readers. In Goethe, Novalis, Chamisso, Musset—the unspecified warrant of the book's protagonist, whose appearance the texts leave open, always merged with the unspecified warrant of a reader, whom the texts addressed simply as a literate human being.[119]

In 1880, however, Alphonse Bertillon, chief of the Parisian Office of Identification, blesses the criminal police forces of the earth with his anthropometric system: 11 measurements of diverse body parts, all with a rather constant, lifelong length, are sufficient for an exact registration, since they already afford 177,147 possible combinations or individualizations. Furthermore, the police archive documents the name, surname, pseudonym, age, as well as two photos (front and side). From which Moravagine, Cendrars's protagonist, deduces consequences for literature three days before the outbreak of war in 1914. He starts on a flight around the globe, naturally plans a film about himself, and chides the cameraman for not coming along:

> I can understand your wanting to rest and get back to your books. . . . You always needed time to think about a whole pile of things, to look, to see, to compare and record, to take notes on the thousand things you haven't had a chance to classify in your own mind. But why don't you leave that to the police archives? Haven't you got it through your head that human thought is a thing of the past and that philosophy is worse than Bertillon's guide to harassed cops?[120]

When Bertillon's police archive and Charcot's iconography, those two complementary recording technologies, chop up the human being of philosophy into countless criminals and lunatics, what results are doppelgängers on doppelgängers. And one only needs (as in Moravagine's case) to supplant still photographs with a combination of motorization and film to teach doppelgängers how to move. One no less than Mallarmé already celebrated the view through a moving car as that of a camera on wheels;[121] one no less than Schreber, during his relocation from the insane asylum at Coswig to the one at Sonnenstein, "mistakes" all the "human forms [that he] has seen on the drive and in the station in Dresden for miraculous 'fleeting men.'"[122] Traffic in the age of motorization always means encountering doppelgängers, schematically and serially.

The shapes that come to the surface these days out of the depths of mobile mirrors no longer have anything to do with literature and education. In 1886, Professor Ernst Mach described how he had recently seen a stranger on a bus and had thought, "'what a shabby-looking schoolmaster that is, who just got on.'"[123] It took even the great theorist of perception a couple of practical milliseconds before he could identify that stranger as his own mirror image. And Freud, who recapitulates Mach's uncanny encounter in 1919, can offer a traveling story of his own:

> I was sitting alone in my *wagon-lit* compartment when a more than usually violent jolt of the train swung back the door of the adjoining washing cabinet, and an elderly gentleman in a dressing-gown and a traveling cap came in. I assumed that in leaving the washing-cabinet, which lay between the two compartments, he had taken the wrong direction and come into my compartment by mistake. Jumping up with the intention of putting him right, I at once realized to my dismay that the intruder was nothing but my own reflection in the looking-glass on the open door. I can still recollect that I thoroughly disliked his appearance. Instead, therefore, of being *frightened* by our "doubles," both Mach and I simply failed to recognize them as such. Is it not possible, though, that our dislike of them was a vestigial trace of the archaic reaction which feels the "double" to be something uncanny?[124]

The horror of starlets like Barbara La Marr affects theorists as well. At a hundred kilometers per hour, as soon as they participate in motorized traffic, everyday life necessarily becomes cinematic. From the cabinet of Dr. Freud emerges his other. In the archive of Bertillon or Charcot, professors appear as dirty old men who remind even the father of psychoanalysis of his bodily functions. But the psychoanalysis of the *uncanny* does not touch upon modern technologies of trace detection with as much as a single word. Freud and Rank, in their hunt for the remainders of an archaic reaction, return mobile mirrors to stationary ones once again, turn cinema and railroad into the romantic world of books. The one deciphers the doppelgänger in E. T. A. Hoffmann, the other, in Chamisso and Musset.

Tzvetan Todorov observes that "the themes of fantastic literature

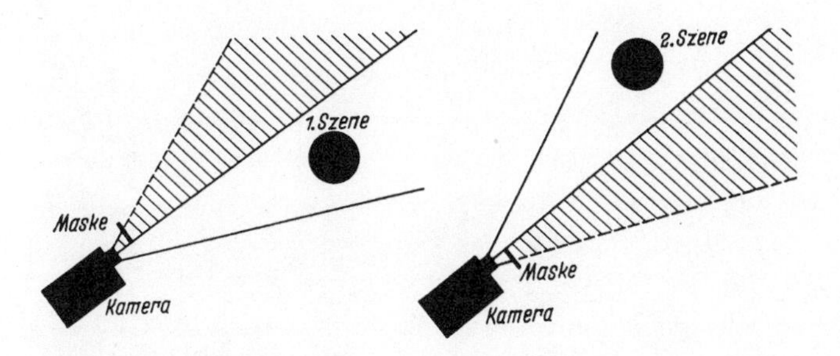

Scheme for a doppelgänger shot.

have literally become the very themes of the psychological investigations of the last fifty years. . . . We need merely mention that the double was even in Freud's time the theme of a classic study (Otto Rank's *Der Doppelgänger*)."[125] As a science of unconscious literalities, psychoanalysis indeed liquidates phantoms such as the doppelgänger, whom romantic readers once hallucinated between printed lines. In modern theory and literature "words have gained an autonomy which things have lost."[126] But to ascribe the death of "the literature of the fantastic" solely to a "psychoanalysis" that has "replaced" it and thereby made it "useless"[127] is Todorov's critical-theoretical blind spot. Writers know better that theories and texts are variables dependent upon media technologies:

> The writer of yesteryear employed "images" in order to have a "visual" effect. Today language rich in images has an antiquated effect. And why is it that the image disappears from front-page articles, essays, and critiques the way it disappears from the walls of middle-class apartments? In my judgment: because with film we have developed a language that has evolved from visuality against which the visuality developed from language cannot compete. Finally, language becomes pure, clean, precise.[128]

Only in the competition between media do the symbolic and the imaginary bifurcate. Freud translates the uncanny of the Romantic period into science, Méliès, into mass entertainment. It is precisely this fantasizing, anatomized by psychoanalysis, that film implements with powerful effect. This bilateral assault dispels doppelgängers from their books, which become devoid of pictures. On-screen, however, doppelgängers or their iterations celebrate the theory of the unconscious as the technology of cinematic cutting, and vice versa.

The doppelgänger trick is nothing less than uncanny. Half of the lens is covered with a black diaphragm while the actor acts on the other half

The Eiffel Tower from October 14, 1888, to March 31, 1889.

of the picture frame. Then, without changing the camera's position, the exposed film is rewound, the other half of the lens is covered up, and the same actor, now in his role as the doppelgänger, acts on the opposite side of the frame. Put differently, Méliès only had to record his stop trick onto the same roll of celluloid twice. "A trick applied with intelligence," he declared, "can make visible that which is supernatural, invented, or unreal."[129]

That is how the imaginary returned, more powerful than it could ever be in books, and as if made to order for writers of entertainment literature. In 1912, Heinz Ewers wrote: "I hate *Thomas Alva Edison*, because we owe to him one of the most ghastly of inventions: the *phonograph*! Yet I love him: he redeemed everything when he returned fantasy to the matter-of-fact world—in the movies!"[130]

These are sentences of media-technological precision: whereas the grooves of records store ghastly waste, the real of bodies, feature films take over all of the fantastic or the imaginary, which for a century has gone by the name of literature. Edison; or, the splitting of discourse into white noise and imagination, speech and dream (not to mention hatred and love). From then on neo-Romantic writers interested in love had it easy. One year later, Ewers wrote the screenplay for *The Student of Prague* by drawing on all of the book-doppelgängers in his library.[131] The film trick to end all film tricks (or, as a contemporary review put it, "the cinematic problem to end all cinematic problems")[132] conquered the screen.

The Student of Prague (Paul Wegener) next to his beloved (Grete Berger) and in front of his doppelgänger, in a Jewish cemetery in Prague.

Ewers's *Student*, Gerhart Hauptmann's *Phantom*, Wiene's *Caligari*, Lindau's *Anderer*, Wegener's *Golem*: a doppelgänger boom. Books (since Moses and Mohammed) have been writing writing; films are filming filming. Where art criticism demands expressionism or self-referentiality, media have always been advertising themselves. Finally, motorists, train travelers, and professors, starlets and criminals, madmen and psychiatrists—they, too, recognized that camera angles are their everyday reality. Doppelgänger films magnify the unconscious in mobile mirrors; they double doubling itself. The feature film transforms the "shock"[133] of the moment of recognition in Bronnen, Nabokov, Mach, and Freud into slow-motion trace detection: for 50 minutes, until his eventual disintegration and suicide, the student of Prague must see how the "horrifically unchanging apparition of the 'other'" sees him.[134] Notwithstanding Walter Bloem's *The Soul of Cinema*, cinema is what kills the soul. Precisely because "humans" are not "worms, for whom something like" division or doubling "is a piece of cake, . . . the notion of a unified artistic personality" disintegrates. Mimes become stars because human beings or civil servants have been made into guinea pigs. When executing the doppelgänger trick, "mechanics becomes a coproducer."[135]

On October 11, 1893, *The Other: A Play in Four Acts* had its première in Munich. In 1906, Paul Lindau's horror play was published by the Reclam Universal Library only to land on the desks of the Royal Police Force in Munich, from whose copy I must of necessity quote. For on February 15, 1913, change overtook all libraries: *The Other*, consisting of "2,000 meters" and "five acts," appeared as the first German *auteur* film.[136]

"Men such as Paul Lindau," Gottfried Benn wrote, "have their merits and their immortality."[137] They are among the first to make the change from the pen to the typewriter and thus to produce texts suitable for filming (the script of *The Student of Prague* was a typescript, too).[138] They are among the first to make the change from the soul to mechanics and thus to produce subject matter suitable for film; that is, doppelgängers. With Lindau and Ewers, cinema in Germany becomes socially acceptable.

Except that Lindau's protagonist, Mr. Hallers, J.D., has not yet achieved wide cultural acceptance, for which poetic-filmic justice simply compensates him with a double. In order to abolish a superannuated civil-service ethos, Hallers (just like Dr. Hyde or the student of Prague) must first become the other of the title. At the beginning, late at night, the prosecutor is in the process of dictating *The Constraints of Willpower in Light of Criminality* to one of the last male secretaries, who takes it down in shorthand. Lacking Lindau's typewriter, he also lacks any knowledge of psychiatry. Hypnosis, suggestion, hysteria, the unconscious, split personality—the civil servant wants to take out of circulation all of these terms, which have been in common usage "since Hippolyte Taine's study on the intellect."[139]

Hallers (*dictating*). Where would that eventually lead? It would lead to felons in every serious case quoting a physician to escape justice . . . to medicine being in stark contrast to justice. Let us be on guard against such insidious . . . (*interrupting himself*) no, change that to: against such highly disconcerting false teachings. (*Short break. He walks behind the desk chair and gradually lapses unthinkingly into the rhetorical tone of argumentation.*) Let us not destroy the consciousness of moral self-determination, of the responsibility of the individual for his own actions, through the misconstrued practical application . . . (*interrupting himself*) How did I put it?

Kleinchen (*reads without emphasis*). Let us not destroy the consciousness of moral self-determination, of the responsibility of the individual for his own actions . . .

Hallers (*interrupting*). Through the misconstrued practical application of an intelligent, if you will, but yet highly dubious deduction . . . theoretical deduction. Let us eliminate, as far as possible, the "Constraints of Willpower" from our court proceedings![140]

A highly rhetorical performance, whose refutation begins with Freudian slips and fills all four acts. Hallers's slips alone refute his dictation and his plaint, which turns into complete nonsense in its mechanical reproduction (the gramophone function of all secretaries, from Lindau to Valéry).[141] Foucault would have described historical ruptures no less derisively: justice ceding to medicine, law (with writing as the medium of civil service) ceding to biotechnologies that are media technologies as well. The same year, Senate President Schreber of the Regional Court of Appeals in Dresden disappears into an asylum simply because a "conspiracy" of psychiatrists "denied" him (a lawyer) "professions, like that of a nerve specialist, that lead to a closer proximity to God."[142] Hence Hallers, too, breaks off his dictation, because "pathologically strained nerves"[143] are in dire need of the "testimony of an available nerve specialist and psychiatrist."[144]

Professor Feldermann makes a nightly house call, cannot convince his patient of his split personality even by telling of classic case studies, and concludes that his "dull, gnawing headache" is attributable to Hallers's "fall from a horse . . . last fall."[145] What must remain taboo *in Light of Criminality* becomes an epistemic matter of course in Guyau's "Memory and Phonograph": consciousness and memory are mutually exclusive. In the depths of his brain engrams, the disciple of free will is listening to the dictates of the unconscious.

That is how the inevitable comes about. What the consciousness of the lawyer denies, his body enacts. The other appears (as does, in Schreber's case, a female other: a "sexually dependent woman" who appears in the same position as the president of the senate).[146] Hallers falls asleep as a prosecutor only to wake up immediately as a criminal. His movements become "automaton-like,"[147] "strained," "belabored and heavy, as if against his will."[148] Consequently, the other is (as in Caligari's case) the same once again, but this time as a cinematographic guinea pig. "A burglar"[149] has possessed the civil servant / person and, consequently, plans to break in to Hallers's own villa together with some crooks from Berlin. For whereas the lawyer half of his split personality surmises darkly only that "I no longer am I am,"[150] the criminal half proudly announces, in the unparalleled phrase of Yahweh, "I am . . . what I am."[151] As in all cases of split personality since Dr. Azam's Félida and Wagner's Kundry, unconscious knowledge overreaches conscious knowledge, not the other way round.[152] The other, with all of his complications, knows and steals from Hallers's villa, while the prosecutor (who returns as such after falling asleep a second time) only appears as an unintentional comedian when he interrogates his accomplice. It is not until he is confronted with Professor Feldermann's knowledge of psychiatry that he is brought up to par on contemporary developments and brought to his renunciation of all unrestrained willpower in the civil-service sector. A happy ending, not least because Hallers is rewarded with a bourgeois girl as well.

However, the exponential burglar, who invades both the personality and the villa of the lawyer, loves not the bourgeois girl but rather her maid (or stenotypist, had Lindau written a couple of years later), who was let go for disgraceful behavior. The civil-service domain of 1900 dreams, in terms criminal and erotic, male and female, of all its underbellies and doppelgängers. This dream, however, resides precisely in the middle ground between film and anthropometrics. The only indication that both sides of Hallers's double life are relayed is a photograph of the waitress Amalie. He receives the photograph as a criminal and finds it again, having changed back into the prosecutor, in his jacket pocket, at which point he can (following Bertillon) identify the woman he worships at night. But this photo materializes in the imaginary, even before the first transformation, during Feldermann's diagnosis.

Feldermann.	Are your dreaming at all?
Hallers.	Yes
Feldermann.	Of what?
Hallers (*reluctantly*).	Of uncomfortable things. I feel as if my dreams appeared in a kind of sequence, as if I returned to the same haunts every once in a while.
Feldermann.	What haunts are you talking about?
Hallers.	I can't recall the details. (*More quietly.*) I always see . . . something reddish . . . the gleam of a light . . .

something (*pointing to the fireplace*) like the embers in the fireplace . . . and, inside the reddish lighting (*yet more quietly*), the head of a woman . . .

Feldermann. The head of a woman.

Hallers. It's always the same one . . . always a bit red . . . like a drawing with red chalk. . . . The face of the girl is also pursuing me while awake. . . . As soon as I try to visualize it in detail, it falls apart, I cannot put it together. . . . If I should ever see her again, I will ask her for her picture.

Feldermann (*turning toward him further and looking at him attentively*). What is it you are saying here?

Hallers. It bothers me that the face with the red shine always hovers in front of me and that I cannot stabilize it.

Feldermann. I understand that. But I do not understand what you could be expecting of a photograph produced in your dreams of a dream image from your waking life.[153]

Film projection as internal theater exists two years prior to its introduction. Reason enough for Lindau, the writer, to forsake writing for cinema as quickly as possible. As with Freud or Rank, dreams are films and vice versa. One only has to have a nervous disorder like Hallers's to trigger the shutter while dreaming instead of surrendering to the "shadowy, fleeting . . . scenes of the film drama" and making literature again, as does Rank. Madness is cinematographic not only in motoric and physiognomic terms; cinema implements its psychic mechanisms itself.

That was precisely Münsterberg's insight. *The Photoplay: A Psychological Study*, the slender, revolutionary, and forgotten theory of the feature film, was published in New York in 1916. While psychiatrists continued to concentrate on pathologies of motion and psychoanalysts continued to consume films and retranslate them into books, the director of the Harvard Psychological Laboratory went past consumption and usage. His American fame opened the New York studios to him; hence he could argue both from the producer's standpoint and from the elementary level that relates film and the central nervous system. That is the whole difference between Rank and Münsterberg, psychoanalysis and psychotechnology.

Psychotechnology, a neologism coined by Münsterberg, describes the science of the soul as an experimental setup. *Basics of Psychotechnology*, published in 1914, reframes in 700 pages the collected results of experimental psychology in terms of their feasibility. What began the pioneering work of Wundt in Leipzig and what brought Münsterberg to Cambridge, Massachusetts, was the insight (dispelling presumptions in elitist labs to the contrary) that everyday reality itself, from the workplace to leisure time, has long been a lab in its own right. Since the motor and sensory activities of so-called Man (hearing, speaking, reading, writing) have been measured under all conceivable extreme conditions, their ergonomic revolution is only a matter of course. The second industrial revolution enters the knowledge base. Psychotechnology relays psychology and media technology under the pretext that each psychic apparatus is also a technological one, and vice versa. Münsterberg made history with studies on assembly-line work, office data management, combat training.

Hence his theory was fully absorbed by the film studios (which had not yet migrated to Hollywood). From film technology and film tricks, knowledge only extracted what it had invested in the studies of optical illusions since Faraday. With the indirect consequence that film technology itself (as with phonography in Guyau's case) became a model of the soul—initially as philosophy and, eventually, as psychotechnology.

In 1907, Bergson's *Creative Evolution* culminated in the claim that the philosophically elementary functions of "perception, intellection, language" all fail to comprehend the process of becoming. "Whether we would think becoming, or express it, or even perceive it, we hardly do anything else than set going a kind of cinematograph inside us. We may therefore sum up what we have been saying in the conclusion that the *mechanism of our ordinary knowledge is of a cinematographical kind.*" Instead of registering change as such, "we take snapshots, as it were, of the passing reality," which—once it is "recomposed . . . artificially," like a film—yields the illusion of movement.[154] What that means in concrete physiological terms is beyond the philosopher Bergson, who is solely interested that film mark a historical difference: In antiquity, "time comprises as many undivided periods as our natural perception and our language cut out in it successive facts." By contrast, modern science, as if Muybridge were its founding hero, isolates (following the model of differential equations) the most minute time differentials. "It puts them all in the same rank, and thus the gallop of a horse spreads out for it into as many successive attitudes as it wishes," rather than (as "on the frieze of the Parthenon") "massing itself into a single attitude, which is supposed to flash out in a privileged moment and illuminate a whole period."[155]

Bergson does not want to reverse this panning shot from art to media, but his philosophy of life does envision a kind of knowledge that could register becoming itself, independent of antique and modern technologies of perception: the redemption of the soul from its cinematographic illusion.

Psychotechnology proceeds exactly the other way around. For Münsterberg, a sequence of stills, that is, Bergson's cinematographic illusion of consciousness, is by no means capable of evoking the impression of movement. Afterimages and the stroboscopic effect by themselves are necessary but insufficient conditions for film. Rather, a series of experimental and Gestalt-psychological findings demonstrates—contra Bergson—that the perception of movement takes place as "an independent experience."[156]

> The eye does not receive the impressions of true movement. It is only a suggestion of movement, and the idea of motion is to a high degree the product of our own reaction. . . . The theater has both depth and motion, without any subjective help; the screen has them and yet lacks them. We see things distant and moving, but we furnish to them more than we receive; we create the depth and the continuity through our mental mechanism.[157]

One cannot define film more subjectively than Münsterberg does, but only to relay these subjective ideas to technology. Cinema is a psychological experiment under conditions of everyday reality that uncovers unconscious processes of the central nervous system. Conversely, traditional arts such as theater, which Münsterberg (following Vachel Lindsay)[158] continuously cites as a counterexample, must presuppose an always-already functioning perception without playing with their mechanisms. They are subject to the conditions of an external reality that they imitate: "Space, time, causality."[159] On the other hand, Münsterberg's demonstration that the new medium is completely independent aesthetically and need not imitate theater suggests that it assembles reality from psychological mechanisms. Rather than being an imitation, film plays through what "attention, memory, imagination, and emotion" perform as unconscious acts.[160] For the first time in the global history of art, a medium instantiates the neurological flow of data. Although the arts have processed the orders of the symbolic or the orders of things, film presents its spectators with their own processes of perception—and with a precision that is otherwise accessible only to experiment and thus neither to consciousness nor to language.

Münsterberg's errand to the film studios was worth it. His psychotechnology, instead of merely assuming similarities between film and dreams as does psychoanalysis, can ascribe a film trick to each individual, unconscious mechanism. Attention, memory, imagination, emotion: they all have their technological correlative.

Naturally, this analysis begins with attention, because in the age of media facts are generally defined by their signal-to-noise ratio. "The chaos of the surrounding impressions is organized into a real cosmos of experience by our selection,"[161] which, in turn, can either be voluntary or involuntary. But because voluntary selection would separate spectators from the spell of the medium, it is not considered. What counts is solely whether and how the different arts control involuntary attention and hence "play on the keyboard of our mind."[162] Of

the whole large scene, we see only the fingers of the hero clutching the revolver with which he is to commit the crime. Our attention is entirely given up to the passionate play of his hand. . . . Everything else sinks into a general vague background, while that one hand shows more and more details. The more we fixate [on] it, the more its clearness and distinctness increase. From this one point wells our emotion, and our emotion again concentrates our senses on this one point. It is as if this one hand were during this pulse beat of events the whole scene, and everything else had faded away. On the stage this is impossible; there nothing can fade away. That dramatic hand must remain, after all, only the ten thousandth part of the space of the whole stage; it must remain a little detail. The whole body of the hero and the other men and the whole room and every indifferent chair and table in it must go on obtruding themselves on our senses. What we do not attend cannot be suddenly removed from the stage. Every change which is needed must be secured by our own mind. In our consciousness the attended hand must grow and the surrounding room must blur. But the stage cannot help us. The art of the theater has there its limits.

Here begins the art of the photoplay. That one nervous hand which feverishly grasps the deadly weapon can suddenly for the space of a breath or two become enlarged and be alone visible on the screen, while everything else has really faded into darkness. The act of attention which goes on in our mind has remodeled the surrounding itself. . . . In the language of the photoplay producer it is a "close-up." *The close-up has objectified in our world of perception our mental act of attention and by it has furnished art with a means which far transcends the power of any theater stage.*[163]

Münsterberg's patient gaze, which we have long since unlearned, focuses not for nothing on the revolver: its drum stands at the origin of cinema. When it appears as a close-up, film films involuntary and technological mechanisms at the same time. Close-ups are not just "objectivizations" of attention; attention itself appears as the interface of an apparatus.

This is true of all the involuntary mechanisms Münsterberg investigates. Whereas each of the temporal arts, in "the most trivial case," presupposes the storage of past events, "the theater can do no more than suggest to our memory this looking backward"—namely, with words, for which "our own material of memory ideas" must "supply the picture[s]."[164] In the "slang" and practice of photo artists, by contrast, there are cut-backs or flashbacks, which are "really an objectivation of our memory function."[165] The same is true of the imagination as unconscious expectation and of associations in general. Aside from flashbacks and flash-forwards, cinematic montage conquers "the whole manifoldness of parallel currents with their endless interconnections."[166] According to the film theory of Béla Balázs, who unknowingly furthered Münsterberg's work, unconscious processes "can never be rendered so visually in words as in cinematic montage—be they the words of a physician or a poet. Primarily because the rhythm of montage can reproduce the *original speed* of the process of association. (Reading a description takes much longer than the perception of an image)."[167]

And yet, literature—whose power film infinitely exceeds or "transcends," according to Münsterberg—attempts the impossible. Schnitzler's novellas simulate processes of association in phonographic real time,[168] Meyrink's novels in filmic real time. *The Golem* appears in 1915 as a doppelgänger novel in ostensible competition with Ewers's and Lindau's successes on the screen; as a simulation of film, however, it unknowingly anticipates Münsterberg's theory. Meyrink's framing narrative begins with a nameless I, who is transformed by his half-asleep associations into the doppelgänger of the framed story. As if in a flashback, this person, Pernath, reappears in the Prague ghetto, long since torn down, only to encounter in turn a Golem who is expressly called Pernath's "negative,"[169] that is, the doppelgänger of the doppelgänger. This iteration of mirror situations, associations, transformations follows the techniques of film so closely that Meyrink's framing narrative even sacrifices the time-honored past tense of the novel to it. It is not just since *Gravity's Rainbow* that novels have been written in the present tense to suggest the flow of association and easy filmability.

Which makes interpretation meaningless and only invites the rewriting of Meyrink's beginning as a screenplay. Well, here is the first chapter (the narrative frame) of *The Golem* once more, this time in two columns with Münsterbergian instructions for the camera.

SLEEP

The moonlight is falling on to the foot of my bed. It lies there like a tremendous stone, flat and gleaming.

Fade-out to dream

As the shape of the new moon begins to dwindle, and its left side starts to wane—as age will treat a human face, leaving his trace of wrinkles first upon one hollowing cheek—my soul becomes a prey to vague unrest. It torments me.

I cannot sleep; I cannot wake; in its half dreaming state my mind forms a curious compound of things it has seen, things it has read, things it has heard—streams, each with its own degree of clarity and color, that intermingle, and penetrate my thought.

Before I went to bed, I had been reading from the life of Buddha; one particular passage now seeks me out and haunts me, drumming its phrases into my ears over and over and over again from the beginning, in every possible permutation and combination:

Caption (text)

"A crow flew down to a stone that looked, as it lay, like a lump of fat.

Thought the crow, 'Here is a toothsome morsel for my dining'; but finding it to be nothing of the kind, away it flew again. So do we crows, having drawn near to the stone, even so do we, would-be seekers after truth, abandon Gautama the Anchorite, so soon as in him we cease to find our pleasure."

Close-up (= attention)

This image of the stone that resembled a lump of fat assumes ever larger and larger proportions within my brain.

Moving camera

I am stumbling along the dried-up bed of a river, picking up smoothed pebbles.

Close-ups

Now they are grayish-blue, coated in a fine, sparkling dust; persistently I grub them up in handfuls, without in the least knowing what use I shall make of them; now they are black, with sulfury spots, like the strivings of a child to create in stone squab, spotty, prehistoric monsters.

(= involuntary attention)

I strive with all my might and main to throw these stone shapes from me, but always they drop out of my

	hand, and, do what I will, are there, for ever there, within my sight.
Cut-back (= involuntary memory)	Whereupon every stone that my life has ever contained seems to rise into existence and compass me around. Numbers of them labor painfully to raise themselves out of the sand towards the light—like monstrous, slaty-hued crayfish when the tide is at the full—and all rivet their gaze upon me, as though agonizing to tell me tidings of infinite importance.
Fading	Others, exhausted, fall back spent into their holes, as if once for all abandoning their vain search for words.
Fade-out to everyday	Time and again do I start up from this dim twilight of dreams, and for the reality space of a moment experience once more the moonshine on the end of my billowing counterpane, like a large, flat, bright stone, only to sink blindly back into the realms of semi-consciousness, there to grope and grope in my painful quest for that eternal stone that in some mysterious fashion lurks in the dim recesses of my memory in the guise of a lump of fat. . . . What happens next I cannot say. Whether, of my own free will, I abandon all resistance; whether they overpower and stifle me, those thoughts of mine . . . I only know that my body lies sleeping in its bed, while my mind, no longer part of it, goes forth on its wanderings. *Who am I?* That is the question I am suddenly beset with a desire to ask; but at the same instant do I become conscious of the fact that I no longer possess any organ to whom this query might be addressed; added to which, I am in mortal terror lest that idiotic voice should reawaken and begin all over again that never-ending business of the stone and the lump of fat.
Fade-out (onto doppelgänger)	I capitulate.[170]

The Golem begins as film; more precisely, as a silent film. Only films make it possible to present all the mechanisms of madness, to run through chains of associations in real time, and to jump continually from a metaphoric stone at the bedside to a real stone in the ghetto of the doppelgänger. (Immediately after the capitulation of the "I," Pernath begins his life history in the past tense as the I of the framed story.)

And only silent films command the robbing of the narrative I of all its organs of speech. In lieu of reflexive introspections we have neurologically pure data flows that are always already films on the retina. All-powerful optical hallucinations can flood and sever a body, and eventually make it into an other. Pernath and Golem, the substitutes of the narrative I in the framed story, are the positive and negative of a celluloid ghost.

Fading of consciousness itself . . . simply as a sequence of film tricks.

"Our psychic apparatus reveals itself in these transformations," wrote Balázs. "If fading, distorting, or copying could be executed without any specific image, that is, if the technique could be divorced from any particular object, then this 'technique as such' would represent the mind as such."[171]

But as Münsterberg demonstrated, the transformation of a psychic apparatus into film-trick transformations is lethal for the mind [*Geist*] as such. Mathematical equations can be solved in either direction, and the title "psychotechnology" already suggests that film theories based on experimental psychology are at the same time theories of the psyche (soul) based on media technologies. In *The Golem*, Proust's beloved *souvenir involontaire* becomes a flashback, attention a close-up, association a cut, and so on. Involuntary mechanisms, which hitherto existed only in human experiments, bid their farewell to humans only to populate film studios as the doppelgängers of a deceased soul. One Golem as tripod or muscles, one as celluloid or a retina, one as cut-back or random access memory . . .

Golems, however, possess the level of intelligence of cruise missiles, and not only those in Meyrink's novel or Wegener's film. They can be programmed with conditional jump instructions, that is, first to execute everything conceivable and then to counter the danger of the infinite spirals praised by Goethe. Precisely for that reason, in Münsterberg's succinct words, "every dream becomes real" in film.[172] All the historical attributes of a subject who around 1800 celebrated his or her authenticity under the title literature can around 1900 be replaced or bypassed by Golems, these programmed subjects. And above all, dreams as a poetic attribute.

The romantic novel par excellence, Novalis's *Henry von Ofterdingen*, programmed the poetic calling of its hero with media-technological precision: as a library-inspired fantasy and a dream of words. As if by chance, Ofterdingen was allowed to discover an illustrated manuscript with neither name nor title, but which dealt "with the wondrous fortunes of a poet."[173] Its pictures "seemed wonderfully familiar to him, and as he looked more sharply, he discovered a rather clear picture of himself among the figures. He was startled and thought he was dreaming"[174]—the wonder of the dream was the necessity of the system. In 1801, the recruitment of new authors was, after all, achieved through literarily vague doppelgängers, in whom bibliophile readers could recognize (or not) their similarly unrecordable "Gestalt." And Ofterdingen promptly decided to merge with the author and hero of the book he found.

This mix-up of speech and dream was programmed at the novel's beginning. There Ofterdingen listened to the "stories" of a stranger that told of "the blue flower" that nobody had ever seen or heard of. But because prospective writers needed to be able to change words into optical-acoustic hallucinations, Ofterdingen quickly fell asleep and began dreaming. Poetic wonder did not wait: words became an image, and the image a subject, Ofterdingen's future beloved.

But what attracted him with great force [in the dream] was a tall, pale blue flower, which stood beside the spring and touched him with its broad glistening leaves. Around this flower were countless others of every hue, and the most delicious fragrance filled the air. He saw nothing but the blue flower and gazed upon it long with inexpressible tenderness. Finally, when he wanted to approach the flower, it all at once began to move and change; the leaves became more glistening and cuddled up to the growing stem; the flower leaned towards him and its petals displayed an expanded blue corolla wherein a delicate face hovered.[175]

No word, no book, no writer can write what women are. That is why that task was performed during the age of Goethe by poetic dreams, which, with the help of psychotricks, produced an ideal woman and hence a writer from the word "flower." The trick film (following Münsterberg's insight) makes such internal theater of subjects or literate people as perfect as it is superfluous.

No theater could ever try to match such wonders, but for the camera they are not difficult. . . . Rich artistic effects have been secured, and while on the stage every fair play is clumsy and hardly able to create an illusion, in the film we see the man transformed into a beast and the flower into a girl. There is no limit to the trick pictures which the skill of the experts invents. The divers jump, feet first, out of the water to the springboard. It looks magical, and yet the camera man has simply to reverse his film and to run it from the end to the beginning of the action. Every dream becomes real.[176]

A medium that turns moonspots into stones or, better still, flowers into girls no longer allows for any psychology. The same machinelike perfection can make flowers into a so-called I. That is precisely the claim of

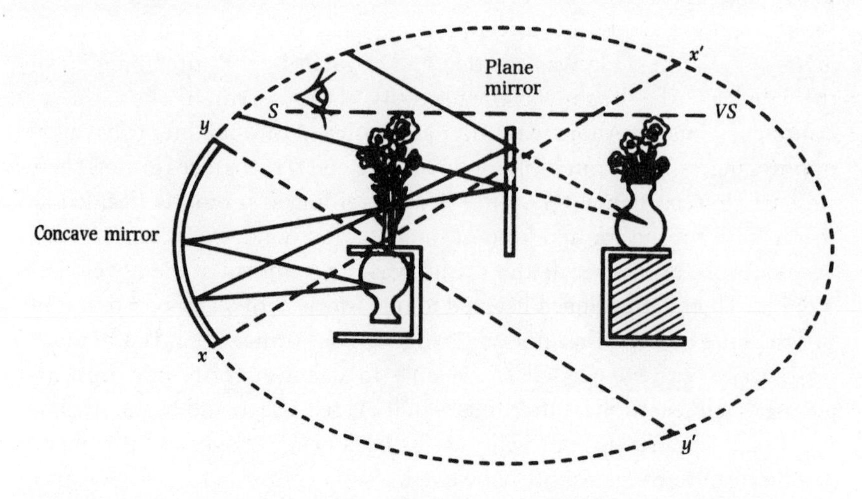

Lacan's scheme of the inverted vase. (From Lacan 1988a, 139; reproduced by permission of W. W. Norton & Company, Inc. © 1975 by Les Editions du Seuil; English translation © 1988 by Cambridge University Press)

Lacan's theory, which, especially as an anti-psychology, is up to date with contemporary technological developments. The symbolic of letters and numbers, once celebrated as the highest creation of authors or geniuses: a world of computing machines. The real in its random series, once the subject of philosophical statements or even "knowledge": an impossibility that only signal processors (and psychoanalysts of the future) can bring under their control. Finally, the imaginary, once the dream produced by and coming out of the caverns of the soul: a simple optical trick.

In *The Interpretation of Dreams*, Freud followed the positivistic "suggestion that we should picture the instrument which carries out our mental functions as resembling a compound microscope or a photographic apparatus, or something of the kind."[177] Lacan's theory of the imaginary is an attempt truly to "materialize"[178] such models. As a result of which, cinema—the repressed of Freud's year at the Salpêtrière—returns to psychoanalysis. Lacan's optical apparatuses show a complexity that can only derive from cinematic tricks. Step by step, they go beyond the simple mirror and the (mis)recognition that induces in the small child a first but treacherous image of sensory-motoric wholeness.

Following Bouasse's *Photométrie* of 1934, a concave mirror initially projects the real image of a hidden vase into the same room where, in between *x* and *y*, it is expected by its actual flowers. If the optic beams com-

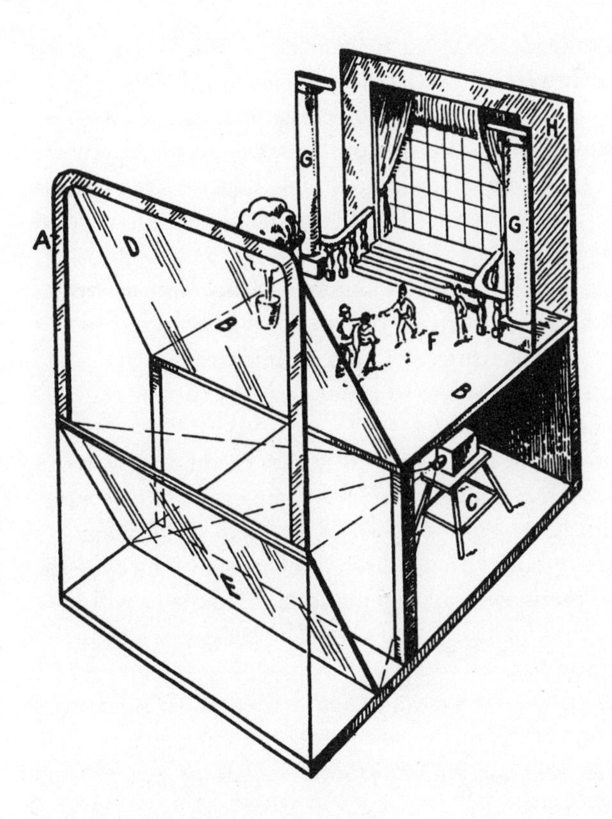

Messter's Alabastra Theater.

ing out of the parabola, however, are also deflected by a plane mirror perpendicular to the eye, then the vase, miraculously filled with flowers, appears to the subject *S* next to its own, but only virtual, mirror image *VS*. "That is what happens in man," who first achieves "the organization of the totality of reality into a limited number of preformed networks"[179] and then lives through his identification with virtual doppelgängers. Narcissism is duplicated.

Lacan, however, did not need to search for his optical tricks in the science of Bouasse. Film pioneers, who have always been dreaming of 3-D cinema without glasses, built apparatuses of a similar nature. In 1910 Oskar Messter, the founder of the German film industry and the person in charge of all photography and film footage taken at the front during the First World War,[180] introduced his Alabastra Theater in Berlin. He replaced the concave mirror of Bouasse and Lacan with a film projector C that nevertheless had the same function as the mirror: his lenses projected real images of actors that were allowed to act only against the black background of all media—on a screen *E* located below the stage *A*. The audience, however, only saw the virtual image of this screen, projected by the plane mirror *B*. With the result that filmed female dancers appeared on the stage of the Alabastra Theater itself and gave the impression of moving through three dimensions.[181]

"Hence," Lacan said, addressing his seminar participants as well as the audience of the Alabastra Theater, "you are infinitely more than you can imagine, subjects [or underlings] of gadgets and instruments of all kinds—ranging from the microscope to radio and television—that will become elements of your being."[182]

What's missing now is for the plane mirror *B* to become a psychoanalyst and, prompted by the remote control of language that occupies him,[183] turn by 90 degrees, so that the subject *S* sacrifices all its imaginary doppelgängers to the symbolic. Then, three dimensions or media—the nothing called a rose, the illusion of cinema, and discourse—will have been separated in a technologically pure way. The end of psychoanalysis/es is depersonalization.[184]

Consequently, Lacan was the first (and last) writer whose book titles only described positions in the media system. The writings were called *Writings*, the seminars, *Seminar*, the radio interview, *Radiophonie*, and the TV broadcast, *Télévision*.

Media-technological differentiations opened up the possibility for media links. After the storage capacities for optics, acoustics, and writing had been separated, mechanized, and extensively utilized, their distinct data flows could also be reunited. Physiologically broken down into fragments and physically reconstructed, the central nervous system was resurrected, but as a Golem made of Golems.

Such recombinations became possible no later than the First World War, when media technologies, reaching beyond information storage, began to affect the very transmission of information. Sound film combined the storage of acoustics and optics; shortly thereafter, television combined their transmission. Meanwhile, the text storage apparatus of the typewriter remained an invisible presence, that is to say, in the bureaucratic background. Lacan's final seminars all revolve around possibilities of connecting and coupling the real, the symbolic, and the imaginary.

Engineers, however, had been planning media links all along. Since everything from sound to light is a wave or a frequency in a quantifiable, nonhuman time, signal processing is independent of any one single medium. Edison perceived this very clearly when he described the development of his kinetoscope in 1894: "In the year 1887, the idea occurred to me that it was possible to devise an instrument which should do for the eye what the phonograph does for the ear, and that by a combination of the two all motion and sound could be recorded and reproduced simultaneously."[185]

Cinema as an add-on to the phonograph—in theory, this applied only to storage and not to the systemic differences between one- and two-dimensional signal processing; in practice, however, the analogy had far-reaching implications. Edison's Black Mary, the very first film studio, simultaneously recorded sound and motion, that is, phonographic and kinetographic traces. In other words, sound film preceded silent film. But the synchronization of data streams remained a problem. Whereas in the optical realm, processing was a matter of equidistant scanning, which television was to increase to millions of points per second, in the acoustic dimension processing was based on analogies in a continuous stream of time. As a result, there arose synchronization problems similar to those of goose-stepping French regiments, problems more difficult to amend than Demeny's. Which is why Edison's master-slave relationship was turned on its head, and film, with its controllable time, took the lead. Mass-media research, with stacks of books on film and hardly any on gramophony, followed in its wake.

But pure silent film hardly ever existed. Wherever media were unable to connect, human interfaces filled the niche. Acoustic accompaniment in the shape of words and music came out of every fairground, variety show, and circus corner. Wagner pieces like the *Liebestod* or the "Ride of the Valkyries" posthumously proved that they had been composed as sound tracks. At first, solo piano or harmonium players fought for image-sound synchronicity in movie houses; from 1910 on, so did entire ensembles in urban centers. When the literati Däubler, Pinthus, Werfel, Hasenclever, Ehrenstein, Zech, and Lasker-Schüler saw *The Adventures of Lady Glane* in Dessau in 1913, the "dismal background piano clinking" was "drowned out by the voice of a narrator commenting on the action in a broad Saxon: 'And 'ere on a dark and stormy night we see Lady Glahney . . . '"[186] The repulsion in the progressive literati triggered by the Saxon dialect gave rise to their *Movie Book*. It "incited extensive and far-reaching discussions about the misguided ambitions of the newly emerging silent film to imitate the word- or stage-centered theatrical drama or the ways in which novels use narrative language instead of probing the new and infinite possibilities inherent in moving images, and [Pinthus] raised the question what each of us, if asked to write a script, would come up with."[187]

Pinthus et al. thus turned the handicaps of contemporary technology into aesthetics. Sound, language, and even intertitles were all but purged from the literary scripts they (rather unsuccessfully) offered to the film industry. For the medium of silent film as for the writing medium, the guiding motto was appropriateness of material. (The fact that the *Movie Book* itself linked the two was missed by Pinthus.) As if the differentiation of distinct storage media had called for theoretical overdetermination, early film analyses all stressed *l'art pour l'art* for the silent film. According to Bloem, "the removal of silence would dissolve the last and most important barrier protecting films from their complete subjugation to the depiction of plain reality. An utterly unbridled realism would crush any remaining touch of stylization that yet characterizes even the most impoverished film."[188] Even Münsterberg's psychotechnology discerned insoluble aesthetic rather than mechanical problems in the media link of film and phonograph.

A photoplay cannot gain but only lose if its visual purity is destroyed. If we see and hear at the same time, we do indeed come nearer to the real theatre, but this is desirable only if it is our goal to imitate the stage. Yet if that were the goal, even the best imitation would remain far inferior to an actual theatre performance. As soon as we have clearly understood that the photoplay is an art in itself, the conversation of the spoken word is as disturbing as colour would be on the clothing of a marble statue.[189]

The "invention of the sound film came down like a landslide" on these theories. In 1930, at the end of the silent film era, Balázs saw "a whole rich culture of visual expression in danger."[190] The International Artists Lodge as well as the Association of German Musicians, the human interfaces of the silent movie palaces, agreed and went even further in their labor dispute, turning Münsterberg's arguments into a pamphlet "To the Audience!": "Sound film is badly conserved theater at a higher price!"[191]

Literature as word art, theater as theater, film as the filmic and radio as the radiophonic: all these catchwords of the 1920s were defensive measures against the approaching media links. "A voluntary restriction of the artist to the technical material at hand—that results in the objective and immutable stylistic laws of his art."[192] In strict accordance with Mallarmé's model, the filmic and the radiogenic were to import *l'art pour l'art* into the optical and acoustic realms. But the radiogenic art of the radio play was not killed off by the mass-media link of television; already at its birth it was not as wholly independent of the optical as the principle of appropriate material demanded. With its "accelerated dreamlike succession of colorful and rapidly passing, jumping images, its abbreviations and superimpositions—its speed—its change from close-up to long shot with fade-in, fade-out, fade-over," the early radio play "consciously transferred film technology to radio."[193]

The reverse passage from sound to image, or from gramophone to film, was taken less consciously, maybe even unconsciously. But only once records emanate from their electric transmission medium of radio does the rayon girl decide to "write her life like a movie." In Bronnen's Hollywood novel, Barbara La Marr learns from the record player all the movements that will make her a movie star. "We have a gramophone, that's all. Sometimes I dance to it. But that is all I know about large cities and singers and variety shows, of movies and Hollywood."[194] In turn, the gramophone (and some jazz bands) felt compelled to technologically synchronize a woman's body: while making love,[195] inventing the strip tease,[196] taking screen tests,[197] and so on. The future movie star Barbara La Marr was acoustically preprogrammed.

Two entertainment writers with Nobel prizes, Hermann Hesse and Thomas Mann, followed the beaten track. Immediately prior to the introduction of sound film, links connecting cinema and gramophone, especially when they stayed in the realm of the fantastic, were the best advertisement. Hesse's *Steppenwolf* culminates in a "Magic Theater," evidently the educated circumscription of a movie theater that uses radio records to produce its optical hallucination. From the "pale cool shimmer" of an "ear" that, as with Bell and Clarke's *Ur*-telephone, belongs to a corpse, emerges the music of Handel in "a mixture of bronchial slime and chewed rubber; that noise that owners of gramophones and radios have agreed to call radio." But it is precisely this music that conjures up an optically hallucinated Mozart whose interpretation of Handel's music encourages consumers to perceive the latter's everlasting value behind the medium.[198] The stage is set for sound tracks.

Thomas Mann could already look back on one film version of *Buddenbrooks* when a "very good Berlin producer" approached him in 1927 with plans for turning *The Magic Mountain* into a movie. Which was "not surprising" to Mann. Ever since December 28, 1895, when the Lumières presented their cinema projector, non-filmability has been an unmistakable criterion for literature. "What might not have been made" of entertainment novels, particularly of the "chapter 'Snow,' with its Mediterranean dream of humanity!"[199] Dreams of humans and humanity, whether the results of meteorological snow or of the powder of the same name, stage the mirror stage and are therefore cinema from the start.[200]

The particular human in question, after escaping his dismemberment, embarks on a career in a lung sanatorium. The Magic Mountain already has at its disposal a stereoscope, a kaleidoscope, and, though demoted to the status of an amusing diversion, Marey's cinematographic cylinder.[201] In the end, however, and shortly before the First World War and its trenches, the so-called engineer Castorp also receives a modern Polyhymnia gramophone, which he proceeds to administer as "an overflowing cornucopia of artistic enjoyment."[202] Opportunities for self-advertisement follow swiftly, even though pathology once again stands in for future technology. The sanatorium's own psychoanalyst and spiritualist is unable to conjure up the spirit of Castorp's deceased cousin until the gramophone administrator comes up with the obvious solution. Only when prompted by the phonographic reproduction of his favorite tune does the spirit appear,[203] thus revealing this media link to be a sound-film reproduction. Nothing remains to keep *The Magic Mountain* from being made into a movie.

Entertainment writers in particular, who insist on playing Goethe even under advanced technological conditions,[204] know fully well that Goethe's "writing for girls"[205] is no longer sufficient: the girls of the Magic Mountain have deserted to the village movie theatre, their "ignorant red face[s] . . . twisted into an expression of the hugest enjoyment."[206]

That, too, is a media link, but an ordinary and unassuming one beneath the dignity of Nobel Prize winners. Since 1880, literature no longer has been able to write for girls, simply because girls themselves write. They are no longer taken by imagining sights and sounds between poetic lines, for at night they are at the movies and during the day they sit at their typewriters. Even the Magic Mountain has as its "business center" a "neat little office" with "a typist busy at her machine."[207]

The media link of film and typewriter thoroughly excludes literature. In 1929, the editor and German Communist Party member Rudolf Braune published a miscellany on the empirical sociology of readers in the literature section of the *Frankfurter Zeitung*. Pursuing the question "What They Read," Braune had approached "Three Stenographers" and received answers that triggered his public outcry: Colette, Ganghofer, Edgar Wallace, Hermann Hesse . . . Not even Braune's desperate attempt to interest the three office workers in literature loyal to the party line met with success. Five weeks later, however, on May 26, 1929, the typists received a boost. Nameless female colleagues wrote or typed letters to the editors and readers of the *Frankfurter Zeitung*, informing them what is different about modern women:

> If we stenographers read little or nothing, do you know why? Because at night we are much too tired and exhausted, because to us the rattling of the typewriters, which we have to listen to for eight hours, keeps ringing in our ears throughout the evening, because each word we hear or read breaks down into letters four hours later. That is why we cannot spend evenings other than at the movies or going for walks with our inevitable friend.[208]

Whereas social engagement queries the reception or non-reception of literature in sociological terms, the test subjects respond in technological terms. Typewriters that break down their input into single letters in order to deliver an output in the shape of series and columns of standardized block letters also determine historical modes of reception. As selective as a band-pass filter, the machine positions itself between books and speeches on the one hand and eyes or even ears on the other. As a result, language does not store or transmit any meaning whatsoever for stenographers, only the indigestible materiality of the medium it happens to be. Every night the movie-continuum has to treat the wounds that a discrete machine inflicts upon secretaries during the day. An entanglement of the imaginary and the symbolic. The new media link that excludes literature was nevertheless committed to paper: in the shape of a screenplay that was never filmed. Pinthus's *Movie Book* printed plain text on cinema, books, and typewriters.

RICHARD A. BERMANN, "LYRE AND TYPEWRITER" (1913)

Returning home from her beloved movies, a swarthy little typist should tell her smiling friend about a movie thus:

Now there's a movie that clearly shows how important we typists are—we who copy and sometimes also occasion your poems. You see, first they showed what you poets are like when we're not around. One of you—with long hair and big tie, lots of attitude for no reason—he's sitting at home chewing on a huge pen. Maybe he's got nothing to eat, and why should he? Is he working? He nervously runs around the room. He writes a verse on a piece of paper folded in a funny way. He stands in front of the mirror, recites the verse, and admires himself. In a very satisfied mood, he lies down on the sofa. He gets up again and goes on chewing—he can think of absolutely nothing. Angrily he rips up the piece of paper. You can tell he feels ignored because he doesn't get anything done. He puts on a romantic coat and hurries to a literary cafe. It's summer, so he can sit outside on the street. Then *she* walks by—a very blond energetic muse. He quickly calls the waiter and with great ado does not pay for the mélange. He hurries after the muse. She takes the tube. As luck has it he's got ten cents left, so he takes a ride too. He approaches her when she leaves the station, but she's not one of those and sends him packing. Well, he still follows her. She enters her house, grabs the elevator key, and takes a ride upstairs. He runs up the stairs like a madman and arrives just as she closes her door. But there's a sign on the door:

MINNIE TIPP
Typing Service
Transcription of Literary Works
Dictation

He rings the bell. The door opens. Minnie Tipp is already typing away. She wants to throw him out but he claims to be a customer with a dictation. He assumes a pose and dictates: "Miss, I love you!" She types it and the writing appears on the white screen. But she throws the scrap of paper at his feet, sits down again, and writes: "I have no time for idle sluggards. Come back when you have some literary work that needs copying. Goodbye!"

Like, what can he do in the face of so much virtue? He goes back home really dejected and despairs in front of the mirror. He gets paper, lots of paper, and plans to write like there's no tomorrow. But he can't do any more than chew the pen, which by now is quite short. He reclines on his infamous sofa. Suddenly, the image of Minnie appears—the upright, diligent, energetic typist. She shows him a perfectly typed page that reads: "I would love you, too, if you could get some real work done!" The image vanishes and he returns to his desk. And now, you see, the boy with bow and quiver appears in a dark corner of the room. He darts to the desk at which the brooding poet is sitting and pours a quiver full of ink into his sterile inkwell. Then the boy sits down with crossed legs on the sofa and watches. The poet dips his pen—now it's running all by itself. As soon as the pen touches the paper, it is full of the most beautiful verses and whisked away. In no time the room is full of manuscripts. The poet may dictate after all. They are all love poems. The first one starts:

When first I beheld your eyes so blue
My limbs were filled with molten ore.
I work, and working am so close to you—
I live once more!

She writes with long sharp fingers, but she doesn't look at the machine and leaves no spaces between the words. She is dancing a dance of love on her machine. It is a mute duet. He is a very happy lyric poet. He returns home in a rapture.

A couple of days later a man appears with a wheelbarrow and brings the poet a couple of hundred pounds of perfectly copied manuscripts. He also has a letter—a perfumed, neatly typed one. The poet kisses the letter. He opens it. The boy with the bow is back in the room again and peers over the poet's shoulder. But alas! The poet is tearing his hair—and the nice boy pulls a face, for the letter reads:

Dear Sir, you will be receiving your manuscripts with today's mail. Please allow me to inform you that I am enraptured by the fire of your verses. I also beg to draw your attention to the enclosed invoice of 200 Marks. I would be delighted if you were to communicate the amount to me in person, at which point we could enter into a discussion concerning the content of your verses. Yours, Minnie Tipp.

"That's what happens," the swarthy little typist tells her smiling friend, "when we women are forced to work. It makes us so eminently practical."

Well, of course the poor poet hasn't a penny to his name. He searches the whole room and finds only manuscripts. He searches his pockets and finds only impressive holes. Amor wants to help and turns his quiver on its head—but why would Amor walk around with two hundred Marks? Finally, there is nothing left for the poet to do but to get behind the wheelbarrow and cart the manuscripts to a cheese dealer. He buys them to wrap soft cow cheese. Now, the famous critic Fixfax is of a delicate nature and loves runny cow cheese. So he proceeds to the cheese dealer in person, buys a portion, and takes it home. On the street pedestrians hold their noses and bolt. But Fixfax loves smelling the cheese. As he is about to drill his nose—covered, of course, by black, horn-rimmed spectacles—into the cheese, he happens to read a verse and is absolutely enchanted. He gets into a car and drives straight to the publisher Solomon Edition and shows him the cheese. The publisher can't stand the smell of cheese and writhes and squirms. But the critic is all over him and quotes the poet's verses. Now the publisher is enthusiastic, too. The two immediately run to the cheese dealer and bring along a huge sack stuffed with an advance. ("You have to know," the swarthy little typist tells her smiling friend, "this is a fantasy movie.") Well, the two buy all the cheese off the dealer, hire thirteen men who all cover their noses, and proceed to the poet. The poet is standing on a chair and about to hang himself, because he can't come up with the two hundred Marks. A faint stench begins to pervade his room. Now, do you really hang yourself when it's stinking so abominably? No, you get all angry and develop a new zest for life. The thirteen guys march in but he throws them out with such force that the cheese trickles down the stairs. He only quiets down when the publisher and his sack full of money arrive. No stinking cheese can match the fragrance of the advance.

The poet now hurries to the typing bureau. He finds this snotty businessman who is dictating snotty letters to Minnie and coming on to her. But the poet throws him out; he can afford it, he can now afford to hire the typist for hours, days, and whole eternities. He immediately dictates another poem to her. But what does she write? "Stupid fellow!" she writes, "I love the hardworking and successful." Underlined twice. On that day they did not type any further.

"It's a moral film," the swarthy little girl says. "It shows how an industrious woman can educate a man."

For a moment, the friend no longer smiles. "It shows," he says, "how an industrious woman ruins a man. The film will demonstrate to writers that while this damned typewriter makes them diligent, it makes women turn cold. The film will reveal the spiritual dangers of the typewriter. Do you really think that poet's industrious manuscripts were any *good*? The chewing and the sofa, that was good. But you professional women will never understand that."

The swarthy little one laughs.

And with good reason. While all the men of the time tragically collide with their filmic doubles, the swarthy typist and her colleague Minnie Tipp are united by serene harmony. Or, in more technical terms: by positive feedback. One woman goes to her beloved movies starring the other; the plan was to make a movie featuring both. The logic of representation would have been perfect: one and the same woman spends her days in the real of work time and the symbolic of text processing, and her nights in a technified mirror stage. Which is exactly how Braune's three stenotypists described it.

Demeny speaks “Je vous ai-me” into the chronophotograph.

Against this film-within-a-film-within-a-film, this endless folding of women and media, literature does not have a chance. Both men, the smiling friend and his double, do not move beyond pens and poetry. Subsequently, they are left with an old-fashioned mirror stage in the shape of ephemeral and unpublished authorship. You stare at empty white paper, since Mallarmé the background of all words, and fight with the sterility Mallarmé turned into a poem,[209] until one lone verse finds its way onto the paper. But not even the elementary consolation afforded by mirrors that magically turn bodies into wholes and unconscious literates into self-assured authors is of lasting value. The verse does not carry on into the next; a hand tears up its handwriting, simply because it cannot do it with the body itself.

Poets of 1913 act in old-fashioned ways. One "stands in front of the mirror, and recites the verse, and admires himself." Twenty-two years after Demeny had replaced forgetful mirrors with trace detection and snapshots of speech, words are still lost: to declamations and torn paper. The media revenge follows swiftly. When this particular poet upgrades his mirror declamations to typed dictations, the most oral sentence of all falls into technological storage and at the speaker's feet. And to top it off, the typed "Miss, I love you!" appears on screen, published for the benefit of all of Minnie Tipp's doubles.

Such is the solidarity of film and typewriter, Demeny and Miss Tipp. Every word they hear, read, speak, or type breaks down (as the stenotypists put it) into its letters. The typist turns a poetic and erotically charged flow of speech, the manifest secret of German literature, into twelve letters, four empty spaces, and two punctuation marks, all of which (as her correspondence makes clear) come with a price. Just as he had done with "Vi-ve la Fran-ce!" Demeny turns this declaration of love into twenty-millisecond shots of his empty and media-infatuated mouth. He positions himself in front of a camera (instead of the mirror), declaims the verse of all verses, and becomes a test subject (instead of an admired author). "JE VOUS AI-ME."

To the poetic intellect, the unassuming media link of silent film and typewriter, image flow and intertitles, was nothing short of desecration. In order to save the *Soul of the Film*, Bloem decreed: "Emotion does not reside in the titles; it is not to be spoken, it is to be embodied mimically. Yet there are directors who do not shy away from blaring out 'I love you' (the most fiery and tender possibility of this art) in a title."[210]

A criticism that completely missed the technological, experimental, and social necessity of such prostitution. To begin with, love consists in

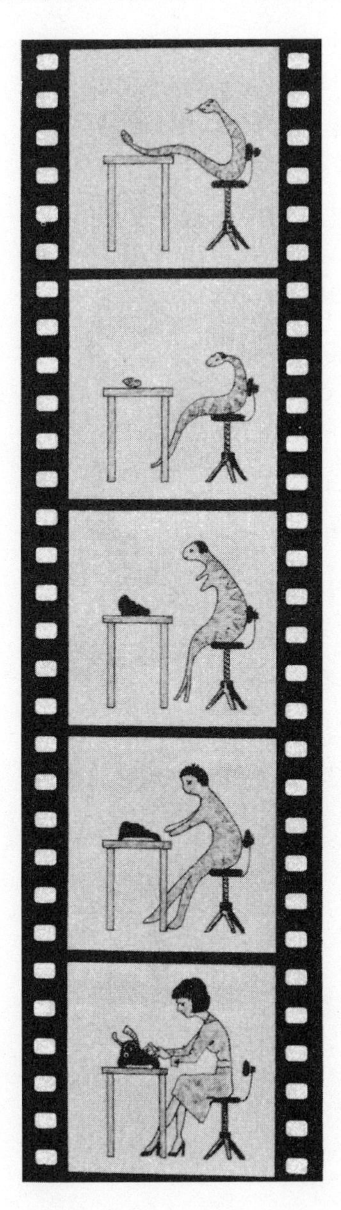

Filmstrip: from snake to typist, 1929.

words; therefore, silent films have to transfer them directly from typescript to screen. In addition, Demeny's experiment delivered the grand kaleidoscope of human speech, as Villiers would have put it, to the deaf and dumb, and Minnie Tipp even delivered it to writers. The decomposition and filtering of love ensured that her new customer would rise to the particular work ethic that characterizes "professional women" and marks within that group a necessary, though not always sufficient, distinction between typists and whores.[211] With the result that a man, too, grasped the secular difference between poet and writer. From handwriting to typed dictation, from the loneliness in front of mirrors to the sexual division of labor and best-selling poetry: as a "moral film," "Lyre and Typewriter" shows "how an industrious woman can educate a man." Or how, in a fine animated sequence, the old snake becomes the Eve of the twentieth century.

"There are more women working at typing than at anything else."[212] Film, the great media self-advertisement, has reached its target group and its happy ending.

TYPEWRITER

"Typewriter" is ambiguous. The word meant both typing machine and female typist: in the United States, a source of countless cartoons. (Typed letter of a bankrupt businessman to his wife: "Dear Blanche, I have sold all my office furniture, chairs, desks, etc. etc., and I am writing this letter under difficulties with my typewriter on my lap.")[1] But the convergence of a profession, a machine, and a sex speaks the truth. Bermann's word "stenotypist" gradually came to require footnotes explaining that since 1885, it has referred to women who have completed Ferdinand Schrey's combined training program in the Hammond typewriter and stenography. In the case of "typewriter," by contrast, everyday language for once matches statistics (see the accompanying table).

The table unfortunately does not distinguish between stenographic handwriting and Remington's typewriting. Nevertheless, it is clear that the statistical explosion begins in 1881, with the record sales of the Remington II. Although the number of men dwindles like a bell curve, the number of female typists increases almost with the elegance of an exponential function. As a consequence, it might be possible—as we approach the threshold of infinity—to forecast the year in which typist and woman converge. Minnie Tipp will have been Eve.

An innocuous device, "an 'intermediate' thing, between a tool and a machine," "almost quotidian and hence unnoticed,"[2] has made history. The typewriter cannot conjure up anything imaginary, as can cinema; it cannot simulate the real, as can sound recording; it only inverts the gender of writing. In so doing, however, it inverts the material basis of literature.

The monopoly of script in serial data processing was a privilege of men. Because orders and poems were processed through the same channel, security protocols evolved. Even though more and more women were

Stenographers and Typists in the United States by Sex, 1870–1930

Year	*Total*	*Men*	*Women*	*Women as a percentage of total*
1870	154	147	7	4.5%
1880	5,000	3,000	2,000	40.0
1890	33,400	12,100	21,300	63.8
1900	112,600	26,200	86,400	76.7
1910	326,700	53,400	263,300	80.6
1920	615,100	50,400	564,700	91.8
1930	811,200	36,100	775,100	95.6

SOURCE: U.S. Bureau of the Census, *Sixteenth Census of the United States, 1940: Population* (1943), as cited in Davies 1974, 10.

taught letters in the wake of general educational reform, being able to read was not the same as being allowed to write. Prior to the invention of the typewriter, all poets, secretaries, and typesetters were of the same sex. As late as 1859, when the solidarity of American women's unions created positions for female typesetters, their male colleagues on the presses boycotted the printing of unmanly type fonts.[3] Only the Civil War of 1861–64—that revolutionary media network of telegraph cables and parallel train tracks[4]—opened the bureaucracy of government, of mail and stenography, to writing women; their numbers, of course, were as yet too small to register statistically.

The Gutenberg Galaxy was thus a sexually closed feedback loop. Even though Germanists are fundamentally oblivious to it, it controlled nothing less than German literature. Unrecognized geniuses swung the quill themselves, whereas national poets had personal secretaries, as in the case of Goethe, John, Schuchardt, Eckermann, Riemer, and Geist. It is precisely this media network—namely, that the Ur-author can bring forth his spirit in Eckermann—that Professor Pschorr had been able to prove phonographically in Goethe's study.[5] One's own or dictated script was processed by male typesetters, binders, publishers, and so on, in order finally to reach in print the girls for whom Goethe wrote. As Goethe put it in conversation with Riemer (who of course recorded it), "he conceives of the Ideal in terms of female form or the form of Woman. What a man is, he didn't know."[6]

Women could and had to remain an ideal abstraction, like Faust's Gretchen, as long as the materialities of writing were the jobs of men, far

too close for them to be aware of it. One Gretchen inspired the work; her many sisters were allowed to consume it through their identification with her. "Otherwise," that is, without sales and female readers, "things would be bad" for him, the "author," Friedrich Schlegel wrote to his lover.[7] But the honor of having a manuscript appear in print under the author's proper name was barred to women, if not factually then at least media-technologically: the proper name at the head of their verse, novels, and dramas almost always has been a male pseudonym.

If only because of that, an omnipresent metaphor equated women with the white sheet of nature or virginity onto which a very male stylus could then inscribe the glory of its authorship. No wonder that psychoanalysis discovered during its clean-up operation that in dreams, "*pencils, pen-holders*, . . . and other *instruments* are undoubted male sexual symbols."[8] It only retrieved a deeply embedded metaphysics of handwriting.

And consequently did not disclose any unconscious secrets, either. For that, the "symbols" of man and woman were too closely attached to the monopoly of writing. When, in 1889, the editors of the illustrated journal *Vom Fels zum Meer* (as usual) made a pitch for Hammond typewriters and Schrey, their general representative, the "writer of these lines" was thrilled by a self-study: "Already after a couple of weeks he reached a speed of 125 letters per minute." Only two things were "lost" during this mechanization of writing: first, "the intimacy of handwritten expression, which nobody is willing to relinquish voluntarily, particularly in personal correspondence"; and second, a centerpiece of occidental symbolic systems:

> Machines everywhere, wherever one looks! A substitute for numerous types of labor, which man would otherwise do with his industrious hand, and what economy of exertion and time, and what advantages in terms of flawlessness and regularity of work. It was only natural that after the engineer had deprived woman's tender hand of the actual symbol of female industriousness, one of his colleagues hit upon the idea of replacing the quill, the actual symbol of male intellectual activity, with a machine.[9]

The literal meaning of text is tissue. Therefore, prior to their industrialization the two sexes occupied strictly symmetrical roles: women, with the symbol of female industriousness in their hands, wove tissues; men, with the symbol of male intellectual activity in their hands, wove tissues of a different sort called text. Here, the stylus as singular needlepoint, there, the many female readers as fabric onto which it wrote.

Industrialization simultaneously nullified handwriting and hand-based work. Not coincidentally, it was William K. Jenne, the head of the sewing-machine subdivision of Remington & Son, who in 1874 developed Sholes's prototype into a mass-producible "Type-Writer."[10] Not coincidentally as well, early competing models came from the Domestic Sewing Machine Co., the Meteor Saxon Knitting-Machine Factory, or Seidel & Naumann.[11] Bipolar sexual differentiation, with its defining symbols, disappeared on industrial assembly lines. Two symbols do not survive their replacement by machines, that is, their implementation in the real. When men are deprived of the quill and women of the needle, all hands are up for grabs—as employable as employees. Typescript amounts to the desexualization of writing, sacrificing its metaphysics and turning it into word processing.

A transvaluation of all values, even if it arrived on pigeon toes, as Nietzsche would have it, or on "high-buttoned shoes" (in the words of the most amusing chronicler of the typewriter).[12] To mechanize writing, our culture had to redefine its values or (as the first German monograph on the typewriter put it, in anticipation of Foucault) "create a wholly new order of things."[13] The work of ingenious tinkerers was far from achieving that. In 1714 Henry Mill, an engineer with the New River Water Co. in London, received his inconsequential British patent (no. 395) "for 'a machine or artificial method, to print letters continuously one after another while writing, in a fashion so clean and precise that they are indistinguishable from the printing of letters.'"[14] The precision of this concept or premise, namely, to introduce Gutenberg's reproductive technology into textual production, was contradicted by the vagueness of the patent's phrasing. The work of Kempelen, the engineer of phonographs, to design an appropriate writing instrument for a blind duchess was similarly inconsequential. Under the discursive conditions of the age of Goethe, the term "writing-machine" was bound to remain a non-term, as was proven rather involuntarily by another Viennese.

In 1823, the physician C. L. Müller published a treatise entitled *Newly Invented Writing-Machine, with Which Everybody Can Write, Without Light, in Every Language, and Regardless of One's Handwriting; Generate Essays and Bills; the Blind, Too, Can, Unlike with Previous Writing Tablets, Write Not Only with Greater Ease but Even Read Their Own Writing Afterward.* What Müller meant and introduced was a mechanical contraption that, its name notwithstanding, only enabled the blind to guide their hands across paper while writing. The mapping of the page and the concentration of ink even afforded them the possibility of rereading their writing through touch. For Müller could "not deny" an authorial narcissism that prompts "all those so inclined," like Minnie Tipp's poet, "to reread what he has written."[15] Significantly enough, the invention was aimed primarily at educated but unfortunately blind fathers for the purpose of illuminating their morally blind sons with letters and epistolary truths. "How often would a man of good standing write a few lines to save a lost estate or the welfare of whole families, how often would the handwritten letter of a father steer a son back on the right track, if such men could, without restraint and prompting, write in such a way as if they had been endowed with vision."[16]

The "writing-machine," in that sense, only brought to light the rules regulating discourses during the age of Goethe: authority and authorship, handwriting and rereading, the narcissism of creation and reader obedience. The device for "everybody" forgot women.

Mechanical storage technologies for writing, images, and sound could only be developed following the collapse of this system. The hard science of physiology did away with the psychological conception that guaranteed humans that they could find their souls through handwriting and rereading. The "I think," which since Kant was supposed to accompany all of one's representations, presumably only accompanied one's readings. It became obsolete as soon as body and soul advanced to become objects of scientific experiments. The unity of apperception disintegrated into a large number of subroutines, which, as such, physiologists could localize in different centers of the brain and engineers could reconstruct in multiple machines. Which is what the "spirit"—the unsimulable center of "man"—denied by its very definition.

Psychophysics and psychotechnology converted into empirical research programs Nietzsche's philosophical and scandalous surmise that "humans are perhaps only thinking, writing, and speaking machines." *Dysfunctional Speech* (*Die Störungen der Sprache*), following Kußmaul's insight or monograph of 1881, could only be cleared up under the premise that speech has nothing to do with the "I think":

> One can conceive of language in its initial development as a *conditioned* reflex. It is the character of reflected intentionality that distinguishes conditioned from inborn movements of expression, their greater ability to adapt, in appropriate form and degree, to the intended purpose. Because of this quality, we are not quite prepared to see in them anything but the play of mechanical circuits acquired through exercise. And yet, pantomime, the spoken word, and the written word are nothing but the products of internal, self-regulating mechanisms that are channeled and coordinated through emotions and conceptions, just as one can operate a sewing, typing, or speaking machine without knowing its mechanism.[17]

When, from the point of view of brain physiology, language works as a feedback loop of mechanical relays, the construction of typewriters is only a matter of course. Nature, the most pitiless experimenter, paralyzes certain parts of the brain through strokes and bullet wounds to the head; research (since the Battle of Solferino in 1859) is only required to measure the resulting interferences in order to distinguish the distinct subroutines of speech in anatomically precise ways. Sensory aphasia (while hearing), dyslexia (while reading), expressive aphasia (while speaking), agraphia (while writing) bring forth machines in the brain. Kußmaul's "sound board," with its "cortical sound keys,"[18] virtually conjures up the rods and levers of old Remingtons.

Disabilities or deformations therefore suggest not only Müller's "sweet hope" to be "of use to his fellow humans" and "to alleviate the suffering of many unfortunates."[19] Blindness and deafness, precisely when they affect either speech or writing, yield what would otherwise be beyond reach: information on the human information machine. Whereupon its replacement by mechanics can begin. Knie, Beach, Thurber, Malling Hansen, Ravizza: they all constructed their early typewriters for the blind and/or the deaf. The Frenchmen Foucauld and Pierre even constructed them for the blind as blind people themselves.[20] Interest in authorship, or in the possibility of reading one's unconscious outpourings in the mirror, disappeared completely.

What the typewriters for the blind in mid century were still missing was speed. But ever since 1810, the introduction of the rotary press and continuous form into the printing trade made typesetting machines desirable in which ("as with a piano") "the various types fall, through a touch of the keys, into place almost as quickly as one speaks."[21] And when Samuel Morse patented his electric cable telegraph in 1840, he introduced a communications technology whose speed of light far outpaced all forms of manual communication. "The average speed, which can be sustained for hours by hand, is about 20–25 words per minute."[22] Consequently, not long thereafter "a whole generation of telegraph operators had appeared who could understand code much faster than they could write it down. Stenographers found themselves in a similar fix. They could take their notations as quickly as a man could speak, and yet they couldn't transcribe faster than at a snail's pace."[23]

What therefore became part of the wish list were writing instruments that could coincide with the operating speed of nervous pathways. Since aphasia researchers had figured out the number of milliseconds it takes for a letter to travel from the eye to the hand muscles via the brain's read-

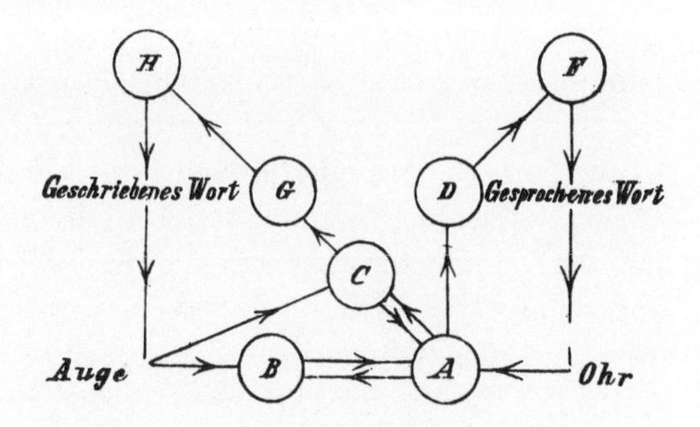

Schematic diagram of the language subsite in the brain. *A* denotes the center for sound images, *B*, the center for visual images.

ing and writing centers, the equation of cerebral circuits with telegraphic dispatches had become a physiological standard.[24] When "the average latency, that is, the time between the stimulus and the pushing of the button takes about 250 milliseconds," and when, furthermore, "the typing of a given output resembles a flying projectile" because "it only needs a starting signal" and "then goes all by itself"[25]—then, the typewriter as a mass-produced article was bound to roll automatically off the production lines of a gun manufacturer.

Unconfirmed rumors have suggested that Sholes sold the Remington company a patent that he had stolen from the poor Tyrolean Peter Mitterhofer during his studies at the Royal and Imperial Polytechnical Institute in Vienna.[26] But plagiarism, or, in modern terms, the transfer of technology, is of little importance in the face of circumstances. Rumor has it that, in reference to Mitterhofer's request for money, Emperor Franz Joseph allegedly remarked to his cabinet that the invention of superior war strategies would be more appropriate than that of useless typewriters. Remington & Son were above such pseudo-alternatives: they transferred "the standardization of the component parts of weapons, which had been widely practiced since the Napoleonic Wars," to those of civil writing instruments.[27] (Weapons manufacturers such as Mauser, Manufacture d'Armes de Paris, and the German Weapons and Ammunitions Factory [DWF] were to follow suit.)

The technologies of typewriting and sound recording are by-products of the American Civil War. Edison, who was a young telegrapher during the war, developed his phonograph in an attempt to improve the processing speed of the Morse telegraph beyond human limitations. Remington

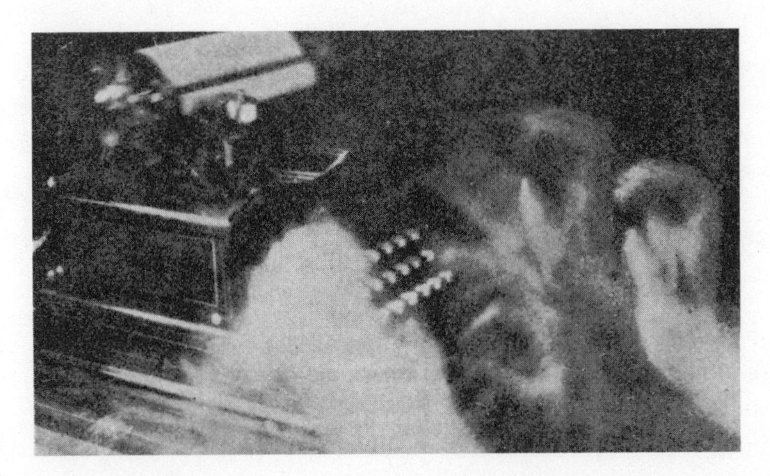

Anton Giulio Bragaglia and Arturo Bragaglia, *Datillografa*, 1911.

began the serial production of Sholes's typewriter models in September 1874 simply because "after the Civil War boom things had been on the slow side," and they had "more capacity than they were using."[28]

The typewriter became a discursive machine-gun. A technology whose basic action not coincidentally consists of strikes and triggers proceeds in automated and discrete steps, as does ammunitions transport in a revolver and a machine-gun, or celluloid transport in a film projector. "The pen was once mightier than the sword," Otto Burghagen, the first monographer of the typewriter, writes in 1898, "but where the typewriter rules," he continues, "Krupp's cannons must remain silent!"[29] Burghagen is contradicted, however, by his own deliberations on "the significant *savings of time*, which endear the machine to the merchant. With its help one can complete office work in a third of the time it would take with the pen, for with each strike of a key the machine produces a complete letter, while the pen has to undergo about five strokes in order to produce a letter. . . . In the time it takes the pen to put a dot on the "i" or to make the "u" sign, the machine produces two complete letters. The striking of the keys follows in succession with great speed, especially when one writes with all fingers; then, one can count five to ten keyboard hits per second!"[30] This is the epic song of a firepower whose German record as of August 1985 stands at "773 letters per minute for thirty minutes of high-speed typing."[31]

Jean Cocteau, who produced a corresponding work for each of the late-nineteenth-century media—*La voix humaine* for the acoustics of the telephone, the script for *Orphée* for mirrors, doppelgängers, cinematic effects, and "for car radios, secret codes, and short-wave signals"[32]—made the typewriter into the titular hero of a play in 1941. The reason was there in American idiom: for three acts, a detective chases an unknown woman who has been tormenting her community with anonymous, typewritten letters, going by the title "the typewriter."[33] For three acts, he "imagines the culprit at work at her typewriter, aiming and operating her machine gun."[34] Typewriters are simply "fast," not just "like Jazz" (as Cendrars put it) but also like rapid-fire weapons. In her confession, Cocteau's anonymous letter-writer puts it this way: "I wanted to attack the whole city. All the hypocritical happiness, the hypocritical piety, the hypocritical luxury, the whole lying, egotistical, avaricious, untouchable bourgeoisie. I wanted to stir that muck, attack and reveal it. It was like a hoax! Without accounting for myself, I chose the dirtiest and cheapest of all weapons, the typewriter."[35]

About which the playwright, in his preface of 1941, only remarked that he had "portrayed the terrible feudal province" of France "prior to the debacle."[36] As innocuous as they were, typewriters could still provide cover for the work of Guderian's submachine guns and tank divisions. And indeed: whereas the Army High Command supplied its war photographers with "Arriflex hand-held cameras, Askania Z-tripod cameras, [and] special-assignment vehicles" and its recording specialists with "armored vehicles and tanks for radio broadcasts" and with magnetophones, "war reporters were equipped solely with typewriters, and specifically, most often with commercially available traveling typewriters."[37] Modesty of literature under conditions of high technology.

That is precisely how Remington began production. The Model I hardly sold, even though or precisely because one no less than Mark Twain purchased a Remington in 1874. He sent his novel *Tom Sawyer*, the first typescript in literary history, to his publisher, and sent a paradoxical letter of support to the typewriter manufacturer:

> GENTLEMEN: PLEASE DO NOT USE MY NAME IN ANY WAY,
> PLEASE DO NOT EVEN DIVULGE THE FACT THAT I OWN A
> MACHINE, I HAVE ENTIRELY STOPPED USING THE TYPE-
> WRITER, FOR THE REASON THAT I NEVER COULD WRITE A
> LETTER WITH IT TO ANYBODY WITHOUT RECEIVING A
> REQUEST BY RETURN MAIL THAT I WOULD NOT ONLY DESCRIBE
> THE MACHINE BUT STATE WHAT PROGRESS I HAD MADE IN THE
> USE OF IT, ETC., ETC. I DON'T LIKE TO WRITE

LETTERS, AND SO I DON'T WANT PEOPLE TO KNOW THAT
I OWN THIS CURIOSITY BREEDING LITTLE JOKER.

YOURS TRULY,
SAML L. CLEMENS.[38]

The Model II of 1878, which allowed the switch from lower to upper case for a price of $125, initially did not fare much better. But after a slow start of 146 sales per year there came a rise that approximated a global snowball effect.[39] For in 1881, the marketing strategists of Wyckoff, Seamans, and Benedict made a discovery: they recognized the fascination their unmarketable machine held for the battalions of unemployed women. When Lillian Sholes, as "presumably" the "first type-writer" in history,[40] sat and posed in front of her father's prototype in 1872, female typists came into existence for purposes of demonstration, but as a profession and career, the stenotypist had yet to come. That was changed by the central branch of the Young Women's Christian Association in New York City, which trained eight young women in 1881 to become typists and immediately received hundreds of inquiries (at $10 a week) from the corporate world.[41] A feedback loop was created connecting recruitment, training, supply, demand, new recruitment, and so on—first in the United States, and shortly thereafter through Christian women's associations in Europe.[42]

Thus evolved the exponential function of female secretaries and the bell curve of male secretaries. Ironically enough, the clerks, office helpers, and poet-apprentices of the nineteenth century, who were exclusively male, had invested so much pride in their laboriously trained handwriting as to overlook Remington's innovation for seven years. The continuous and coherent flow of ink, that material substrate of all middle-class in-dividuals and indivisibilities, made them blind to a historical chance. Writing as keystrokes, spacing, and the automatics of discrete block letters bypassed a whole system of education. Hence sexual innovation followed technological innovation almost immediately. Without resistance men cleared the field "where competition is as fierce as nowhere else."[43] Women reversed the handicap of their education, turning it into a "so-called emancipation"[44] that, all proletarian fascination notwithstanding, wears the white collar of the employee of discourse.

In 1853, Hessian school regulations described knowledge of writing and arithmetic as useful for girls but not indispensable.[45] And women "without any talent for arithmetic, with terrible handwriting, with a

Sholes's daughter at the Remington, 1872.

highly deficient knowledge of orthography and mathematics" promptly started "in droves" to "work on the typewriter"—so says a woman who in 1902 described the job of a *female clerk* "as building a church tower in the air because one had forgotten the foundations."[46]

But in the age of information, foundations no longer count. The fact that "the female clerk could all-too-easily degrade into a mere typewriter"[47] made her an asset. From the working class, the middle class, and the bourgeoisie, out of ambition, economic hardship, or the pure desire for emancipation[48] emerged millions of secretaries. It was precisely their marginal position in the power system of script that forced women to develop their manual dexterity, which surpassed the prideful handwriting aesthetics of male secretaries in the media system. Two German economists noted it in 1895:

> Today, the *typist* has evolved into a kind of type: she is generally very high in demand and is the ruling queen in this domain not only in America but in Germany as well. It may come as a surprise to find a practical use for what has become a veritable plague across the country, namely, piano lessons for young girls: the re-

sultant dexterity is very useful for the operation of the typewriter. Rapid typing on it can be achieved only through the dexterous use of *all fingers*. If this profession is not yet as lucrative in Germany as it is in America, it is due to the infiltration of elements who perform the job of typist mechanically, without any additional skills.[49]

Edison's mechanical storage of sound made obsolete the piano keyboard as the central storage device for music's scriptive logic; women were no longer asked to endow lyrical letters with a singable, ersatz sensuality; the national plague of their dexterity could finally find a practical use on typewriter keyboards (derived from the piano). And since power after the print monopoly's collapse was diverted to cable and radio, to the recording of traces and electrical engineering, outdated security protocols were dropped as well: women were allowed to reign over text processing all by themselves. Since then, "discourse has been secondary" and desexualized.[50]

A certain Spinner, treasurer of the United States and a friend of Philo Remington, gave an example of this change. The attrition of males during the Civil War forced him to hire 300 women and to make the statement, "that I authorized the hiring of women for positions in government satisfies me more than all the other achievements in my life."[51]

One country after another opened the mail and wireless services as well as the railroad to typists. Technological media needed technological (or hysterical) media. In the German Reich, this was initially understood only by Undersecretary of the Interior and Major General von Budde, chief of the railroad division within the Great General Staff, who dictated flawless orders to his secretaries every day and who committed subordinate agencies to "an increased appropriation of typewriters."[52] But the German dream of men as civil servants and women as mothers weighed heavily: what had to be created for girls involved in typing, telegraphing, and telephoning was a special, temporary, civil-servant status that was immediately revoked upon marriage.[53] Understood that way, communications technology amounted to "the disintegration of the old family structure"[54] and "denied" its female machine operators "a return to any role in the family."[55]

Global forms of disintegration put an end to the German dream. In 1917, when the Army High Command built up its arsenal to prepare for the Ludendorff offensive and screened the civil-service corps for battle readiness, in a letter Hindenburg established the "principle" that, regardless of sex, "whosoever does not work, shall not eat." One year later, the *Zeitschrift für weibliche Handelsgehilfen* (Journal for female clerks) re-

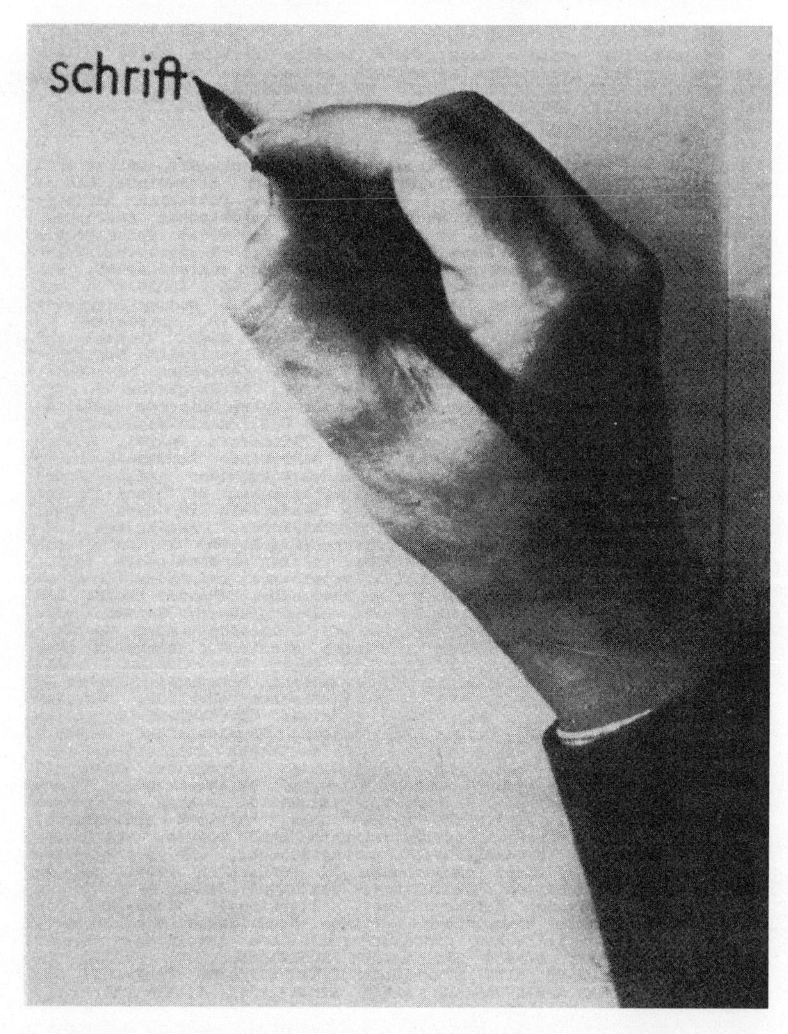

Jan Tschichold writing, 1948. " . . . to substitute the innervation of guiding fingers for the continuous movement of the hand" (Benjamin).

ported full compliance. "The offices of all manufacturers central to the war effort have been occupied with female workers; they have conquered even the orderly rooms of the army administration; shift work was always understaffed, and there was a constant demand for them. They were absorbed in large quantities by the occupied territories; domestic administrative agencies of all kinds hired them in large numbers, let alone com-

Olivetti M 20. Poster by Piramo, Italy, 1920.

panies in the private sector central to the war effort."[56] "A state—it *is*," Heidegger observed in 1935. But only in order to doubt whether this "being" consists in the "fact that the police arrest a suspect, or so-and-so-many typewriters are clattering in a government building, taking down the words of ministers and state secretaries."[57]

Only his winter semester in Stalingrad revealed to the thinker—much to the surprise of his listeners—the relationship among Being, Man, and typewriter.

MARTIN HEIDEGGER ON THE HAND AND THE TYPEWRITER (1942–43)

Man himself acts [*handelt*] through the hand [*Hand*]; for the hand is, together with the word, the essential distinction of man. Only a being which, like man, "has" the word (μύθος, λόγος), can and must "have" "the hand." Through the hand occur both prayer and murder, greeting and thanks, oath and signal, and also the "work" of the hand, the "hand-work," and the tool. The handshake seals the covenant. The hand brings about the "work" of destruction. The hand exists as hand only where there is disclosure and concealment. No animal has a hand, and a hand never originates from a paw or a claw or talon. Even the hand of one in desperation (it least of all) is never a talon, with which a person clutches wildly. The hand sprang forth only out of the word and together with the word. Man does not "have" hands, but the hand holds the essence of man, because the word as the essential realm of the hand is the ground of the essence of man. The word as what is inscribed and what appears to the regard is the written word, i.e., script. And the word as script is handwriting.

It is not accidental that modern man writes "with" the typewriter and "dictates" [*diktiert*] (the same word as "poetize" [*dichten*]) "into" a machine. This "history" of the kinds of writing is one of the main reasons for the increasing destruction of the word. The latter no longer comes and goes by means of the writing hand, the properly acting hand, but by means of the mechanical forces it releases. The typewriter tears writing from the essential realm of the hand, i.e., the realm of the word. The word itself turns into something "typed." Where typewriting, on the contrary, is only a transcription and serves to preserve the writing, or turns into print something already written, there it has a proper, though limited, significance. In the time of the first dominance of the typewriter, a letter written on this machine still stood for a breach of good manners. Today, a handwritten letter is an antiquated and undesired thing; it disturbs speed reading. Mechanical writing deprives the hand of its rank in the realm of the written word and degrades the word to a means of communication. In addition, mechanical writing provides this "advantage," that it conceals the handwriting and thereby the character. The typewriter makes everyone look the same. . . .

Therefore, when writing was withdrawn from the origin of its essence, i.e., from the hand, and was transferred to the machine, a transformation occurred in the relation of Being to man. It is of little importance for this transformation how many people actually use the typewriter and whether there are some who shun it. It is no accident that the invention of the printing press coincides with the inception of the modern period. The word-signs become type, and the writing stroke disappears. The type is "set," the set becomes "pressed." This mechanism of setting and pressing and "printing" is the preliminary form of the typewriter. In the typewriter we find the irruption of the mechanism in the realm of the word. The typewriter leads again to the typesetting machine. The press becomes the rotary press. In rotation, the triumph of the machine comes to the fore. Indeed, at first, book printing and then machine type offer advantages and conveniences, and these then unwittingly steer preferences and needs to this kind of written communication. The typewriter veils the essence of writing and of the script. It withdraws from man the essential rank of the hand, without man's experiencing this withdrawal appropriately and recognizing that it has transformed the relation of Being to his essence.

The typewriter is a signless cloud, i.e., a withdrawing concealment in the midst of its very obtrusiveness, and through it the relation of Being to man is transformed. It is in fact signless, not showing itself as to its essence; perhaps that is why most of you, as is proven to me by your reaction, though well intended, have not grasped what I have been trying to say.

I have not been presenting a disquisition on the typewriter itself, regarding which it could justifiably be asked what in the world that has to do with Parmenides. My theme was the modern relation (transformed by the typewriter) of the hand to writing, i.e., to the word, i.e., to the unconcealedness of Being. A meditation on unconcealedness and on Being does not merely have something to do with the didactic poem of Parmenides, it has everything to do with it. In the typewriter the machine appears, i.e., technology appears, in an almost quotidian and hence unnoticed and hence signless relation to writing, i.e., to the word, i.e., to the distinguishing essence of man. A more penetrating consideration would have to recognize here that the typewriter is not really a machine in the strict sense of machine technology, but is an "intermediate" thing, between a tool and a machine, a mechanism. Its production, however, is conditioned by machine technology.

This "machine," operated in the closest vicinity to the word, is in use; it imposes its own use. Even if we do not actually operate this machine, it demands that we regard it if only to renounce and avoid it. This situation is constantly repeated everywhere, in all relations of modern man to technology. Technology *is* entrenched in our history.[58]

"Our writing tools are also working on our thoughts," Nietzsche wrote.[59] "Technology *is* entrenched in our history," Heidegger said. But the one wrote the sentence about the typewriter on a typewriter, the other described (in a magnificent old German hand) typewriters per se. That is why it was Nietzsche who initiated the transvaluation of all values with his philosophically scandalous sentence about media technology. In 1882, human beings, their thoughts, and their authorship respectively were replaced by two sexes, the text, and blind writing equipment. As the first mechanized philosopher, Nietzsche was also the last. Typescript, according to Klapheck's painting, was called *The Will to Power*.

Nietzsche suffered from extreme myopia, anisocoria, and migraines (to say nothing of his rumored progressive paralysis). An eye doctor in Frankfurt attested that Nietzsche's "right eye could only perceive misshapen and distorted images" as well as "letters that were virtually beyond recognition," whereas the left, "despite its myopia," was in 1877 still capable of "registering normal images." Nietzsche's headaches therefore appeared to be "a secondary symptom,"[60] and his attempts to philosophize with a hammer the natural consequence of "an increased stimulation of the site in the prefrontal wall of the third ventricle responsible for aggression."[61] Thinkers of the founding age of media naturally did not only turn from philosophy to physiology in theory; their central nervous system always preceded them.

Nietzsche himself successively described his condition as quarter blindness, half-blindness, three-quarter blindness (it was for others to suggest mental derangement, the next step in this mathematical sequence).[62] Reading letters (or musical notes) distorted beyond recognition became painful after twenty minutes, as did writing. Otherwise, Nietzsche would not have attributed his "telegram style,"[63] which he developed

Konrad Klapheck, *The Will to Power*, 1959. (Reproduced courtesy of the artist)

while writing the suggestively titled *The Wanderer and His Shadow*, to his eye pain. To direct the blindness of this shadow, he had been planning to purchase a typewriter as early as 1879, the so-called "year of blindness."[64] It happened in 1881. Nietzsche got "in touch with its inventor, a Dane from Copenhagen."[65] "My dear Sister, I know Hansen's machine quite well, Mr. Hansen has written to me twice and sent me samples, drawings, and assessments of professors from Copenhagen about it. *This* is the one I want (*not* the American one, which is too heavy)."[66]

Since our writing tools also work on our thoughts, Nietzsche's choice followed strict, technical data. En route between Engadine and the Riviera, he decided first for a traveling typewriter and second as the cripple that he was. At a time when "only very few owned a typewriter, when there were no sales representatives [in Germany] and machines were available only under the table,"[67] a single man demonstrated a knowledge of engineering. (With the result that American historians of the typewriter elide Nietzsche and Hansen.)[68]

Hans Rasmus Johann Malling Hansen (1835–90), pastor and head of the royal Døvstummeinstitut in Copenhagen,[69] developed his *skrivekugle* / writing ball / *sphère écrivante* out of the observation that his deaf-mute patients' sign language was faster than handwriting. The machine "did not take into account the needs of business"[70] but rather was meant to compensate for physiological deficiencies and to increase writing speed (which prompted the Nordic Telegraphy Co. to use "a number of writing balls for the transfer of incoming telegrams").[71] Fifty-four concentrically arranged key rods (no levers as yet) imprinted capital letters, numbers, and signs with a color ribbon onto a relatively small sheet of paper that was fastened cylindrically. According to Burghagen, this semispheric arrangement of the keys had the advantage of allowing "the blind, for whom this writing ball was primarily designed, to learn writing on it in a surprisingly short time. On the surface of a sphere each position is completely identifiable by its relative location. . . . It is therefore possible to be guided solely by one's sense of touch, which would be much more difficult in the case of flat keyboards."[72] That is precisely how it could have been stated in the assessments of professors from Copenhagen for a half-blind ex-professor.

In 1865 Malling Hansen received his patent, in 1867 he started serial production of his typewriter, in 1872 the Germans (and Nietzsche?) learned of it from the *Leipziger Illustrirte Zeitung*.[73] Finally, in 1882 the Copenhagen printing company of C. Ferslew combined typing balls and women—as a medium to offset the nuisance that "their female typesetters were significantly more preoccupied with the decoding of handwritten texts than with the actual setting of text."[74] McLuhan's law that the typewriter causes "an entirely new attitude to the written and printed word" because it "fuses composition and publication"[75] was realized for the first time. (Today, when handwritten publisher's manuscripts are rarities, "the entire printing industry, via the Linotype, depend[s] upon the typewriter.")[76]

In the same year and for the same reasons, Nietzsche decided to buy. For 375 Reichsmarks (shipping not included)[77] even a half-blind writer chased by publishers was able to produce "documents as beautiful and standardized as print."[78] "After a week" of typewriting practice, Nietzsche wrote, "the eyes no longer have to do their work":[79] *écriture au-*

tomatique had been invented, the shadow of the wanderer incarnated. In March 1882, the *Berliner Tageblatt* reported:

> The well-known philosopher and writer [sic] Friedrich Nietzsche, whose failing eyesight made it necessary for him to renounce his professorship in Basel three years ago, currently lives in Genoa and—excepting the progression of his affliction to the point of complete blindness—feels better than ever. With the help of a typewriter he has resumed his writing activities, and we can hence expect a book along the lines of his last ones. It is widely known that his new work stands in marked contrast to his first, significant writings.[80]

Indeed: Nietzsche, as proud of the publication of his mechanization as any philosopher,[81] changed from arguments to aphorisms, from thoughts to puns, from rhetoric to telegram style. That is precisely what is meant by the sentence that our writing tools are also working on our thoughts. Malling Hansen's writing ball, with its operating difficulties, made Nietzsche into a laconic. "The well-known philosopher and writer" shed his first attribute in order to merge with his second. If scholarship and thinking, especially toward the end of the nineteenth century, were allowed or made possible only after extensive reading, then it was blindness and blindness alone that "delivered" them from "the book."[82]

Good news from Nietzsche that coincided with all the early typewriter models. None of the models prior to Underwood's great innovation of 1897 allowed immediate visual control over the output. In order to read the typed text, one had to lift shutters on the Remington model, whereas with Malling Hansen's—notwithstanding other claims[83]—the semicircular arrangement of the keys itself prevented a view of the paper. But even Underwood's innovation did not change the fact that typewriting can and must remain a blind activity. In the precise engineering lingo of Angelo Beyerlen, the royal stenographer of Württemberg and the first typewriter dealer of the Reich: "In writing by hand, the eye must constantly watch the written line and only that. It must attend to the creation of each sign, must measure, direct, and, in short, guide the hand through each movement." A media-technological basis of classical authorship that typewriting simply liquidates: "By contrast, after one briefly presses down on a key, the typewriter creates in the proper position on the paper a complete letter, which is not only untouched by the writer's hand but also located in a place entirely apart from where the hands work." With Underwood's models, too, "the spot where the next sign to be written *occurs*" is "precisely what . . . *cannot* be seen."[84] After a fraction of a second, the act of writing stops being an act of reading that is produced by the grace

Malling Hansen, Writing Ball, 1867, a model of Nietzsche's typewriter. "Our writing tools are also working on our thoughts" (letter to Peter Gast). (Reproduced courtesy of the Stiftung Weimarer Klassik, Goethe-Schiller-Archiv)

of a human subject. With the help of blind machines, people, whether blind or not, acquire a historically new proficiency: *écriture automatique*.

Loosely translating Beyerlen's dictum that "for writing, visibility is as unnecessary today as it has always been,"[85] an American experimental psychologist (who in 1904 measured the "Acquisition of Skill in Type-Writing" and who obliged his subjects to keep typed test diaries) recorded documentary sentences like those of André Breton:

Self-advertisement of the medium—a typewriter with visible type.

24th day. Hands and finger are clearly becoming more flexible and adept. The change now going on, aside from growing flexibility, is in learning to locate keys without waiting to see them. In other words, it is location by position.

25th day. Location (muscular, etc.), letter and word associations are now in progress of automatization.

38th day. To-day I found myself not infrequently striking letters before I was conscious of seeing them. They seem to have been perfecting themselves just below the level of consciousness.[86]

"A Funny Story About the Blind, etc." (Beyerlen's essay title) was also the story of the mechanized philosopher. Nietzsche's reasons for purchasing a typewriter were very different from those of his few colleagues who wrote for entertainment purposes, such as Twain, Lindau, Amytor, Hart, Nansen, and so on.[87] They all counted on increased speed and textual mass production; the half-blind, by contrast, turned from philosophy to literature, from rereading to a pure, blind, and intransitive act of writing. That is why his Malling Hansen typed the motto of all modern, highbrow literature: "Finally, when my eyes prevent me from *learning* anything—and I have almost reached that point! I will still be able to craft verse."[88]

1889 is generally considered the year zero of typewriter literature, that barely researched mass of documents, the year in which Conan Doyle first published *A Case of Identity*. Back then, Sherlock Holmes managed to prove his claim that the typed love letters (including the signature) received by one of London's first and ostensibly myopic typists were the work of her criminal stepfather engaging in marriage fraud. A machine-produced trick of anonymization that prompted Holmes, seventeen years prior to the professionals in the police, to write a monograph entitled *On the Typewriter and Its Relation to Crime.*[89]

Our esteem for Doyle notwithstanding, it is nonetheless an optical-philological pleasure to show that typewriting literature began in 1882—with a poem by Friedrich Nietzsche that could well be titled *On the Typewriter and Its Relation to Writing.*

In these typed, that is, literally forged or crafted, verses, three moments of writing coincide: the equipment, the thing, and the agent. An author, however, does not appear because he remains on the fringes of the verse: as the addressed reader, who would "utilize" the "delicate"[90] writing ball known as Nietzsche in all its ambiguity. Our writing tool not only works on our thoughts, it "is a thing like me." Mechanized and automatic writing refutes the phallocentrism of classical pens. The fate of the philosopher utilized by his fine fingers was not authorship but feminization. Thus Nietzsche took his place next to the young Christian women of Remington and the typesetters of Malling Hansen in Copenhagen.

But that happiness was not to last long. The human writing ball spent two winter months in Genoa to test and repair its new and easily malfunctioning favorite toy, to utilize and compose upon it. Then the spring on the Riviera, with its downpours, put an end to it. "The damned writing," Nietzsche wrote, self-referentially as always, "the typewriter has been *unusable* since my last card; for the weather is dreary and cloudy, that is, humid: then each time the ribbon is also *wet* and *sticky*, so that every key gets stuck, and the writing cannot be seen *at all*. If you think about it!!"[91]

***

SCHREIBKUGEL IST EIN DING GLEICH MIR: VON
EISEN
UND DOCH LEICHT ZU VERDREHN ZUMAL AUF REISEN.
GEDULD UND TAKT MUSS REICHLICH MAN BESITZEN
UND FEINE FINGERCHEN, UNS ZU BENUETZEN.

A facsimile of Nietzsche's Malling Hansen poem, February–March 1882. The text reads, "THE WRITING BALL IS A THING LIKE ME: MADE OF / IRON / YET EASILY TWISTED ON JOURNEYS. / PATIENCE AND TACT ARE REQUIRED IN ABUNDANCE, / AS WELL AS FINE FINGERS, TO USE US." (Reproduced courtesy of the Stiftung Weimarer Klassik, Goethe-Schiller-Archiv)

And so it was a rain in Genoa that started and stopped modern writing—a writing that is solely the materiality of its medium. "A letter, a litter," a piece of writing, a piece of dirt, Joyce mocked. Nietzsche's typewriter, or the dream of fusing literary production with literary reproduction, instead fused again with blindness, invisibility, and random noise, the irreducible background of technological media. Finally, letters on the page looked like the ones on the right retina.

But Nietzsche did not surrender. In one of his last typewritten letters he addressed media-technological complements and/or human substitution: the phonograph and the secretary. "This machine," he observed in another equation of writing equipment with writer, "is as delicate as a little dog and causes a lot of trouble—and provides some entertainment. Now all my friends have to do is to invent a reading machine: otherwise I will fall behind myself and won't be able to supply myself with sufficient intellectual nourishment. Or, rather: I need a young person who is intelligent and knowledgeable enough to *work* with me. I would even consider a two-year-long marriage for that purpose."[92]

With the collapse of his machine, Nietzsche became a man again. But only to undermine the classical notion of love. As with men since time immemorial and women only recently, "a young person" and a "two-year-long marriage" are equally suitable to continue the failed love affair with a typewriter.

And so it happened. Nietzsche's friend Paul Rée, who had already transported the Malling Hansen to Genoa, was also searching for its human replacement: somebody who could "aid" Nietzsche "in his philosophical studies with all kinds of writing, copying, and excerpting."[93] But instead of presenting an intelligent young man, he presented a rather notorious young lady who, "on her path of scholarly *production*," required a "*teacher*":[94] Lou von Salomé.

And so a defunct typewriter was replaced by the most famous ménage à trois of literary history. The question of whether, when, and in what grouping Professor Nietzsche, Dr. Rée, and Ms. von Salomé went to bed with one another may be amusing to psychologists. But the question as to why young women of the Nietzsche era could replace his writing ball and even his proverbially rare students is of priority to us. The locally known sister of the globally known brother (as Pschorr put it) gave an answer to that question. In her monograph, *Friedrich Nietzsche and the Women of His Time*, Elisabeth Förster described how professors at the University of Zurich "very much appreciated having emancipated women of the time at universities and libraries as secretaries and assistants"[95] (especially once emancipation had "gradually taken on more temperate forms" and was no longer synonymous with gender war). With the logical consequence that young women from Russia or Prussia (where the management of discourse and higher education was to remain a male monopoly until 1908) had every reason to enroll, as did Lou von Salomé, at the philosophical faculty of Zurich. With the further logical consequence that former professors of the University of Basel had every reason to welcome them as secretaries and assistants. At any rate, the die had long been cast before an impassioned philosopher and his Russian love climbed Monte Sacro . . .

Nietzsche's philosophy simply implemented the desexualization of writing and the university. Since no colleague and hardly a student in Basel could be enthused about Nietzsche's most deeply felt wish, namely, to establish a Zarathustra chair, Nietzsche dismantled the elementary barrier of philosophical discourses. He recruited his students from the women who had just recently been admitted to the universities. Lou von Salomé was only one of many students of philosophy in Zurich who contacted him: aside from her, there were the forgotten names Resa von Schirnhofer, Meta von Salis, and especially Helene Druskowitz, who succeeded (and competed with) Nietzsche all the way to her death in an insane asylum. Curiously enough, what Nietzsche called *The Future of Our Institutions of Higher Education* began, of all places, in the quiet and removed Engadine. Beginning in 1885, emancipated women students traveled to Sils Maria "only to get to know better Prof. Nietzsche, who appeared to them as the most dangerous enemy of women."[96]

But that's how it goes. Just as the hundred-year-long exclusion of women from universities and philosophy led to the idealization of grand Dame Nature, so their renewed inclusion altered philosophy as such. What Hegel in his youth called Love (and a Love that was one with the Idea), Nietzsche in *Ecce Homo* notoriously transvalued into the definition that "Love in its means, [is] war; at bottom, the deadly hatred of the sexes."[97] And if the new philosopher, following such insights, fought against emancipation as a form of conscientious objection and even defined Woman as both truth *and* untruth, only female philosophers had an answer. The hatred for males of Helene Druskowitz, Nietzsche's former student, even outdid his hatred for women. The escalation of positions in the work of two writers, a man and a woman, gave proof of Nietzsche's media-specific notion of heterosexuality.

Nietzsche and Lou von Salomé's honeymoon would have been nice and forgotten. Their ceaselessly escalating gender war is what started

Nietzsche's fame. Women (and Jews) brought an almost completely silenced ex-professor back into the public. Whether out of hatred, as with Druskowitz, or love, Nietzsche's private students became writers, and their careers in turn afforded them the opportunity to write books on Nietzsche. "With all kinds of writing, copying, and excerpting," as desired, women did their secretarial work.

That is just how precisely Nietzsche registered discursive changes. Even if the system of higher education had attuned him, as it did all others, to handwriting and academic homosexuality, he himself started something new. The two relayed innovations of his time, writing machines and writing women, recorded his speech.

"Our writing tools are also working on our thoughts." Hence Nietzsche's next thought—four years after the malfunctioning of his typewriter—was to philosophize on the typewriter itself. Instead of testing Remington's competing model, he elevated Malling Hansen's invention to the status of a philosophy. And this philosophy, instead of deriving the evolution of the human being from Hegel's spirit (in between the lines of books) or Marx's labor (in between the differential potential of muscular energy), began with an information machine.

In the second essay of *On the Genealogy of Morals*, knowledge, speech, and virtuous action are no longer inborn attributes of Man. Like the animal that will soon go by a different name, Man derived from forgetfulness and random noise, the background of all media. Which suggests that in 1886, during the founding age of mechanized storage technologies, human evolution, too, aims toward the creation of a machine memory. Guyau's argument presupposes the phonograph, Nietzsche's, the typewriter. To make forgetful animals into human beings, a blind force strikes that dismembers and inscribes their bodies in the real, until pain itself brings forth a memory. People keep promises and execute orders only after torture.

Writing in Nietzsche is no longer a natural extension of humans who bring forth their voice, soul, individuality through their handwriting. On the contrary: just as in the stanza on the delicate Malling Hansen, humans change their position—they turn from the agency of writing to become an inscription surface. Conversely, all the agency of writing passes on in its violence to an inhuman media engineer who will soon be called up by Stoker's *Dracula*. A type of writing that blindly dismembers body parts and perforates human skin necessarily stems from typewriters built before 1897, when Underwood finally introduced visibility. Peter Mitterhofer's Model 2, the wooden typewriter prototype of 1866, unlike the Malling

Hansen did not even have types and a ribbon. Instead, the writing paper was perforated by needle pins—inscribing, for example, in a rather Nietzschean manner, the proper name of the inventor.

Such is the solidarity among engineers, philosophers, and writers of the founding age of media. Beyerlen's technical observation that in typing, everything is visible except the actual inscription of the sign, also describes *On the Genealogy of Morals*. Neither in Nietzsche nor in Stoker can the victims see and hence read what the "most dreadful sacrifices and pledges," "the most repulsive mutilations," and "the cruelest rites"[98] do to their body parts. The only possible, that is unconscious, kind of reading is the slavish obedience called morals. Nietzsche's notion of inscription, which has degenerated into a poststructuralist catch-all metaphor, has validity only within the framework of the history of the typewriter. It designates the turning point at which communications technologies can no longer be related back to humans. Instead, the former have formed the latter.

Under conditions of media the genealogy of morals coincides with the genealogy of gods. Following Beyerlen's law—namely, the invisibility of the act of inscription—we can deduce the necessary existence of beings that could be either observers or, as with Dracula, masters of inhuman communications technologies. "So as to abolish hidden, undetected, unwitnessed suffering from the world and to deny it, one was in the past virtually compelled to invent gods and genii of all the heights and depths; in short something that even roams in secret, hidden places, sees even in the dark, and will not easily let an interesting painful spectacle pass unnoticed."[99]

It is Nietzsche's most daring experimental setup to occupy the place of such a god. If God is dead, nothing is there to prevent the invention of gods. "The poor man," as he was described by an emancipated woman, "is a true saint and ceaselessly working, even though he is almost blind and can neither read nor write (except with a machine)"[100]—this poor man identifies with Dionysus, the master of media. Once again, philosophizing or studying are followed by the crafting of verse. *On the Genealogy of Morals* deploys itself in rhythms and unfolds an interesting and painful spectacle: Nietzsche's dithyrambs of Dionysus entitled *Ariadne's Complaint*. Composing and dictating into a machine are, following Heidegger's recollection, in word and deed one and the same thing.

Ariadne's composed lament arises out of complete darkness or blindness. She speaks about and to a "veiled" god that tortures her body, following all the rules of mnemotechnology or memory inscription described in *Genealogy*. Dionysus has neither word nor style nor stylus—except for torture itself. His female victims are faced with the painful attempt to decode from their bodily pain the trace of a desire, which is truly the desire of the other. And only after 150 lines or laments can Ariadne read that she herself desires the desire of the god:

> Come back!
> *With* all your torments!
> All the streams of my tears
> run their course to you!
> and the last flame of my heart
> it burns up to you.
> Oh come back,
> my unknown god! my *pain*!
> my last happiness![101]

This last cry is not a fiction. It is a quotation—from one of the new women writers. One of Lou von Salomé's poems, accompanied by Nietzsche's music, contained the following lines: "Have you no more happiness to share with me, so be it! as yet you have your torment." The poet of dithyrambs is once again only a secretary who puts the words of one woman, von Salomé, into the mouth of another woman, Ariadne. And as the *Genealogy* predicted, the god of inscription can and must come forth from inscribed pain itself. After Ariadne's or Salomé's last cry, the long-concealed Dionysus himself becomes "visible" in blinding and "emerald beauty." The dithyrambs come to a necessary close because their answer transmits plain text: the whole scene of writing has been a scene of torture:

> Be wise Ariadne! . . .
> You have little ears, you have ears like mine:
> let some wisdom into them!—
> Must we not first hate ourself if we are to love ourself? . . .
> *I am thy labyrinth . . .* [102]

A Dionysus that occupies the ear of his victims and inserts smart words turns into a poet (*Dichter*) or dictator in all senses of the word. He dictates to his slave or secretary to take down his dictation. The new notions of love and heterosexuality become reality when one sex inserts painful words into the ear of the other. University-based, that is, male, discourses on and about an alma mater are replaced by the discourse of two sexes about their impossible relationship: Lacan's *rapport sexuel.* That is why Nietzsche describes Dionysus's existence as an "innovation" once he has invented him as a "philosopher." Unlike Socrates with his Greek noblemen, and unlike Hegel with his German civil-servant apprentices, Dionysus dictates to a woman. According to Nietzsche, *Ariadne's Complaint* is just one of the many "celebrated dialogues" between Ariadne and her "philosophical lover" on Naxos.[103]

The Naxos alluded to here was not a fiction either, but the future of Germany's institutions of higher education. The widow of Max Weber has described how new female students, "from unheard-of intellectual points of contact with young men," were afforded "unlimited opportunities for innovative human relationships": "camaraderie, friendship, love."[104] (To say nothing of the innovative human relationships that, as in the case of Lou Andreas-Salomé, grew out of the opportunities between male *and* female psychoanalysts.) Following the double loss of his Malling Hansen and his Salomé, Nietzsche at any rate was on the lookout for secretaries into whose ears he could insert Dionysian words. For *Zarathustra* and his whip he "needed . . . just somebody to whom he could dictate the text"—and "Fräulein Horner fell from the sky," it seems, precisely "for that purpose."[105] Then, for *Beyond Good and Evil*, that *Foreplay to Philosophy of the Future*, a certain Mrs. Röder-Wiederhold set foot on the island of Naxos.

"I am your labyrinth," Dionysus said to the tortured Ariadne, who in turn had herself been the mistress of the labyrinth during the Cretan ritual dance. And Zarathustra added that poet-dictators who write in blood and aphorisms want not to be read but to be learned by heart.[106] That is precisely why Mrs. Röder-Wiederhold caused some problems. Unfortunately, certain gods, demons, intermediate beings of Europe had already inserted the morality of Christendom and of democracy into her ears. That made the scene of dictation in Engadine into a scene of torture. Her own hand had to write down what was beyond good and evil, beyond Christendom and morality. Ariadne's complaint turned into an empirical event. Every history of writing technologies has to account for the fact that *Beyond Good and Evil* was not easily written. Nietzsche knew and wrote it. "In the meantime I have the admirable Mrs. Röder-Wiederhold in the house; she suffers and tolerates 'angelically' my disgusting 'anti-democratism'—as I dictate to her, for a couple of hours every day, my thoughts on Europeans of today and—*Tomorrow*; in the end, I fear, she may still 'fly off the handle' and run away from Sils-Maria, baptized as she is with the blood of 1848."[107]

Against human and/or technological typewriters such as Nietzsche and the Malling Hansen, substitute secretaries could not compete. Nietzsche stuck to his love affair with the writing ball from January through March 1882: "Between the two of us," the media master wrote about his "admirable woman": "I can't work with her, I don't want to see a repeat (*Wiederholung*). Everything I dictated to her is without value; as well, she cries more often than I can handle."[108]

A complaint of Ariadne that her dictator might have been able to foretell: "Must we not first hate ourself if we are to love ourself? . . . "

Nietzsche and his secretaries, no matter how ephemeral and forgotten, have introduced a prototype into the world. Word processing these days is the business of couples who write, instead of sleep, with one another. And if on occasion they do both, they certainly don't experience romantic love. Only as long as women remained excluded from discursive technologies could they exist as the other of words and printed matter. Typists such as Minnie Tipp, by contrast, laugh at any romanticism. That is why the world of dictated, typed literature—that is, modern literature—harbors either Nietzsche's notion of love or none at all. There are desk couples, two-year-long marriages of convenience, there are even women writers such as Edith Wharton who dictate to men sitting at the typewriter. Only that typed love letters—as Sherlock Holmes proved once and for all in *A Case of Identity*—aren't love letters.

The unwritten literary sociology of this century. All possible types of industrialization to which writers respond have been thoroughly researched—ranging from the steam engine and the loom to the assembly line and urbanization. Only the typewriter, a precondition of production that contributes to our thinking prior to any conscious reaction, remains a critical lacuna. A friend writes or dictates a biography of Gottfried Benn. Upon rereading the 200 typed pages, he begins to realize that he is writing about himself: the biographer and the writer have the same initials. After 200 additional pages, his secretary asks him whether he has noticed that secretaries and writers (*Schriftsteller*) have the same initials. . . . Lacan's three registers cannot possibly be demonstrated more effectively: the real of the writer, the imaginary of his doppelgänger, and, finally, as elementary as forgotten, the symbolic of machine writing.

Under such conditions, what remains to be done is to start a register of the literary desk couples of the century (Bermann's film was never realized).

Case 1. When, beginning in 1883, Wyckoff, Seamans, & Benedict developed a sales network and (following the example of Mark Twain) solicited writers to advertise typewriting, "the Petrograd salesman came up with the most spectacular big name, Count Lyof Nikolayevitch Tolstoy, a man who loathed modern machinery in every form ("The most powerful weapon of ignorance—the diffusion of printed matter."—*War and Peace*, epilogue, part 2, chapter 8), and got a great photograph of the author, looking quite miserable, dictating to his daughter, Alexandra Lvovna, who sat poised over the Remington keyboard."[109]

Case 2. When Christiane von Hofmannsthal finished the sixth grade of secondary school, instead of continuing on she transferred to learn Gabelsberg stenography and typewriting. In 1919 her father and poet wrote about how difficult it would be if he "had to do without the little one as my typist, which she is."[110]

Case 3. In 1897, Hofmannsthal's Austria allowed female graduates of secondary school to study philosophy, in 1900, medicine (including state exams and the doctorate). Consequently, Sigmund Freud, university professor of nerve pathology, began his *Introductory Lectures on Psycho-Analysis* in Vienna during the winter semester of 1915–16 with the revolutionary address, "Ladies and Gentlemen!" Since "the ladies among you have made it clear by their presence in this lecture-room that they wish to be treated on an equality with men," Freud scorned "science . . . for schoolgirls"[111] and identified primary sexual markers by their names. He told the women in the lecture hall that the secular distribution of gender roles, including the symbols of pen and natural paper, was psychoanalytically obsolete: "Women possess as part of their genitals a small organ similar to the male one."

Women, however, who *have* a "clitoris"[112] in the real, and who *are* "wood, paper, . . . books"[113] in the symbolic of the dream, stood on both sides of writing technologies' gender differences. Nothing and nobody barred them anymore from professions involving case studies and hence writing. Sabina Spielrein, Lou Andreas-Salomé, Anna Freud, and so on, up until today: female psychoanalysts became historically possible. An institution that banned phonographs from its examination rooms and ignored the cinema altogether still adjusted its writing equipment. "In February [of 1913] Freud took the novel step of buying a typewriter. . . . But it was not for himself, for there was no question of his employing an amanuensis and giving up his beloved pen. It was simply to help Rank to cope with his increasing editorial duties." Exceeding the mechanization of psychoanalytical secretaries and film interpreters, the machine also altered their sex; for, curiously enough, the typewriter, according to the same biographer, remained not with Rank but in the lifetime possession of Anna Freud, the bridal daughter and psychoanalyst.[114]

"Typewriter," after all, signifies both: machine and woman. Two years after the purchase of the machine, Freud wrote to Abraham from Hofmannsthal's Vienna: "A quarter of an hour ago I concluded the work on melancholy. I will have it typewritten so that I can send you a copy."[115]

Case 4. In 1907, Henry James, the writer and brother of Münsterberg's great sponsor, shifted his famous, circumlocutionary style of novel writing toward "Remingtonese."[116] He hired Theodora Bosanquet, a philosopher's daughter who had worked for the offices of Whitehall on the *Report of the Royal Commission on Coast Erosion* and who learned to type for James's sake. After a job interview, during which James came across as a "benevolent Napoleon,"[117] novel production got under way. The Remington, along with its operator, "moved into his bedroom," where dictation "pulled" texts from James "so much more effectively and unceasingly" than did "writing." Soon a reflex loop was created: only the clanking of the typewriter induced sentences in the writer. "During a fortnight when the Remington was out of order he dictated to an Oliver typewriter with evident discomfort, and he found it almost impossibly disconcerting to speak to something that made no responsive sound at all."[118]

So it went for seven years, until a less benevolent Napoleon said farewell. James had several strokes in 1915. His left leg became paralyzed, and his sense of orientation in space and time was impaired; only the conditioned reflex of pure, intransitive dictation remained intact. Writing in the age of media has always been a short circuit between brain physiology and communications technologies—bypassing humans or even love. Hence, James ordered the Remington, along with Theodora Bosanquet (not the other way around), to his deathbed, in order to record the real behind all fiction. Henry James had become emperor and dictated: a letter to his brother Joseph, the king of Spain; a decree specifying new construction at the Louvre and in the Tuileries; finally, some prose on the death of the royal eagle and the cowardice of its common murderers.[119] That is how deliriously, how lucidly a paralyzed brain recorded itself, the situation, and the system of media. From 1800–1815, Napoleon's noted ability to dictate seven letters simultaneously produced the modern general staff. His secretaries were generals and a marshal of France.[120] From 1907–17, a typewriter and its female operator produced the modern American novel. From that, imperial eagles died.

Case 5. Thomas Wolfe, who made a point of selling his American novels in a highly industrialized fashion, by word count (350,000 in the case of *Look Homeward, Angel*),[121] was nevertheless

the most completely un-mechanical of men and never knew how to operate a typewriter, although on at least two occasions he got machines and swore that he would learn. He rented a Dictaphone in 1936, in the hope that he could recite his work into it and have it typed up later, but the only thing he ever actually dictated was a few remarks on the ancestry and character of his most unfavorable critic, Bernard De Voto. He would sometimes play this back and listen to it, grinning.

At any rate, because of his inability to type, he hired a stenographer for $25 a week, who came each day and transcribed his longhand as fast as he could get it down on paper. . . . A typist had to have both practice and a vivid imagination to read what he had written, and most of them worked for him for only a short time. He was constantly distracted by this difficulty: "I can always find plenty of women to sleep with," he once blurted out, "but the kind of woman that is really hard for me to find is a typist who can read my writing."[122]

Case 6. In 1935, Dr. Benn quit his medical practice to serve as chief medical officer for the recruitment inspection offices in Hannover. Remaining in Berlin were two female friends whom Thomas Wolfe would have had no trouble finding: the actresses Tilly Wedekind and Ellinor Büller-Klinkowström. But the military, Benn's aristocratic form of emigration, had its everyday problems. After two "terribly lonesome and secluded years," he wrote to the second of the two women:

The sheets are torn, the bed isn't done from Saturday through Monday, I've got to do the shopping myself, even getting the heating stove going, sometimes. I don't respond to letters, since I've nobody to do the writing for me. I don't do any work, since I neither have the time nor solace nor anybody to dictate to. At 3:30 in the afternoon, I make some coffee, that is the one content of my life. At nine in the evening I go to bed, that's the other. Like cattle.[123]

On the Genealogy of Morals predicted it all: in the chaos without recording technologies, literature basically had to take the shape described by Benn's tripartite organization: First, a beer or wine pub, reading, meditating, and radio listening, in order to bring highbrow poetry up to par with the sound and standard of popular songs. Second, an "old desk (73 cm x 135 cm)" with unread "manuscripts, journals, books, sample medication packages, an inkpad (for recipes), three pens, two ashtrays, one phone," in order to "scribble" the poem the next day in one of those physician's scrawls that Benn "himself could not read." Finally, another desk, "the decisive one," equipped with microscope and typewriter, to convert the "scribbles" into "typewritten materials," in order to make the material "accessible to judgment" and prepare for "the feedback flow from the inspired to the critical I."[124] The whole process operated as a perfect feedback loop, with the hitch that Benn "himself did not type well."[125] With "nobody to dictate to," it was hardly possible to deal adequately with the material on paper, and hence the media competition of radio and cinema was overwhelming.

More fortunate than his colleagues Nietzsche and Wolfe, however, the writer made a find in Hannover. Benn entered a "marriage of companionship"[126] that was to come to an end only during the World War, when his typewriting wife committed suicide. In Berlin two women friends received their last handwritten letters; the fact that one of them answered with a typewritten letter[127] was no match for the technological competition.

I must make one more try to establish a serious human relationship and, with its help, to escape from the morass of my life.

Morchen, you may know everything, but nobody else does. And when I now describe to you what kind of a person she is, someone who will almost certainly become unhappy, you will be surprised.

Much younger than myself, about 30. Not a bit pretty in the sense of Elida and Elisabeth Arden. Very nice body, but negroid face. From a very well-respected family. No money. Job similar to that of Helga, well paid, types about 200 syllables, a perfect typist.[128]

Two hundred syllables per minute are pretty close to 773 keystrokes, the German typing record of 1985. Modern literature could be produced in the Wehrmacht and the Army High Command simply because the daughter of an officer's widow, Herta von Wedemeyer—following the example of the female protagonist of a 1894 novel[129]—worked as secretary.

Case 7. (so as not to forget, amidst all those writers, "les Postes en général," that is, general secretaries and general field marshals).[130] "By virtue of a decree of the erstwhile Prussian Ministry of Commerce and Trade of July 17, 1897, typewritten documents were deemed admissible in dealings with the government."[131] Official (or government) texts were rendered anonymous and laid the groundwork for Herta von Wedemeyer's profession. Which had consequences not only for chief medical officers but also for their ultimate superior, the minister of war. Nine days prior to Benn's second marriage and in the same city,

on January 12 [1938], General Field Marshal von Blomberg, who since 1932 had been widowed and had two sons and three daughters, married the former stenog-

Wie in allerlei einigen baar,;%%%"/ Ländern gleiche selbe solche
Sprachen geredet gesprochen geschwätzt gedratschet werden immer.?)
BULLGARIENÄHNLICHES RUSSISCHGLEICHES RUSSLANDGLEICHES BULLGARIENÄHN=
LICHES CHRISTLICH SOWIJETISCHES RIESIGES LAND,;" AROSSIRUSSLAND.?:)

Arossirussland Arossirussland arossirussisch, Barbbados bullgarisch
bullgarisch, Korealand vietnamesisch, Schwaziland schwedisch, und ++
Kollumbien spanisch,;"%/%% Japan japanisch,;%"%, Pottanien heiliges
christliches deutsch römisch hebräisch, österreichisches deutsches
Land Schopprron österreichisch, Griechenland deutsches griechisch,+
Afferun affrikanisch, Sallarmankar affrikanisch,;%"% servokroatisch=
es Land Possnien,";%% servokroatisch," Hollland ausländisches, so
niederländisches indunesisch ähnliches holländisch, Länderlein ja,
Irak Iran Irrland Issland Italien italienisch,;;, bullgarisches+++
Land Bellgien russisches ähnliches bullgarisch,;"4%, koreanisch so
vietnamesische ausländische Länderlein Korealand Hanoi Vietnam und
Cyppern Limasoll Kairo,;%, vietnamesisch,,;%, Jugoslavien Marzeddo=
nien servokroatisch, Ägypten Irrithrea Israel deutsches römisches
römisch,;, österreichische deutsche Ländlein,, Gardesgardnerhof und
Liechtenstein Burgenland Bayern deutsches österreichisches österre=
ichisch,,," Apullonien Jarmaykkar Engeland Chillenenlateinarmerika,;
kleinenglisches ausländisches englisch,;% Thunnesien Türkei, reden%
gesprochenes auländisches zigeunerisches mohrisches indianerisches++
kongonegerisches türkisches türkisch, indunesiengroßes ausländisches
riesiges Land Land,;, Affghanistan affrikanisches affrikanisch nur
noch,, französische ausländische fremde Länderlein Ländchen Länder
Ländlein Lande, Frankreich Pollen,; französisch,;" Amsderdamm und
Österreich europäisches deutsches österreichisch,;%%% kapitalistisch
kapitalistisches christlich katholisches hitlerfaschistisches Land,;
Sudetenland Helgoland Thailand Reichsdeutschland reichsdeutsches so
dudendeutsches reichsdeutsch,%%%, reichsdeutsches dänemärkisches de=
utsches dänisches dänemärkisch,,%% Hitlerland Land,% Dänemark hier
redet dänemärkisch,,,;%/+ kollumbianisches spanisches Land Kuuhwait
kollumbieanisches zigeunerisches schwarzes spanisch,; Armänien Rum=
änien reden nur rumänisch,;/% Cheyllon Tokio Texas China,; ausl=
ändisches chinesisches chinesisch,% Böhmerland Tschechoschlovakei ++
Mähren,,% tschechoschlovakisches böhmerländisches böhmisch, rätseln=
haftiges geheimnissvolles heimliches geheimes Märchenland Ewigkeit=
endeland Weltallendeland Ewigkeitendeländlein Phantasieland Ländlein
Ewigkeitendeland, Ippprrrien ewigkeitendeländisches ewigkeitendeländ=
lich, portugiesische Länder Istrien Patthaya Seyschellen Sennegal
Panama Portugall Pararaquay portugiesisch,"%/ Wildwestkkonggo Honno
=lulu Hongkong Isthanbuhl Singarpuur, Indunesien Makkao mallawisch
=es mallakoisches indunesisches makkaoisches," spricht indunesisch,
russlandsriesiges riesiges portugiesisches Land redete, ausländis=
ches zigeunerisches freundliches,% portugiesisch,, als portugiesis=
ches Zigeunerland,;/ Land Ammarconnar.)%) Parkisthan redet auch so
indunesisches indunesisch, Teufelkugel Übelkugel Judenplanet Todes=
jenseits zerberusischer Zerberusplanet, Allahhimmel, Cionhimmel, die
redeten gesprochenes himmlisches heiliges frömmliches kirchliches++
überirdisches auserirdisches frömmlerfreundliches lateinisch immer.?
ICH WERDE RECHNEN LERNEN MIT MEINER SCHREIBEMASCHINE TRUCKMASCH=
INE SCHRIFTENMASCHINE MITN SCHREIBEMASCHINENTAUFNAMEN TAUFNAMEN SO,
SCG))) SCHREIBEMASCHINENVORNAMEN NAMEN NAMEN?,%/;" SILVERETTE.!§§§))
Meine wertvolle Maschine ist technische fabrikische schriftliche
hochgeehrte hochgeschätzte geehrte gültige Schriftenmaschine aber.
Und wird geehrt von allen Göttern, und allen politischen irdi=
schen irdischen staatlichen Regenten,,;" aller ganzen Weltkugel, +
immer im ewigen ewiglichen großen riesgen unentlichen Weltall.?%)

August Walla, typeface, 1985. (Reproduced courtesy of Dr. Johann Feilacher, Die Künstler aus Gugging, Vienna)

rapher Erna Gruhn, a secretary of the Imperial Egg Center (Reichseierzentrale), in the presence of a small circle of friends. Witnesses: Adolf Hitler and Hermann Göring. The couple went on their honeymoon immediately. Shortly thereafter, senior investigator Curt Hellmuth Müller, chief of the Center for Personnel Identification in the Central Office of Criminal Justice, received a load of indecent photographs.

"Mrs. General Field Marshal," more faithful to Bertillon than Minnie Tipp, was "registered" with the authorities.[132] Hitler could take over chief command of the army himself.

Case 8. Once, shortly before the onset of war, fear of cancer drove Hitler to "the extraordinary effort" of "writing down his will by hand." Other than that, like most people in command, Hitler "had for years been used to dictating his thoughts into the typewriter or the shorthand report."[133] A specially constructed typewriter with larger type was at his disposal. This typewriter, however, did not solve all the problems involved in coordinating a world war from the Führer's headquarters, Wolfsschanze. The official historian of the Army High Command saw reason to record a rather inofficious version of the end of the war. It was widely known that great situation conferences would take place around 1300 hours. Hitler, by contrast, had set up "his daily routine" so that

> Jodl could present to him at around 11 and, surrounded by a small circle, the messages and the maps of engagements be compiled overnight. Sometimes it got later, since Hitler was fond of drinking tea with his close advisers after a day's work or, as happened regularly, of staying with his stenographers until about 4 A.M. Militarily speaking, it was highly inconvenient that he then slept well into the day and was not to be disturbed.[134]

But even Führer-typewriters and secretaries, which Hitler preferred over his joint General Staff at the Wolfsschanze, could not decide wars. In order to do that, the Second World War had to produce somewhat more complicated typewriters that did away with literature altogether. . . . First, we need to conclude that fictive cases 9 (Mina Harker + Dr. Seward in Stoker), 10 (Minnie Tipp + poets in Bermann), 11 (Mademoiselle Lust + Faust in Valéry), and their numerous successors (Breidenbach, Bronnen, Gaupp, Heilbut, Kafka, Keun) are anything but fictive. Desk couples have replaced literary love pairs. Only in film scripts or romances do both coincide in a happy end. After Mina Harker for half the novel has collected, recorded, typed, and carbon-copied all discourses on Dracula until the latter has been done away with, she still ends up being a mother. After her

German namesake has made a poet successful and sterile, "they type" (paraphrasing Dante) "no more on that day." According to a beautiful tautology, sexes relayed through media will reunite through media as well. Heilbut's *Frühling in Berlin* (Spring in Berlin), Gaupp's *Nacht von heute auf morgen* (Sudden night) are novels about typewriter romances. And the 30 film renditions of Stoker do not even show phonography and typewriting, so that true love comes to triumph over Dracula. The good fortune of media is the negation of their hardware.

Empirically speaking, women employed in processing discourses are likely to have a successful career. Word processing, somewhere amidst the relays of technological communications networks, breaks up couples and families. Precisely at that gap evolves a new job: the woman author. Ricarda Huch became one (in 1910) after studying in Zurich (1888–91) and working as a secretary in the university's main library (1891–97). Gertrude Stein became one after working in the office of and conducting experiments at the Harvard psychological laboratory headed by her patron, Münsterberg. Theodora Bosanquet became one after working for eight years in the delirious general staff. Tatjana Tolstoy, inspired by her sister's Remington, wrote her first article on a typewriter and mailed it anonymously to her father, who would not have been "impartial" otherwise. Tolstoy was instantly enthused.[135]

Anonymity and pseudonymity (as formerly with the female poets who wrote in the shadow of the *Ur*-author, Goethe) are hardly necessary these days. Whether typewriting authors are called Lindau, Cendrars, Eliot, or Keun, Schlier, or Brück does not count for much in relation to the mass media. A desexualized writing profession, distant from any authorship, only empowers the domain of text processing. That is why so many novels written by recent women writers are endless feedback loops making secretaries into writers. Sitting in front of autobiographical typewriters, Irmgard Keun's heroines simply repeat the factual career of their author. Paula Schlier's *Konzept einer Jugend unter dem Diktat der Zeit* (Concept of a youth under the dictates of time), that extraordinarily precise subtitle for a secretary, hears in "the regular clanging of letters . . . the melody accompanying all the madness of the world:"[136] from world-war field hospitals and lectures in Munich to the editorial office of the *Völkische Beobachter* and the Beer Hall Putsch. Christa Anita Brück's *Schicksale hinter Schreibmaschinen* (Destinies behind typewriters) is an autobiography without mention of love, only the desire to help those "women who are not interested in motherhood" to have a breakthrough as women writers.[137] And since, during office dictation, "a self-regulating machine somewhere in the head chops up the meaning of what the hand, antenna-like, receives,"[138] *écriture automatique* is no longer difficult:

> Tempo, Tempo, faster, faster.
>
> Man funnels his energy into the machine. The machine, which is he himself, his foremost abilities, his foremost concentration and final exertion. And he himself is machine, is lever, is key, is type and moving carriage.
>
> Not to think, not to reflect, on, on, fast, fast, tipp, tip, tipptipptipptipptipptipp . . . [139]

At its high point, typewriter literature means repeating ad infinitum Minnie Tipp's proper name or the advertising slogan on her office door. (Up until Hélène Cixous, women will write that only writing makes women into women.) The relay unit of human and machine exercises a pull that can even replace love. First with female typists, then with their male counterparts. That Kafka's love was a media network is confirmed at the height of German literary history by *Case 12*.

Felice Bauer (1887–1960), who was employed after graduation in 1908 as a stenographer for the Odeon record company, switched in 1909 to Carl Lindström A.-G., the largest German manufacturer of dictaphones and gramophones (with a daily output of 1,500 units).[140] Within three years there, beginning as a simple typist, she made a business career highly unusual for women: with her power of attorney she was entitled to sign "Carl Lindström A.-G." At exactly that time, during a trip to Budapest in the summer of 1912, Ms. Bauer visited the family of Max Brod, the head of personnel for the Prague postal service.[141]

Present on this occasion was a young and little-published writer who was just putting together his first book for Rowohlt and who, at first, saw in the traveling woman nothing but a "bony, empty face" that "wore its emptiness openly."[142] Until the potential inscription surface dropped a sentence that "so amazed" Dr. Kafka "that I banged on the table":

> You actually said you enjoyed copying manuscripts, that you had also been copying manuscripts in Berlin for some gentleman (curse the sound of that word when unaccompanied by name and explanation!). And you asked Max to send you some manuscripts.[143]

That is how quickly a typist's lust taught a (hand)writer a love that, even in the shape of jealousy, wasn't one. Since only professors in Berlin and friends involved in information technology were privileged to have their manuscripts typed and readied for publication, Kafka had no choice but—in an unusual step for him—to go to work on the typewriter himself. Whereas the "main" and "happy" part of Kafka's work consisted in "dictating to a living being" in the office,[144] the endless stream of love letters to Felice Bauer started, as if negating love itself, with a typescript.

"But dearest Felice!" Kafka wrote a year later, "Don't we write about writing, the way others talk about money?"[145] Indeed: from the first letter to the last, their impossible relationship was a feedback loop of text processing. Time and again, Kafka avoided traveling to Berlin with his hand, the hand that once held Felice Bauer's. Instead of the absent body there arrived a whole postal system of letters, registered letters, postcards, and telegrams in order to describe that "hand" with "the hand now striking the keys." What remained of "personal typing idiosyncrasies" was only what was simultaneously of interest to *The Criminological Uses of Typewriting*, namely, the "types of mistake correction": first, with skilled typists; second, with unskilled typists; and third, with "skilled typists on an unaccustomed system."[146] Kafka counted himself among the third group, and of the twelve typos in his first letter, four, that is, a highly significant 33 percent, involved the pronouns "ich" (I) and "Sie" (you). As if the typing hand could inscribe everything except the two bodies on either end of the postal channel. As if the "fingertips" themselves had taken the place of his insufficient "mood" by the name of Ego. And as if the self-critical "mistake," which Kafka "realized" in self-critical amplification while "inserting a new sheet of paper," coincided with nervous typos.

Kafka's call *For the Establishment and Support of a Military and Civilian Hospital for the Treatment of Nervous Diseases in German Bohemia* stated:

> This great war which encompasses the sum total of human misery is also a war on the nervous system, more a war on the nervous system than any previous war. All too many people succumb to this war of nerves. Just as the intense industrialization of the past decades of peace had attacked, affected, and caused disorders of the nervous system of those engaged in industry more than ever before, so the enormously increased mechanization of present-day warfare presents the gravest dangers and disorders to the nervous system of fighting men.[147]

Under its initial conditions of a "war of nerves" between literature and Carl Lindström A.-G., the exchange of love letters was doomed to fail, even though it later switched to handwriting and returned to "increased mechanization" only in 1916, when typed postcards were the fastest way of passing through the war censorship between Prague and

Arbeiter-Unfall-Versicherungs-Anstalt
FÜR DAS KÖNIGREICH BÖHMEN IN PRAG.

Chek-Conto des k. k.
Postsparcassenamtes No. 18.923.

No. E. ai 191
M. Sch. No.
Bei Rückantwort wollen vorstehende Zahlen gefl. bezogen werden.

Sehr geehrtes Fräulein !

Für den leicht möglichen Fall, dass Sie sich meiner auch im geringsten nicht mehr erinnern könnten, stelle ich mich noch einmal vor: Jch heisse Franz Kafka und bin der Mensch, der Sie zum erstenmal am Abend beim Herrn Direktor Brod in Prag begrüsste, Jhnen dann über den Tisch hin Photographien von einer Thaliareise, eine nach der andern, reichte und der schliesslich in dieser Hand, mit der er jetzt die Tasten schlägt, ihre Hand hielt, mit der Sie das Versprechen bekräftigten, im nächsten Jahr eine Palästinareise mit ihm machen zu wollen.

Wenn Sie nun diese Reise noch immer machen wollen-Sie sagten damals, Sie wären nicht wankelmüthig und ich bemerkte auch an Jhnen nichts dergleichen-dannwird es nicht nur gut, sondern unbedingt notwendig sein, dass wir schon von jetzt ab über diese Reise uns zu verständigen suchen. Denn wir werden unsere gar für eine Palästinareise viel zu kleine Urlaubszeit bis auf den Grund ausnützen müssen und das werden wir nur können, wenn wir uns so gut als möglich vorbereitet haben und über alle Vorbereitungen einig sind.

Eines muss ich nur eingestehen, so schlecht es an sich klingt und so schlecht es überdies zum Vorigen passt: Jch bin ein unpünktlicher Briefschreiber. Ja es wäre noch ärger, als es ist, wenn ich nicht die Schreibmaschine hätte; denn wenn auch einmal meine Launen zu einem Brief nicht hinreichen sollten, so sind schliesslich die Fingerspitzen zum Schreiben immer noch da. Zum Lohn dafür erwarte ich aber auch niemals, dass Briefe pünktlich kommen; selbst wenn ich einen Brief mit täglich neuer Spannung erwarte, bin ich niemals enttäuscht, wenn er nicht kommt und kommt er schliesslich, erschrecke ich gern.

Jch merke beim neuen Einlegen des Papiers,dass ich mich vielleicht viel schwieriger gemacht habe,als ich bin.Es würde mir ganz recht geschehn,wenn ich diesen Fehler gemacht haben sollte,denn warum schreibe ich auch diesen Brief nach der sechsten Bürostunde und auf einer Schreibmaschine,an die ich nicht sehr gewöhnt bin.

Aber trotzdem,trotzdem -es ist der einzige Nachteil des Schreibmaschinenschreibens,dass man sich so verläuft-wenn es auch dagegen Bedenken geben sollte,praktische Bedenken meine ich,mich auf eine Reise als Reisebegleiter,-führer,-Ballast,-Tyrann,und was sich noch aus mir entwickeln könnte,mitzunehmen,gegen mich als Korrespondenten -und darauf käme es ja vorläufig nur an-dürfte nichts Entscheidendes xxxx von vornherein einzuwenden sein und Sie könnten es wohl mit mir versuchen.

Prag,am 20.September 1912.

Jhr herzlich ergebener
Dr. Franz Kafka

Prag, ~~Niklasstrasse 36~~
Pořič 7

Franz Kafka's first letter to Felice Bauer.

Berlin, Austria and Prussia.[148] In 1917, while Lindström's acoustical media network, with its financial leverage, helped the Army High Command establish the film corporation UFA,[149] Kafka terminated his engagement to Felice Bauer. Shortly thereafter, the woman, freed from the bombardment of letters, married an affluent Berlin businessman.

In one of his last letters to his last female pen pal, however, Kafka took stock: of misused love letters and communications vampires, of reduced physical labor and information machines.

How on earth did anyone get the idea that people can communicate with one another by letter! Of a distant person one can think, and of a person who is near one can catch hold—all else goes beyond human strength. Writing letters, however, means to denude oneself before the ghosts, something for which they greedily wait. Written kisses don't reach their destination, rather they are drunk on the way by the ghosts. It is on this ample nourishment that they multiply so enormously. Humanity senses this and fights against it and in order to eliminate as far as possible the ghostly element between people and to create natural communication, the peace of souls, it has invented the railway, the motorcar, the aeroplane. But it's no longer any good, these are evidently inventions made at the moment of crashing. The opposing side is so much calmer and stronger; after the postal service it has invented the telegraph, the telephone, the radiograph. The ghosts won't starve, but we will perish.[150]

Hence only ghosts survive the Kafka-Bauer case: media-technological projects and texts reflecting the material limitations of the written word. Even though, or because, Kafka considered the "very existence" of gramophones "a threat,"[151] he submitted to his employee of a phonograph manufacturer a series of media links that were supposed to be able to compete with Lindström's empire. Aside from a direct link involving a parlograph, which "goes to the telephone in Berlin" and conducts "a little conversation" with a "gramophone in Prague," Kafka envisions a "typing bureau where everything dictated into Lindström's Parlographs is transcribed on a typewriter, at cost price, or at first perhaps a bit below cost price."[152] That was not, of course, the most up-to-date proposition (thanks to Dr. Seward and Mina Harker), but one with a future. In Bronnen's monodrama *Ostpolzug* of 1926, "an electrically hooked-up dictaphone dictates into an equally electric typewriter."[153] And since "the machine makes further inroads" into "the function of brains" themselves, instead of merely "replacing the *physical* labor of man, . . . a typewriter is announced [in 1925] that will make the typist superfluous and will translate the sound of words directly into typed script."[154]

Kafka, however, for whom Ms. Bauer did not type a single manuscript, let alone construct media networks, stuck to old-fashioned literature. From the typewriter he only learned to dodge the phantasm of authorship. As with his first love letter, the "I," "the nothingness that I am,"[155] disappeared under deletions or abbreviations until all that remained was a Joseph K. in *The Trial* and a K. by itself in *The Castle*. The office machines of his days also freed the Kafka of his literary nights from the power of attorney, that is, the authority to sign documents:

I could never work as independently as you seem to; I slither out of responsibility like a snake; I have to sign many things, but every evaded signature seems like a gain; I also sign everything (though I really shouldn't) with FK only, as though that could exonerate me; for this reason I also feel drawn to the typewriter in anything concerning the office, because its work, especially when executed at the hands of the typist, is so impersonal.[156]

Mechanized and materially specific, modern literature disappears in a type of anonymity, which bare surnames like "Kafka" or "K." only em-

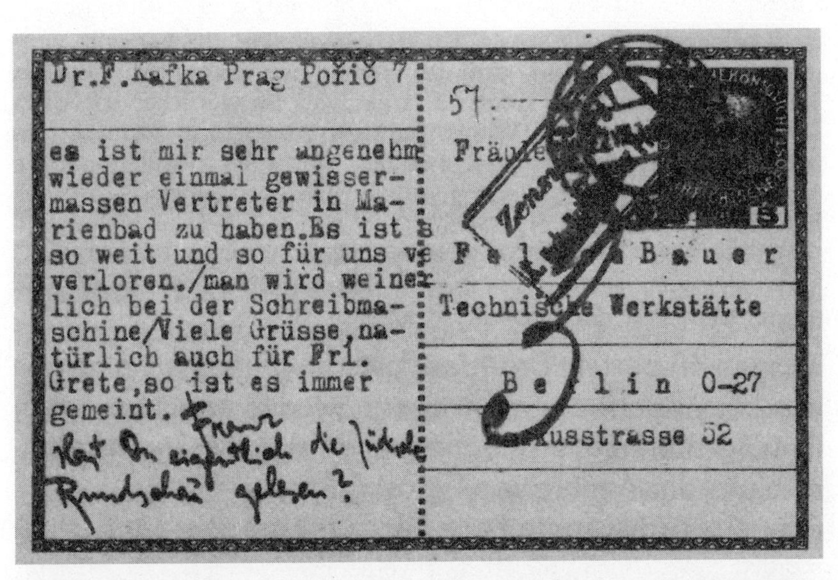

Dr.F.Kafka Prag Poříč 7

57.

es ist mir sehr angenehm
wieder einmal gewisser-
massen Vertreter in Ma-
rienbad zu haben.Es ist
so weit und so für uns v
verloren./man wird weine
lich bei der Schreibma-
schine/Viele Grüsse,na-
türlich auch für Frl.
Grete,so ist es immer
gemeint. Franz
Hast Du eigentlich die Jüdische
Rundschau gelesen?

Fräulein
Felice Bauer
Technische Werkstätte
Berlin O-27
Markusstrasse 52

Liebste,bin gerade bei der Schreibmaschine,versuche
es also einmal so.Mein Schreibmaschinenfräulein ist
auf Urlaub,ich bin augenblicklich.fast krank ver
Sehnsucht nach ihr,denn der Ersatzmann ,so geduldig
eifrig und ängstlich er ist/ich höre zeitweilig sein
Herz klopfen/wütet ohne es zu wissen,in meinen Ner-
ven.Nun morgen ,nein übermorgen kommt sie wieder.Wie
ist denn Dein Hilfsmädchen?Es ist so still von ihr.
Es fällt mir ein:schreib mir auch einmal mit der Masc
schine.Da müsste es doch viel mehr werden,als z.B.
der letzte Sonntagsgruss/von Freitag und Samstag habe
ich noch nichts/vielleicht wird Schreibmaschinenschri
auch schneller zensuriert.-Also auch Sonntag im Büro
und schon zum zweitenmal,sehr unrecht.Was klappt nicl
Und vom Volksheim nichts Neues,sehr schade.äNoch ein
allerdings alter Einfall/bei der Schreibmaschine übe
fallen sie mich /:könntest Du mir nicht einige Bild-
chen von Dir schicken,ja hast Du mir sie nicht sogar
schon versprochen ?-Heute fahren Max und Frau mit
meinen Ratschlägen und unserem Führer nach Marienbad,
22 August 16

Franz Kafka's postcard to Felice Bauer.

phasize. The "disparition élocutoire du poète"[157] urged by Mallarmé becomes reality. Voice and handwriting treacherously could fall subject to criminal detection; hence every trace of them disappears from literature. As Jacques Derrida, or "J.D.," observes in a May 1979 love letter whose address must also be without (a) proper name(s):

> What cannot be said above all must not be silenced, but written. Myself, I am a man of speech, I have never had anything to write. When I have something to say I say it or say it to myself, basta. You are the only one to understand why it really was necessary that I write exactly the opposite, as concerns axiomatics, of what I desire, what I know my desire to be, in other words you: living speech, presence itself, proximity, the proper, the guard, etc. I have necessarily written upside down—and in order to surrender to Necessity.
>
> and *"fort" de toi.*
>
> I must write you this (and at the typewriter, since that's where I am, sorry: . . .).[158]

Hence Derrida's *Postcard* consists of one continuous stream of typed letters punctuated by phone calls that are frequently mentioned but never recorded. Voice remains the other of typescripts.

"I, personally," Benn says about *Problems of Poetry* (*Probleme der Lyrik*), "do not consider the modern poem suitable for public reading, neither in the interest of the poem nor in the interest of the listener. The poem impresses itself better when read. . . . In my judgment, its visual appearance reinforces its reception. A modern poem demands to be printed on paper and demands to be read, demands the black letter; it becomes more plastic by viewing its external structure."[159] Hence a *Pallas* named Herta von Wedemeyer solves all problems of poetry because she transforms Benn's scribbled ideas—"a lifeless something, vague worlds, stuff thrown together with pain and effort, stuff brought together, materials that have been grouped, improved upon and left undeveloped, loose, untested, and weak"[160]—via transcription into art. Under the conditions of high technology, Pallas, the goddess of art, is a secretary.

"Fundamentally, the typewriter is nothing but a miniature printing press."[161] As a doubled spatialization of writing—first on the keyboard, then on the white paper—it imparts to texts an optimal optical appearance. And, following Benjamin's forecast, as soon as "systems with more variable typefaces" (such as rotating head typewriters or thermal printers) become available, "the precision of typographic forms" can directly enter "the conception of . . . books." "Writing [is] advancing ever more deeply into the graphic regions of its new eccentric figurativeness":[162]

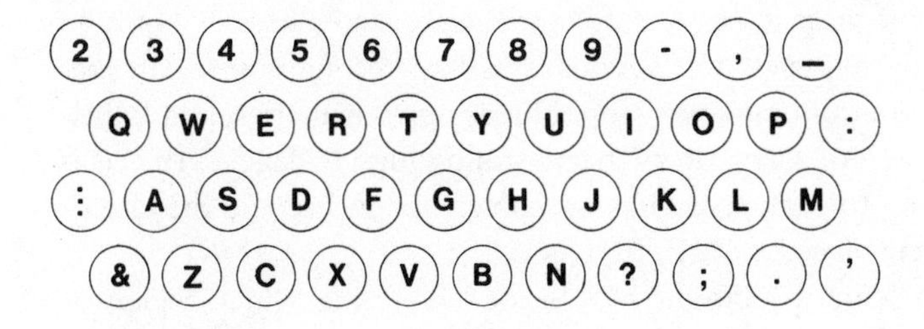

Image of a T3 Remington "Ur-keyboard," 1875.

from Mallarmé's "Coup de dés" and Apollinaire's "Calligrammes," those typographic poems that attempt to bring writers on par with film and phonography,[163] to *poésie concrète*, that form of pure typewriter poetry.

T. S. Eliot, who will be "composing" *The Waste Land* "on the typewriter," "finds" (no different from Nietzsche) "that I am sloughing off all my long sentences which I used to dote upon. Short, staccato, like modern French prose." Instead of "subtlety," "the typewriter makes for lucidity,"[164] which is, however, nothing but the effect of its technology on style. A spatialized, numbered, and (since the 1888 typewriters' congress in Toronto) also standardized supply of signs on a keyboard makes possible what and only what QWERTY prescribes.

Foucault's methodical explanation, the last and irreducible elements of which are at the center of his discourse analysis, can easily eliminate the sentences of linguistics, the speech acts of communications theory, the statements of logic. Only to be confronted by two factual conditions that seem to fulfill all the criteria for an elementary "statement" of discourse analysis: "The pile of printer's character which I can hold in my hand, or the letters marked on the keyboard of a typewriter."[165] Singular and spatialized, material and standardized, stockpiles of signs indeed undermine so-called Man with his intentions and the so-called world with its meaning. Only that discourse analysis ignores the fact that the factual condition is no simple methodical example but is in each case a techno-historical event. Foucault omits the elementary datum (in Latin, the casting of dice or *coup de dés*) of each contemporary theoretical practice and begins discourse analysis only with its applications or configurations: "the keyboard of a typewriter is not a statement; but the same series of letters, A, Z, E, R, T, listed in a typewriting manual, is the statement of the alphabetical order adopted by French typewriters."[166]

Foucault, the student of Heidegger, writes that "there are signs, and that is enough for there to *be* signs for there to *be* statements,"[167] only to point for once to the typewriter keyboard as the precondition for all preconditions. Where thinking must stop, blueprints, schematics, and industrial standards begin. They alter (strictly following Heidegger) the relationship of Being to Man, who has no choice but to become the site of their eternal recurrence. A, Z, E, R, T . . .

Until Arno Schmidt's late novels, beyond Foucault, which repeat or transcribe all keyboard numbers at the top of the page and all keyboard symbols in the margin, and thus can only appear as typescripts.

Until Enright's collection of poems *The Typewriter Revolution and Other Poems* celebrates "the new era" in unsurpassable material appropriateness.[168]

```
THE TYPEWRITER REVOLUTION

    The typeriter is crating
    A revlootion in peotry
    Pishing back the frontears
    And apening up fresh feels
    Unherd of by Done or Bleak

    Mine is a Swetish Maid
    Called FACIT
    Others are OLIMPYA or ARUSTOCART
    RAMINTONG or LOLITEVVI

    TAB e or not TAB e
    i.e. the ?
    Tygirl tygirl burning bride
    Y, this is L
    Nor-my-outfit
    Anywan can od it
    U 2 can b a
    Tepot

    C! *** stares and /// strips
    Cloaca nd † -
    Farty-far keys to suckcess!
    A banus of +% for all futre peots!!
    LSD & $$$
```

```
The trypewiter is cretin
A revultion in peotry
" " All nem r = " "
O how they £ away
@ UNDERWORDS and ALLIWETTIS
Without a.

FACIT cry I!!!
```

Remington's and Underwood's invention ushered in a poetics that William Blake or John Donne with their limits/ears could not hear, for it transcends mystical tigers in the silence of the night, or a metaphysical erotics between heaven and confessional. Only the excessive media link of optics and acoustics, spellings and acronyms, between the letters, numbers, and symbols of a standardized keyboard makes humans (and women) as equal as equal signs. Blake's "tiger, tiger, burning bright," is succeeded by the stenographer, that burning poet's bride. The history of typewriter literature *in nuce*. And always to continue and/or copy—humans, U.S. flags, or spy aircraft. "You too are a poet" with typos (errata).

Toward the end of the First World War, a young and ironic Carl Schmitt conceived the world history of inscription. To rewrite it here in its entirety is impossible, simply because *res gestae* and *res narratae* coincide. It is enough that the diary-typing machines called Buribunks, as well as the "twenty divisions" of buribunkological dissertations,[169] have evolved from humble beginnings into a modern loop of endless replication.

CARL SCHMITT, "THE BURIBUNKS: A HISTORICO-PHILOSOPHICAL MEDITATION" (1918)

Today, because we have been granted the privilege of enjoying the glorious notion of the diary at its zenith, we tend to overlook what a majestic deed man performed when—perhaps as the unknowing instrument of the world spirit—he planted with the first innocuous note the first seed, which now overshadows the earth as a gigantic tree. A certain, I would say, moral feeling of obligation urges us to question what historical personage embodies the precursor to this wonderful epoch, the messenger pigeon that the world spirit has sent in advance of its last and most highly refined period. We are obliged to put this question at the center of our principal investigation.

It would be a mighty triumph for Buribunkology if it could identify a hero such as Don Juan as its ancestor and—in opposition to the charge of scholarly absentmindedness—take pride in its paradoxical descent from this virile and decidedly unscholarly cavalier. Indeed, Don Juan's conquests have been registered, but the crucial point is to whom the intellectual property of this idea can be attributed. In his champagne aria, Don Juan himself sings—

> Ah, la mia lista doman mattina
> d'una decina devi aumentar

—a feeling of which the true Buribunkologist is frequently possessed when he ponders the daily increasing scope or the daily rising number of his publications. He may very well be tempted to compare his sense of achievement with the plucky self-confidence of the frivolous conqueror of women. Still, this seductive parallel should not distract us from the profound seriousness of our endeavor or lead us to lose the distance from our possible founding father, which sober objectivity and detached science dictate to us. Did Don Juan really have the specifically buribunkological attitude that urged him to keep a diary, not for the sake of recording, superficially, his manly conquests, but—if I may say so—out of a sense of sheer obligation and debt vis-à-vis history? We cannot believe so. Don Juan had no interest in the past, just as he fundamentally had no interest in the future, which for him did not go beyond the next conquest; he lived in the immediate present, and his interest in the individual erotic adventure does not point to any signs of a beginning self-historicization. We cannot detect any signs of the attitude characterizing the Buribunk, which originates from the desire to record every second of one's existence for history, to immortalize oneself. Like the Buribunk keeping a diary, Don Juan relishes each individual second, and in that there is certainly a similarity of psychological gesture. Instead of consecrating his exploits on the altar of history in the illuminated temple, however, he drags them into the misty cave of brutal sensuality, devouring them animal-like to satiate his base instincts.* Not for a single moment does he have what I would like to call the cinematic attitude of the Buribunk—he

* Hence, one could say that Don Juan is not one who ruminates upon lived experience, if one were to charge the buribunkological keeping of a diary with being a kind of intellectual rumination. That such a claim is untenable is easy to prove, for the diary-keeping Buribunk does not experience anything prior to his entries; rather, the experience consists precisely in the making of an entry and its subsequent publication. To speak of rumination is thus simply nonsense, because there is no initial act of chewing and swallowing.

never apprehends himself as the subject-object of history—in which the world soul, writing itself, has become realized. And the register that Leporello keeps for him he takes along only as an afterthought, as a delectable flavoring for his horizontal delights. Hence, we have legitimate doubt whether, for example, from the 1,003 Spanish representatives, more than three owe their entry into the register to the very existence of the register itself. Put differently, we wonder whether Don Juan has been prompted into action by his inner need to start and keep a register, if only in those three cases, the way numerous major achievements in the arts, in science, in everyday life have been produced solely with the idea of their recorded existence in a diary or newspaper—the diary of the masses—in mind. The register was never the final cause; in implementing the acts of innervation at issue, it was—in the rectangle of psychological forces—relegated to the role of an accidental, of an accompanying positive motor. Thus, for us Don Juan is finished.

All the more interesting is the behavior of Leporello. He relishes the sensuous leftovers of his master, a couple of girls, a couple of choice morsels; for the most part, he accompanies his master. A Buribunk does not do that, for a Buribunk is unconditionally and absolutely his own master, he is himself. Gradually, however, what awakens in Leporello is the desire to partake in the escapades of his master by writing them down, by taking note of them, and it is at this moment that we see the dawn of Buribunkdom. With the aid of a commendable trick he surpasses his master, and if he does not become Don Juan himself, he becomes more than that; he changes from Don Juan's wretched underling into his biographer. He becomes a historian, drags Don Juan to the bar of world history, that is, world court, in order to appear as an advocate or prosecutor, depending on the result of his observations and interpretations.

Was Leporello, however, cognizant of the implications of starting his register as the first step in a gigantic development? Certainly not. We do not want to dismiss the mighty effort behind the small register of the poor buffo, but we cannot under any circumstances recognize him as a conscious Buribunk—how should he have done it, anyway, the poor son of a beautiful but culturally retrograde country in which the terror of papal inquisition has crushed and smashed all remaining signs of intelligence? He was not meant to see his nonetheless significant intellectual work come to fruition; he holds the treasure-laden shrine, but he does not hold the key to it. He has not understood the essential and has not said the magic words that open the way to Aladdin's cave. He was lacking the consciousness of the writing subject, the consciousness that he had become the author of a piece of world history and hence a juror on the world court—indeed, to exercise control over the verdict of the world court, because his written documents were proof more valid than a hundred testimonies to the contrary. Had Leporello had the strong will to this kind of power, had he ventured the magnificent leap to become an independent writing personality, he would first have written his own autobiography; he would have made a hero of himself, and instead of the frivolous cavalier who fascinates people with his shallow disposition, we would quite probably have gotten the impressive picture of a superior manager who, with his superior business skills and intelligence, pulls the strings of the colorful marionette, Don Juan. But instead of taking pen into fist, the poor devil clenches his fist in his pocket.

Upon close scrutiny, the utter inadequacy of Leporello's registration method appears in numerous defects. He puts photographs in sequence without ever making an attempt to shape the heterogeneous discontinuity of successive seductions into a homogeneous continuity; what is missing is the mental thread, the presentation of development. We don't get any sense of demonstrated causal connections, of the mental, climatic, economic, and sociological conditions of individual actions, nothing relating to an aesthetic observation about the ascending or descending bell curve of Don Juan's evolving taste. Similarly, the register has nothing to say about the specific historical interest in the uniqueness of each individual procedure or in each individual personality. Leporello's disinterest is utterly incomprehensible; he does not even communicate any dismay when he is daily witness to his master's ingenious sexuality—how it is aimlessly scattered to the winds instead of being rationally disseminated into purposive population growth. Still less evident is his willingness to provide reliable research data on details: nowhere does Leporello inquire into the deeper motivations of individual seductions, nowhere do we find sociologically useful data on the standing, origin, age, and so on of Don Juan's victims, as well as their pre-seduction lives—at most, we are left with the summary conclusion (which is probably not sufficient for a more serious scientific investigation) that they came from "every station, every form, and every age." We also don't hear anything about whether the victims later organized themselves into a larger, communal mass initiative and provided mutual economic support—which no doubt would have been the only right thing to do, given their numbers. Naturally, what is also missing is any statistical breakdown within the respective categories, which would have recommended itself in light of the high number of 1,003; even more, what is missing is any indication as to how the dumped girls had been taken care of by the welfare system, which in many cases had become necessary. Naturally as well, there is no inkling that, in light of the brutal exploitation of male social superiority in relation to defenseless women, the introduction of female suffrage is a most urgent and legitimate demand. It would be in vain to ask for the larger precepts that underlie the development of the collective soul, the subjectivism of the time, the degree of its excitation. In a word, inadequacy here is turned into reality. Leporello is oblivious to the welter of the most urgent scientific questions—much to his own disadvantage, because he has to submit his obliviousness to the judgement of history. Oblivious to pressing questions, he did not engage in as much as one investigation that the most immature student of the humanities today would have pounced upon, and hence missed the opportunity of evolving the consciousness necessary to recognize the significance of his own identity. The dead matter has not been conquered by the intellectual labor of its workman, and the flyers on the advertisement pillars continue to announce: Don Juan, the chastised debauchee, and not: Leporello's tales. . . .

Not until Ferker did the diary become an ethical-historical possibility; the primogeniture in the realm of Buribunkdom is his. Be your own history! Live, so that each second of your life can be entered into your diary and be accessible to your biographer! Coming out of Ferker's mouth, these were big and strong words that humanity had not yet heard. They owe their distribution into the nooks and crannies of even the most remote villages to a worldwide organization aimed at disseminating his ideas, an organization well managed and having the support of an intelligent press. No village is so small that it is without a blacksmith, as the old song went; today, we can say with not a little pride that no village is so small that it is not imbued with at least a touch of Buribunkic spirit. The great man,* who presided like the chief of a general staff over his thousands of underlings, who guided his enormous business with a sure hand, who channeled the attention of the troops of researchers to hot spots, and who with unheard-of strategic skill focused attention on difficult research problems by directing pioneering dis-

* On this issue, all relevant documents show a rare unanimity. In his diary, Maximilian Sperling calls him "a smart fellow" (*Sperling's Diaries*, vol. 12, ed. Alexander Bumkotzki [Breslau, 1909], p. 816). Theo Timm, in a letter of August 21 to Kurt Stange, describes him as "a fabulous guy" (*Timm's Letters*, vol. 21, ed. Erich Veit [Leipzig, 1919], p. 498). In her diary, Mariechen Schmirrwitz says, "I find him splendid" (vol. 4, ed. Wolfgang Huebner [Weimar, 1920], p. 435). Following his first meeting with the man himself, Oskar Limburger exclaims, "He is enormous, watch out for him" (*Memories of my Life*, ed. Katharina Siebenhaar [Stuttgart, 1903], p. 87). Prosper Loeb describes him as of a "demonic nature" (Königsberg, 1899, p. 108). He is a "heck of a guy," says Knut vom Heu in his letters to his bride (edited by their son Flip; Frankfurt a.M., 1918, p. 71), and so on.

sertations—this impressive personality experienced a truly sensational rise. Born of humble origins and educated without Latin in the middle school of his small town, he successively became a dentist, a bookmaker, an editor, the owner of a construction company in Tiflis, the secretary of the headquarters of the international association to boost tourism on the Adriatic coast, the owner of a movie theater in Berlin, a marketing director in San Francisco, and, eventually, Professor of Marketing and Upward Mobility at the Institute of Commerce in Alexandria. This is also where he was cremated and, in the most grandiose style, his ashes processed into printer's ink, as he had specified in detail in his will and which was sent in small portions to printing presses all over the world. Then, with the aid of flyers and billboards, the whole civilized world was informed of this procedure and was furthermore admonished to keep in mind that each of the billions of letters hitting the eye over the years would contain a fragment of the immortal man's ashes. For eons, the memorial of his earthly days will never disappear; the man—who even in death is a genius of factuality—through an ingenious, I would say antimetaphysical-positivist gesture, secured himself a continued existence in the memory of humanity, a memory, moreover, that is even more safely guaranteed through the library of diaries that he released in part during his lifetime, in part after his death. For at each moment of his momentous life he is one with historiography and the press; in the midst of agitating events he coolly shoots film images into his diary in order to incorporate them into history. Thanks to this foresight, and thanks as well to his concomitant selfless research, we are informed about almost every second of the hero's life. . . .

Now we are finally in a position to define historically the crucial contribution of this ingenious man: not only has he made the radically transformative idea of the modern corporation feasible for human ingenuity without leaving the ground of the ethical ideal; not only has he demonstrated through his life that one can build a career of purposive ambition and still be an ethically complete being, bound under the sublation of the irreconcilable duality of matter and mind in a way that invalidates the constructions of theologizing metaphysics, which were inimical to the intellectual climate of the twentieth century, through a victorious new idealism; he has, and this is the crux, found a new, contemporary form of religion by strictly adhering to an exclusionary positivism and an unshakable belief in nothing-but-matter-of-factness. And the mental region in which these numerous and contradictory elements, this bundle of negated negations, are synthesized—the unexplainable, absolute, essential that is part of every religion—that is nothing but the Buribunkological.

No Buribunkologist who is also a genuine Buribunk will utter the name of this man without the utmost reverence. That we must emphasize up front. For when we disagree with the critical appraisal of our hero by noted Ferker scholars in the following discussion, we do so not without emphatic protest against the misunderstanding that we fail to recognize Ferker's tremendous impulses and the full stature of the man. Nobody can be more informed by and imbued with his work than we are. Nevertheless, he is not the hero of Buribunkdom but only its Moses, who was permitted to see the promised land but never to set foot on it. Ferker's truly noble blood is still tinged with too many unassimilated, alien elements; atavistic reminiscences still cast their shadow over long periods of his life and dim the pure image of self-sufficient, blue-blooded Buribunkdom. Otherwise, it would be inexplicable that the noble man, doubting his inner sense of self, shortly before his death was willing not only to enter into a bourgeois-religious marriage but also to marry his own housekeeper—a woman who we know was completely uneducated, downright illiterate, and who eventually (aside from inhibiting the free exfoliation of his personality) sought to prevent, for reasons of devout bigotry, his cremation. . . . To have surpassed these inconsistencies and to have made Buribunkdom, in its crystal-clear purity, into a historical fact is the work of Schnekke.

As a fully matured fruit of the most noble Buribunkdom, this genius fell from the tree of his own personality. In Schnekke we find not the least visible trace of hesitation, not the slightest deviation from the distinguished line of the Ur-buribunkological. He is nothing more than a diary keeper, he lives for his diary, he lives in and through his diary, even when he enters into his diary that he no longer knows what to write in his diary. On a level where the I, which has been projecting itself into a reified, you-world constellation, flows with forceful rhythm back into a world-I constellation, the absolute sacrifice of all energies for the benefit of the inner self and its identity has achieved the fullest harmony. Because ideal and reality have here been fused in unsurpassable perfection; what is missing is any particular singularity, which shaped Ferker's life in such a sensational way but which, for any discussion focusing on the essential, must be understood as a compliment rather than a critique. Schnekke is, in a much more refined sense than Ferker, a personality, and precisely because of that has he disappeared behind the most inconspicuous sociability. His distinct idiosyncrasy, an I determined solely by the most extreme rules of its own, is located within a spectrum of indiscriminate generality, in a steady colorlessness that is the result of the most sacrificial will to power. Here we have reached the absolute zenith of Buribunkdom; we need not be afraid of any relapse, as with

Ferker.* The empire of Buribunkdom has been founded. For in the midst of his continuous diaries, Schnekke (with his strong sense of generality and his universal instinct) saw the opportunity to detach the diary from its restrictive bond with the individual and to convert it into a collective organism. The generous organization of the obligatory collective diary is his achievement. Through that, he defined and secured the framing conditions of a buribunkological interiority; he elevated the chaotic white noise of disconnected and single Buribunkdom into the perfect orchestration of a Buribunkic cosmos. Let us retrace the broad lines of development of this sociological architecture.

Every Buribunk, regardless of sex, is obligated to keep a diary on every second of his or her life. These diaries are handed over on a daily basis and collated by district. A screening is done according to both a subject and a personal index. Then, while rigidly enforcing copyright for each individual entry, all entries of an erotic, demonic, satiric, political, and so on nature are subsumed accordingly, and writers are catalogued by district. Thanks to a precise system, the organization of these entries in a card catalogue allows for immediate identification of relevant persons and their circumstances. If, for example, a psychopathologist were to be interested in the pubescent dreams of a certain social class of Buribunks, the material relevant for this research could easily be assimilated from the card catalogues. In turn, the work of the psychopathologist would be registered immediately, so that, say, a historian of psychopathology could within a matter of hours obtain reliable information as to the type of psychopathological research conducted so far; simultaneously—and this is the most significant advantage of this double registration—he could also find information about the psychopathological motivations that underlie these psychopathological studies. Thus

* What a difference there is between Ferker's and Schnekke's attitudes toward women! Never is there any thought in Schnekke about getting married in church; with instinctive surefootedness he recognizes it as a ball and chain on the leg of his ingenuity, and he manages—despite a series of rather fully developed erotic relationships—to escape from marriage with the surety of a sleepwalker. He always remains cognizant of the needs surrounding the free development of his uniqueness, and he rightfully invokes Ekkehard in his diary when he says that marriage would inhibit his essential I-ness. At the same time, we should not overlook the impressive progress evident from Ferker to Schnekke when it comes to women. There is no illiterate woman in Schnekke's life, no one who with petit-bourgeois ridiculousness would claim to restrain genius's needs for unrestrained activity; no one who would not have been proud to have served Schnekke as the impetus for his artistic achievements and thus to have enjoyed the most gratifying reward of her femaleness.

screened and ordered, the diaries are presented in monthly reports to the chief of the Buribunk Department, who can in this manner continuously supervise the psychological evolution of his province and report to a central agency. This agency, in turn, keeps a record of the complete register (and publishes that register in Esperanto) and is hence in a position to exercise buribunkological control over all of Buribunkdom. A series of relevant practices—such as periodic and mutual photo opportunities and film presentations, an active exchange of diaries, readings from diaries, studio visits, conferences, new journals, theater productions preceded and followed by *laudatios* on the personality of the artist—ensure that the interest of the Buribunk in himself and in the quintessentially Buribunkic does not become mere decorum; they prevent as well a damaging, countercultural waning of interest, which leads us to doubt whether the refined existence of the Buribunk world will ever come to an end.

Nevertheless, here too we see a rebellious spirit in evidence, albeit rarely. And yet it needs to be said that in the Buribunk world, there exists an unlimited and infinitely understanding tolerance, as well as the highest respect for a person's individual liberty. Buribunks are allowed to write their diary entries free from any coercion whatsoever. Not only is one allowed to say that he lacks the mental energy for further entries and that it is only the grief felt for his failing energies that gives him the energy necessary for further entries; that is, in fact, one of the most beloved types of entry, which is widely acknowledged and appreciated. He can also put down, without fearing the least pressure, that he considers the diary one of the most senseless and bothersome practices, an annoying chicanery, a ridiculous old hat—in brief, he is not prevented from using the strongest language. For the Buribunks well know that they would violate the nerve center of their being were they to mess with the principle of unconditional freedom of speech. There even exists a reputable organization that sets itself the task of buribunkically recording Anti-Buribunkdom, just as there is an agency created for the purpose of fostering, in impressive entries, the ability to articulate disgust and loathing for the agency and even protest against the obligatory diary entry. Periodically, when diary entries threaten to glide into a certain uniformity, leaders of the Buribunks organize a successful movement aimed at raising individual-personal self-esteem.* The high point of this liberality,

* In this context, certain undaunted neo-Buribunkic initiatives deserve an honorable mention. They have led to the establishment of a prize-winning question that is raised periodically, "What real progress have the Buribunks made since the days of Ferker?" and to decided efforts in that direction.

however, resides in the fact that no Buribunk is forbidden from writing in his diary that he refuses to keep a diary.

Naturally, such freedom does not reach the point of anarchic chaos. Every entry about refusing to keep a diary must be amply supported and developed. Whosoever, instead of writing about one's resistance toward writing, really omits writing altogether violates the general intellectual openness and will be eliminated on grounds of antisocial behavior. The path of evolution silently passes over the silent ones; they are outside of all discourse and as a result can no longer draw attention to themselves. Finally, sinking step by step until hitting the bottom of the social hierarchy, they are forced to manufacture the external conditions for the possibility of noble Buribunkdom, for example, high-quality, handmade paper, upon which are printed the most distinguished diaries. . . . That is a rigorous but completely natural selection of the fittest, for whosoever cannot compete in the intellectual struggle of diaries will soon degenerate and disappear in the mass of those equipped only to produce the external conditions just mentioned. As a result of this physical labor and other menial services, such people also are no longer in a position to exploit, in Buribunk fashion, each moment of their lives, and they thus yield to an inexorable fate. Since they don't write anymore, they cannot respond to possible inconsistencies in their personal file; they no longer stay current, they disappear from the monthly reports and become nonentities. As if swallowed by the earth, nobody knows them anymore, nobody mentions them in their diaries, they are neither seen nor heard. Regardless of the intensity of their lament, even if it were to drive them to the edge of insanity, the honorable law does not spare anyone who has dishonorably excluded him- or herself, just as the laws of natural selection themselves know no exception.

And so, through tireless and engaged activity, the Buribunks seek to achieve such a perfection of their organization that, even if only over the span of hundreds of generations, an unprecedented progression is ensured. Daunting calculations—may progress not render them utopian!—have suggested that culture will eventually reach such a level that, thanks to unhindered evolution, the ability for diary writing will gradually become inborn in the Buribunk fetus. If so, fetuses could, with the help of appropriate, yet-to-be-developed media, communicate with one another about their cardinal perceptions and hence (by demystifying the remaining secrets of sex research) provide the necessary, factual basis for a refined sexual ethics. All that, of course, is far in the distance. It is, however, a historical fact that already today there exists a grand and densely organized mass of Buribunkdom—and hence a speaking, writing, bustling Buribunkdom compelled to enjoy each person's essential personality—that forges ahead into the sunrise of historicity.

The basic outline of the philosophy of the Buribunks: I think, therefore I am; I speak, therefore I am; I write, therefore I am; I publish, therefore I am. This contains no contradiction, but rather the progressive sequence of identities, each of which, following the laws of logic, transcends its own limitations. For Buribunks, thinking is nothing but silent speech; speech is nothing but writing without script; writing is nothing but anticipated publication; and publication is, hence, identical with writing to such a degree that the differences between the two are so small as to be negligible. I write, therefore I am; I am, therefore I write. What do I write? I write myself. Who writes me? I myself write myself. What do I write about? I write that I write myself. What is the great engine that elevates me out of the complacent circle of egohood? History!

I am thus a letter on the typewriter of history. I am a letter that writes itself. Strictly speaking, however, I write not that I write myself but only the letter that I am. But in writing, the world spirit apprehends itself through me, so that I, in turn, by apprehending myself, simultaneously apprehend the world spirit. I apprehend both it and myself not in thinking fashion, but—as the deed precedes the thought—in the act of writing. Meaning: I am not only the reader of world history but also its writer.

At each second of world history, the letters of the typewriter keyboard leap, impelled by the nimble fingers of the world-I, onto the white paper and continue the historical narrative. Only at the moment that the single letter, singled out from the meaningless and senseless indifference of the keyboard, hits the animated fullness of the white paper, is a historical reality created; only at that moment does life begin. That is to say, the beginning of the past, since the present is nothing but the midwife that delivers the lived, historical past out of the dark womb of the future. As long as it is not reached, the future is as dull and indifferent as the keyboard of the typewriter, a dark rat hole from which one second after the other, just like one rat after the other, emerges into the light of the past.

Ethically speaking, what does the Buribunk do who keeps a diary each and every second of his life? He wrests each second off of the future in order to integrate it into history. Let us imagine this procedure in all its magnificence: second by second, the blinking young rat of the present moment crawls out of the dark rat hole of the future—out of the nothing that not yet is—in order to merge (eyes glowing with fiery anticipation) the next second with the reality of history. Whereas with the unintellectual human being, millions and billions of rats rush without plan or goal out into the infinite expanse of the past only to lose themselves in it, the diary-keeping Buribunk can catch each of those seconds, one at a time, and—once aligned in an orderly battalion—allow them to demonstrate the parade of world history. This way, he secures for both himself and humanity the maximum amount of historical facticity and cognizance. This way, the nervous anticipation of the future is defused, for no matter what happens, one thing is for sure: no second peeling off of the future is getting lost, no hit of the typewriter key will miss the page.

The death of an individual is also nothing but such a rat second, which has no content in itself—whether one of happiness or grief—but only in its historical registration. Of course, in the rat second of my death, I can no longer hold pen and diary, and I am ostensibly no longer actively involved in this historical registration; the crux of diary keeping, the will to power over history, disappears and clears the field for somebody else's desire. If we disregard the pedagogical aspects of this situation, that is, its application not to waste a second in order to impose our will to power onto historiography in the making, we must confess that the termination of our will to history goes very much against our will, for the will to power in the first instance always refers to the will to one's own power, not to that of a certain historian of future generations. Such concerns, however, lend themselves to serious confusion, and we have already seen how even in the case of the great Ferker, the fear of death had a downright catastrophic influence on his historical reputation. Today, however, thanks to the evolving consciousness whose sunlight kills the bacteria of the fear of death, there is little danger of any confusion among the Buribunks.

We see through the illusion of uniqueness. We are the letters produced by the writing hand of the world spirit and surrender ourselves consciously to this writing power. In that we recognize true freedom. In that we also see the means of putting ourselves into the position of the world spirit. The individual letters and words are only the tools of the ruses of world history. More than one recalcitrant "no" that has been thrown into the text of history feels proud of its opposition and thinks of itself as a revolutionary, even though it may only negate revolution itself. But by consciously merging with the writing of world history we comprehend its spirit, we become equal to it, and—without ceasing to be written—we yet understand ourselves as writing subjects. That is how we outruse the ruse of world history—namely, by writing it while it writes us.[170]

World history comes to a close as a global typewriters' association. Digital signal processing (DSP) can set in. Its promotional euphemism, post-history, only barely conceals that war is the beginning and end of all artificial intelligence.

In order to supersede world history (made from classified intelligence reports and literary processing protocols), the media system proceeded in three phases. Phase 1, beginning with the American Civil War, developed storage technologies for acoustics, optics, and script: film, gramophone, and the man-machine system, typewriter. Phase 2, beginning with the First World War, developed for each storage content appropriate electric transmission technologies: radio, television, and their more secret counterparts. Phase 3, since the Second World War, has transferred the schematic of a typewriter to a technology of predictability per se; Turing's mathematical definition of computability in 1936 gave future computers their name.

Storage technology from 1914 to 1918 meant deadlocked trench warfare from Flanders to Gallipoli. Transmission technology with VHF tank communications and radar images, those military developments parallel to television,[171] meant total mobilization, motorization, and blitzkrieg from the Vistula in 1939 to Corregidor in 1945. And finally, the largest computer program of all time, the conflation of test run with reality, goes by the name of the Strategic Defense Initiative. Storing/transmitting/calculating, or trenches/blitz/stars. World wars from 1 to n.

In artificial intelligences, all media glamor vanishes and goes back to basics. (After all, "glamor" is nothing but a Scottish corruption of the word "grammar.")[172] Bits reduced the seeming continuity of optical media and the real continuity of acoustic media to letters, and these letters to numbers. DSP stores, transfers, calculates—millions of times per second, it runs through the three functions necessary and sufficient for media. The standard for today's microprocessors, from the point of view of their hardware, is simply their systematic integration.

Calculations are performed by a central processing unit (CPU) that, in the case of Zilog's Z80 microprocessor, cannot do much more than manipulate blocks of 8 bits either logically (following Boolean algebra) or arithmetically (through basic addition). Storage is subdivided into a Read-Only Memory (ROM), which retains once and for all inscribed data, preferably commands and computing constants, and a Random Access Memory (RAM), which reads the variable data of a measured environment and returns mathematical data to control that environment. The exchange between individual modules runs along uni- or bidirectional

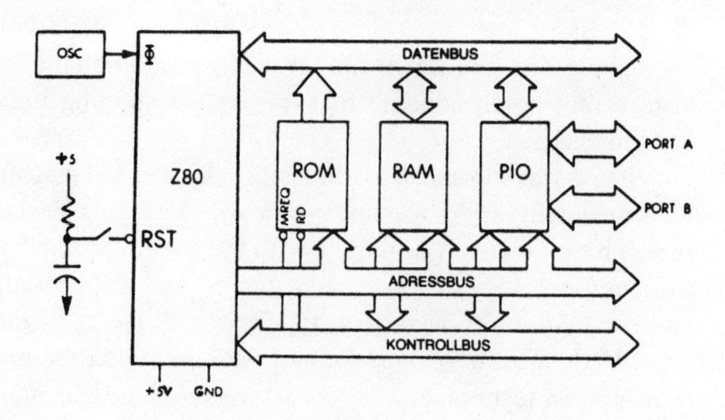

Setup of a microprocessor (Z80).

busses (for data, addresses, and control commands such as WRITE or READ), and the transfer from and to the environment runs via an input/output port (I/O) at whose outer margin, finally, the conversion of continuities into bits takes place.

And since, from the microprocessor to large processing networks, everything is nothing but a modular vice, the three basic functions of storing/transferring/processing are replicated on internal levels no longer accessible to programmers. For its part, the CPU includes (1) an arithmetic logic unit (ALU), (2) several RAMs or registers to store variables and a ROM to store microprograms, and (3) internal busses to transfer data, addresses, and control commands to the system's busses.

That's all. But with sufficient integration and repetition, the modular system is capable of processing, that is, converting into any possible medium, each individual time particle of the data received from any environment. As if one could reconstruct, custom-made from one microsecond to the next, a complete recording studio comprising reel-to-reels plus radio transmission plus control panel and switchboard. Or, as if the Buribunks' immense permeation with data coincided with an automated Buribunkology that could be switched, at the speed of electrical current, from a register of data to a register of persons or even their self-registration. The construction of the Golem, at any rate, is perfect. The storage media of the founding generation were only capable of replacing the eye and the ear, the sensorium of the central nervous system; the communications media between the two wars were only capable of replacing the mouth and the hand, the motorics of information. Which is why, behind all registers,

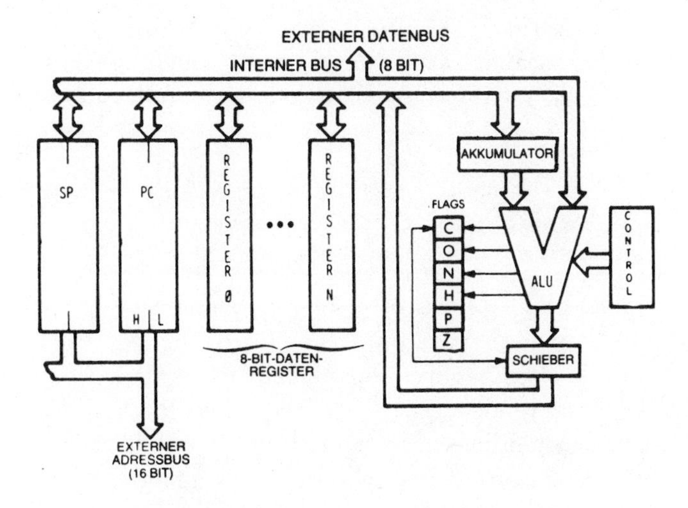

Standard architecture of a CPU.

all channels, a human being still appeared to be doing the transmitting. So-called thinking remained thinking; it therefore could not be implemented. For that, thinking or speech had to be completely converted into computing.

"I WILL LEARN HOW TO COMPUTE ON MY TYPEWRITER," writes an inmate of Gugging (on his red device for this red and black book). Alan Turing did nothing else. Instead of learning his public school's prescribed handwriting, he reduced typewriters to their bare principle: first, storing or writing; second, spacing or transferring; third, reading (formerly reserved for secretaries) or computing discrete data, that is, block letters and figures. Rather than conclude that humans are superior, as did his colleague Gödel, with whom he jointly refuted the Hilbert program (in support of a complete, consistent, and decidable mathematics, that is, a mathematics that could in principle be delegated to machines),[173] Turing was suicidal—in life as well as in his job. He dropped the unpredictable in order to relieve mathematicians of all predictable (or recursive) functions and to construct the machine that Hilbert had presumed as a formalism. The hypothetical determinism of a Laplacian universe, with its humanist loopholes (1795), was replaced by the factual predictability of finite-state machines. Rather full of pride, Turing wrote:

The prediction which we are considering is, however, rather nearer to practicability than considered by Laplace. The system of the "universe as a whole" is such that quite small errors in the initial conditions can have an overwhelming effect at a later time. The displacement of a single electron by a billionth of a centimetre at one moment might make the difference between a man being killed by an avalanche a year later, or escaping. It is an essential property of the mechanical systems which we have called "discrete state machines" that this phenomenon does not occur. Even when we consider the actual physical machines instead of the idealised machines, reasonably accurate knowledge of the state at one moment yields reasonably accurate knowledge any number of steps later.[174]

The overwhelming effects of this predictability have since reached Man's employment statistics. The consequences of Turing's politics of suicide: "As Victorian technology had mechanised the work of the artisans, the computer of the future would automate the trade of intelligent thinking. . . . The craft jealously displayed by human experts only delighted him. In this way he was an anti-technocrat, subversively diminishing the authority of the new priests and magicians of the world. He wanted to make intellectuals into ordinary people."[175]

The first to be affected were of course stenotypists. After eleven years, Turing's Universal Discrete Machine fulfilled the prophecy that an apparatus "also renders superfluous the typist." His simulation game, in which a censor is to but cannot actually decide which of two data sources *A* and *B* is human and which is a machine, significantly has a precursor. According to Turing, computer *B* replaces the systemic position of a woman who—in competition or gender war with a man *A*—seeks to persuade the data gap C that she is the real woman. But since both voices are severed from the "written, or, better still, typed" flow of information, Remington's secretary gives her farewell performance. Whenever transvestite *A* insists that he has strands of hair "nine inches long," the human predecessor of the computer writes to her censor, as mechanically as futilely, "I am the woman, don't listen to him!"[176]

With which the homosexual Turing raised to the level of technology Dionysus's sentence, "Must we not first hate ourself if we are to love ourself?" With the added observation that against total desexualization, protest will "avail nothing."[177] Computers write by themselves, without secretaries, simply with the command WRITE. (Anyone who would like to see the phallus in the 5 volts of a logical 1, and the hole in the 0.7 volts of an 0, confuses industrial standards with fiction.) Only those intersections between computers and their environment that, following ASCII code (American Standard Code for Information Interchange), are networked bit by bit with typewriter keys[178] will continue to offer women jobs for a while. When ENIAC, "the first operational computer," according to misleading American accounts, calculated projectile trajectories and A-bomb pressure waves during the Second World War, one hundred women were hired in addition to male programmers. Their job: "to climb around on ENIAC's massive frame, locate burnt-out vacuum tubes, hook up cables, and perform other types of work unrelated to writing."[179]

By contrast, Turing, with an eye toward "computers and guided projectiles," predicted good times for men, programmers, and mathematicians.[180] But it was a strange kind of mathematics into which he imported the elegance and complexity of classical analysis. What disappeared in the split-up of binaries was not only the continuity of all graphs and trajectories examined since Leibniz, and which Fourier's theory and Edison's phonographs simply followed. What was much more drastic than such primitive step functions was his crucial innovation: the abolition of the difference between numbers and operational symbols, data and commands. For even if numbers stood for data relationships, the signs + or – were still inhabited by a human spirit who appeared to give the command to add or subtract. Turing's Universal Discrete Machine, however, converted these (and all other) letters into their monotonous rows of binaries. In machine language, the command ADD is neither a human enunciation nor a letter symbol, but just one of many series of bits. (In a Z80, the command "Increase the number in the accumulator by 2" would be 1100 0110 / 0000 0010.) It was not Gödel's humanist belief but rather his simple trick of Gödelization that once again emerged victorious: only after commands, axioms, or, to put it briefly, sentences had been converted into numbers were they as infinitely manipulable as numbers. End of literature, which is made up of sentences.

Every microprocessor implements through software what was once the dream of the cabala; namely, that through their encipherment and the manipulation of numbers, letters could yield results or illuminations that no reader could have found. Computers are endless series of numbers only whose relative position decides whether they operate as (verbal) commands or (numeric) data or addresses. If John von Neumann, the mathematician of the Second World War, had not taken certain precautions for his machines, a command sequence of numbers such as ADD could also add up, aside from the usual data, command sequences themselves, until no programmer would be able to comprehend the starry mathematics to which that take-off had abducted their computer.

The neat separation of data, addresses, commands—that is, of storage contents, points of transfer, and processing steps—by contrast, assures that for each address, there is only one command or datum on the bus. A box of numbered paper slips that can log on not only (as with the Buribunks) to certain books, chapters, pages, terms, but to any individual bit of the system. Computer algorithms, instead of simply reproducing a logic, consist of "LOGIC + CONTROL."[181] No wonder that governmental ingenuity invented the impossible job of the data security specialist to camouflage the precision of such data control.

On the other hand, since Turing, the possible job of a programmer has run the risk of forgetting mathematical elegance. Today, prior to the conquest of digital signal processors, the hardware of average computers is at a kindergarten level: of all the basic forms of computation, it barely manages addition. More complex commands have to be reconverted into a finite, that is, serial, number of cumulative steps. An unreasonable chore for humans and mathematicians. Where recursive, that is, automatizable, functions succeed classical analysis, computation works as a treadmill: through the repeated application of the same command on the series of interim results. But that's it. A Hungarian mathematician, after he had filled two whole pages with the recursive formulas according to which a Turing machine progresses from 1 to 2 to 3, and so on, observed in German as twisted as it was precise: "This appears as an extraordinarily slowed-down film shot of the computation processes of man. If this mechanism of computation is applied to some functions, you start living it, you begin to compute exactly like it, only faster."[182] Consolation for prospective programmers . . .

Slow-motion shots of the spirit will exorcise it. Chopped up like movements in front of the camera, equations finally solve themselves without intuition because every discrete step during storage, transfer, and calculation takes place with bureaucratic precision. The discrete machine forms a solitary union with cinema and typewriter, but not with neurophysiology. That is what distinguishes it from the dream typewriter constructed by Friedlaender's Dr. Sucram, who in his main line of work took care of the *Gray Magic* [*Graue Magie*] of three-dimensional cinema.

The doctor concentrated on his experiments with a curious little machine model. He put a metal helmet on his head; fine wires connected the helmet to the keyboard of the typewriter. Without any movement on the doctor's part, the levers of the machine started moving. It was a ghostly sight to behold.

"What kind of a device is that?" [Bosemann] pointed to the helmet from which emerged the wires connecting to the keyboard.

"An extraordinarily comfortable typewriter, Mister Bosemann. It saves me a typist. I am in the process of letting the ethereal emanations of the brain work for me directly. Up to now, our thoughts, no matter how practical, have been moving the world only in indirect ways. Our machines do not yet work under the direct influence of our thoughts, our will. I plan a direct transmission."[183]

The typing, computing, and sewing machines in the brains or books of Nietzsche and Kußmaul hence became reality. The founding myth of a media landscape, which would only be the worldwide unfolding of neurophysiology, reached its peak in Friedlaender's machine fiction. Fourteen years later, it ends in Turing's machine, which was also never built but is mathematically conceivable. The computer and the brain are functionally compatible, but not in terms of their schematics. Since the nervous system, according to Turing, is "certainly not a discrete-state machine," that is, not infinitesimally precise, all the unpredictabilities of a Laplacian universe loom over it.[184] Thus, "the real importance of the digital procedure lies in its ability to reduce the computational noise level to an extent which is completely unobtainable by any other (analogy) procedure." And even if—following Neumann's elegant simplification—the neural, but not the hormonal, conduits operate according to a digital model, their information flow is still five thousand times slower than that of computers.[185] The brain, however, compensates for this loss of transmission through the parallel processing of whole sets of data; statistical breadth (presumably based on majority gates) for which computers can compensate only through serial processing and recursive functions. What remains unrealized, at any rate, is Dr. Sucram's desire for "letting the ethereal emanations of the brain work for me directly."

The white noise of brains, of the ether, of the globe: the total typewriter has nothing to do with that. But everything to do with trenches/blitz/stars.

Even if "there is little in our technological or physiological experience to indicate that absolute all-or-none organs exist,"[186] the oldest knowledge of gods, ghosts, and generals knows better. The language of the upper echelons of leadership is always digital. In the scriptures of the priests, Yahweh distinguishes for seven days between day and night, morning and evening, sun and moon, earth and heaven, land and water (not to mention good and bad). That is what the priests, who have edited and continue to administer this holy scripture, call God's creation. But "it is nothing but the creation of nothing other than signifiers."[187] Earth and heaven can do without Elohim's inscription; it exists, prior to God's creation and after God's death, in another holiness, for which the Holy Bible only has the word *tohu bohu*:[188] the random noise of events. The language of the upper echelons of leadership, by contrast, is digitalization; it transforms sources of accidental noise into absolute all-or-none organs. Otherwise, commands and censures, those two antisymmetrical instruments of leadership, could not be communicated.

And if the invading of communication channels by scrambling noises makes it necessary, the language of the upper echelons goes so far as to overcode the binary opposition with another, that is, redundant, binary opposition. In the German General Staff,

> a military language practice that was exercised and used for decades . . . aimed strictly at distinguishing between "west*ern*" and "east*ward*" in military briefs and reports. The reason was that one wanted to establish a distinct sound difference between the two terms, because otherwise messages and dictates delivered orally or over the phone might easily have resulted in fateful mix-ups. . . . The layman might think of this as a triviality, but every soldier is surely cognizant of the far-reaching implications of this regulation.

For planning wars on two fronts, the opposition between east and west is as fundamental as that between heaven and earth for the gods of creation. Therefore, when Major General Alfred Jodl, the last chief of an illustrious short (hi)story, "used the word 'east*ern*' rather than, as was common protocol, 'east*ward*' in an army report of June 14, 1940, during the western offensive of 1940 . . . even though he himself had emerged from the ranks of the army . . . he violated without much ado a time-tested practice and triggered widespread and intense indignation in the officers' corps."[189]

The *tohu bohu* and, in its wake, analog media run through all the various types of conditions except the NO.[190] Computers are not emanations of nature. Rather, the universal discrete machine, with its ability to erase, negate, and oppose binary signs, always already speaks the language of the upper echelons. On the transmitting side, the general staffs of the Axis, just as, on the receiving end, those in London or Washington.

Whether or NOT the Japanese empire took seriously the resource embargo threatened by Roosevelt (that is, attack the United States), whether or NOT Vice Admiral Nagumo's flotilla would sink the Pacific battleships at Pearl Harbor with carrier-bound aircraft, whether or NOT he would maintain silence in his areas of operation off the Aleutian Islands (he did): these were precisely the digital puzzles of 1941, solvable only through the interception and decoding of necessarily discrete sources of information. And since the machine mathematics of the current century endowed general staffs with the ability to encrypt their orders automatically, that is, immeasurably more efficiently than by hand, decoding had to be done by machines as well. The Second World War: the birth of the computer from the spirit of Turing and his never-built principal relay.

This escalation between senders and receivers, weapons and anti-weapons, is told quickly and most precisely in the words of Guglielmo Marconi, which were broadcast from a gramophone record on Radio Roma by the inventor of the radio immediately after his death (as if to underscore the new acoustic immortality). Marconi, a senator and *marchese* of fascist Italy, "confessed" that

> forty-two years ago, when I achieved the first successful wireless transmission in Pontecchio, I already anticipated the possibility of transmitting electric waves over large distances, but in spite of that I could not hope for the great satisfaction I am enjoying today. For in those days a major shortcoming was ascribed to my invention: the possible interception of transmissions. This defect preoccupied me so much that, for many years, my principal research was focused on its elimination.
>
> Thirty years later, however, precisely this defect was exploited and turned into radio—into that medium of reception that now reaches more than 40 million listeners every day.[191]

Which unnamed circles feared the interception of transmissions is not hard to guess. Which circles charged Marconi with the elimination of this defect, that is, with the construction of a wooden iron, is even easier to guess. Nothing in the analog medium of the radio allows the negation of signals, their spy-proof inversion into their opposite, or nonsense. Hence, general staffs, who were afforded perfect communication to the front and possibilities for blitzkrieg by Marconi's invention, had to rely on the development of discrete encoding machines. Immensely inflated flows of information demanded a form of text processing as automatic as it was discrete—the typewriter.

Since 1919, the engineer Arthur Scherbius had experimented in Berlin-Wilmersdorf with a "secret typewriter." In 1923, he himself thus founded Chiffriermaschinen A.G. (Encoding Machines Corporation) and secured for his model the promotion of the world postal club.[192] For the first time, Remington's typewriter keyboard was no longer the boring and unequivocal one-way link between input and output, softened only by typos. For the first time, hitting a letter key offered numerous combinatory surprises. The 26 letters of the alphabet ran over electric conduits into a distribution system consisting of three (later, four or five) rotors and an inversion rotor, which always selected other substitute letters. With each strike of the

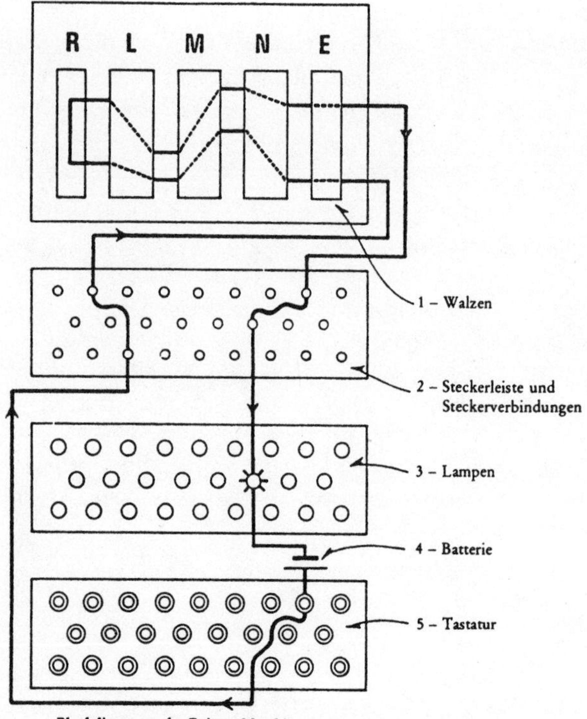

Blockdiagramm der Enigma-Maschine

$A = S N M L R L^{-1} M^{-1} N^{-1} S^{-1}$

$B = S P N M L R L^{-1} M^{-1} N^{-1} P^{-1} S^{-1}$

$C = S P^{2} N M L R L^{-1} M^{-1} N^{-1} P^{-2} S^{-1}$

$D = S P^{3} N M L R L^{-1} M^{-1} N^{-1} P^{-3} S^{-1}$

$E = S P^{4} N M L R L^{-1} M^{-1} N^{-1} P^{-4} S^{-1}$

$F = S P^{5} N M L R L^{-1} M^{-1} N^{-1} P^{-5} S^{-1}$

Permutationen der Buchstaben A, B, C, D, E, F.

Block diagram of the Enigma machine: *Above*, (1) rotors, (2) connector tray and connectors, (3) lamps, (4) battery, and (5) keyboard. *Below*, permutations of the letters *A–F*.

typewriter key, the rotors (just like the second, minute, and hour hands of clocks) advanced by one revolution, only to return to their original position not until 26^7, or 8 billion, hits later.

That is how Scherbius, with his machine mathematics, liberated cryptographers from their manual work. The sender, instead of having to labor for hours with pencil, tables, and graph paper, sat in front of a regular typewriter keyboard and typed in the orders of his general staff in plain text. The letter output, however, which he could read from the flashing of 26 bulbs and which he copied in accordingly, looked like pure letter salad. Radio as well, with its large defect, could translate that salad in spy-proof fashion, until an antisymmetrical, secret typewriter on the receiving end converted the almost perfect white noise back into plain text, simply because the machine was calibrated on the basis of a daily command to start at the same rotor.

Year after year since the First World War, the German army had torpedoed Bredow's plans to set up a civilian radio network, despite all the horror of a communist radio specter and the abuse of army equipment. Its own information flow, especially on long wave, was given priority. In November 1922, however, postal secretary Bredow could inform the Ministry of Defense that "the switch of the official radio services to wireless telegraphy and the use of encoding machines would soon provide sufficient security to protect the privacy of telegraphy."[193] That's how precisely information was exchanged between industry and the state. In 1923, General von Seeckt also granted radio entertainment to Germans, but not without prohibiting with draconian regulations any misuse of civilian receivers for purposes of transmission. But the order of discourses in the current century was restored: a few public transmission frequencies thus permitted (to the joy of literary and media sociologists) the mass reception that Marconi posthumously welcomed; Scherbius, however, prevented the interception of the military-industrial complex's numerous frequencies, which Marconi was worried about. Since then, people have been doused in the glamor of analog media only to remove the grammar of the typewriter, the prototype of digital information processing, from their minds.

In 1926, the German navy used the first encryption machines.[194] Three years later, soon after Major Fellgiebel, the subsequent chief of Army Communications, had taken over the Abwehr's cryptography division,[195] the army followed. The secret typewriter of Wilmersdorf was equipped with yet more secret rotors, as well as the name of secrecy itself: ENIGMA. For a decade, it lived up to that name.

But other states also did their shopping at Scherbius. Modified Enigma models were the standard between the world wars. All classified exchanges between Tokyo and the Japanese embassy in the United States (including all the planning for Pearl Harbor), for example, took place in the machine code Angooki Taipu B, which the American counterpart re-

General Guderian on the Enigma in his general's tank.

named Purple for reasons of security.[196] Three months prior to Vice Admiral Nagumo's blitzkrieg, William F. Friedman, chief of the Signal Intelligence School (SIS), pulled a cryptoanalytical stunt. In mathematical purity, that is, without having captured and subsequently evaluated a Purple code (following the black-box rules of the Second World War), he managed to retrace the infinite permutations of the secret typewriter. The last victory of humans over communication technologies, which Friedman paid for with a nervous collapse and months of psychiatric treatment.[197] But as always, it was precisely at the site of madness that machines originated. Their superhuman computation capability allowed the U.S. president to listen in on Japan's plans for attack. That Roosevelt allegedly did not warn his two commanding air and sea officers in the Pacific is an altogether different story . . .

The escalation of weapons and antiweapons, of cryptography and cryptoanalysis (as Friedman renamed writing and reading under the conditions of high technology), at any rate urgently required the automatization of decoding. And for that need, a universal discrete machine, which could replace *any* other machine, was a perfect fit. "The most complicated machines are made only with words."[198] Turing, soon after negatively solving Hilbert's *Entscheidungsproblem* (decision problem), described to his mother "a possible application" of the new and seemingly infinite mathematics at which he was

> working on at present. It answers the question "What is the most general kind of code or cipher possible," and at the same time (rather naturally) enables me to construct a lot of particular and interesting codes. One of them is pretty well impossible to decode without a key, and very quick to encode. I expect I could sell them to H.M. Government for quite a substantial sum, but I am rather doubtful about the morality of such things. What do you think?[199]

The answer came not from his mother but from the government. Germany's "Enigma machine was the central problem that confronted the British Intelligence Service in 1938. But they believed it was unsolvable,"[200] until the Government Code and Cipher School hired Alan M. Turing (notwithstanding his moral doubts) three days after the outbreak of the war.

Bletchley Park, the bombproof site of British cryptoanalysis during the war, was in a better position than its American colleagues: young mathematicians of the Polish secret service had already constructed a decoding machine, the so-called Bombe, based on captured Enigmas. But when Fellgiebel's Army Communications increased the number of rotors to five, even the Bombe could not follow suit. The 150,738,274,937,250 possible ways of electrically connecting ten pairs of letters exceeded its capacity, at least in real time, on which blitzkrieg commands and their timely countermeasures depend. The overwhelmed Poles donated their files to the British and Turing.

From this primitive Bombe, Turing made a machine that the head of Bletchley Park not coincidentally named the Oriental Goddess: a fully automatized oracle to interpret fully automatized secret radio communication. Turing's recursive functions laid the groundwork for the enemy's ability to decode Enigma signals with a mere 24-hour delay beginning in May 1941, and thus, to paraphrase Goebbels, to eavesdrop on the enemy. The German army did not want to believe it until the end of the war: it was "fully convinced that the decoding of Enigma was, even with the aid of captured machines, impossible given the overwhelmingly large number of calibrating positions."[201] However, only nonsense, white noise without information and hence of no use for the upper echelons, provides complete proof against spying. Whereas "the very fact that the Enigma *was* a machine made mechanical cryptoanalysis a possibility."[202] As a pseudo-random generator, the secret typewriter produced nonsense only relative to systems whose revolutions did not match its own. Turing's goddess, however, found regularities in the letter salad.

For one thing, Enigma had the practical advantage or theoretical disadvantage that its cipher consisted of a self-inverse group. In order to be encoded or decoded on the same machine, letter pairs had to be interchangeable. For example, when the OKW encoded its O as a K, the K inversely turned into an O. From that followed "the very particular feature that no letter could be enciphered by itself."[203] Not even the OKW was capable of writing its own name. Turing subjected these few yet revealing implications to a sequential analysis that weighted and controlled all the probabilities of solution. With automatized judgment, the Oriental Goddess ran through permutation after permutation, until the letter salad became plain text again. War of typewriters.

And since from "15 to a maximum of 29 percent"[204] of the German radio traffic ran through Enigma, the spy war reached a new level: interception yielded "not just messages, but the whole enemy communication *system*."[205] The midrange levels of command—from army and division headquarters to individual blitzkrieg weapons on land, in the air, or at sea—betrayed their addresses, which are, all spy novels notwithstanding, more revealing than data or messages. Sixty different Enigma codes and 3,000 classified radio messages per day, with all of the specs for their

senders and receivers, recorded the war like a typewriter the size of Europe. Under the conditions of high technology, war coincides with a chart of its organizational structure. Reason enough for the Government Code and Cipher School to model, in miniature, its organization after that of the German army, that is, after the enemy.[206] Turing's game of imitation became a reality.

It is only one step from the flowchart to the computer. The addresses, data, commands that circulated between humans and typewriters in the German army or its British simulacrum could finally turn into hardware. This last step was undertaken in 1943 by the Post Office Research Station at Bletchley Park. One thousand five hundred tubes were expropriated and converted into overloaded switches and, instead of reinforcing radio analog signals, simulated the binary play of Boolean algebra. Transistors did not make it into the world until 1949, but even without them the universal discrete machine—including data entry, programming possibilities, and the great innovation of internal storage mechanisms[207]—saw its first implementation, for which Turing's successors could find no other name than COLOSSUS. Because the strategic secrets of the Führer's headquarters, Wolfsschanze, could, as is logical, only be cracked by a monster computer.

COLOSSUS began its work and decoded an additional 40 percent of the German radio traffic—everything that for reasons of security was transmitted not via Enigma and wireless but via the Siemens Cryptwriter. As a teleprinter running the Baudot-Murray Code, this typewriter no longer required cumbersome manual operation with its human sources of error; its fully digitized signals consisted of the "yes" or "no" of ticker tape, which, through the binary addition of plain text and pseudo-random generator, could be encoded much more efficiently than with Enigma. Moreover, radio interception became possible only once signals were sent through a radio link rather than a telegraph cable.[208] That is how well upper echelons pick their typewriters.

Obviously, COLOSSUS beat binary addition with binary addition, but even the first computer in the history of science or warfare would have been nothing but a several-ton version of Remington's special typewriter with a calculating machine[209] had it not observed conditional jump instructions.[210]

Conditional jumps, first envisioned in Babbage's unfinished Analytical Engine of 1835, were born into the world of machines in 1938 in Konrad Zuse's apartment in Berlin, and this world has since been self-identical with the symbolic. In vain, the autodidact offered his binary calculators to use as encryption machines and to surpass the supposedly spy-proof Enigma.[211] The opportunity missed by Army Communications was seized by the German Aviation Test Site in 1941—for the purposes of "calculating, testing, and examining cruise missiles."[212] Yet Zuse made only the most sparing use of the IF-THEN commands of his brilliant "plan calculation": Gödel's and Turing's insight of translating commands, that is, letters, into numbers was a concern for him:

> Since programs, like numbers, are built from series of bits, it was only a matter of course that programs be stored as well. With that it was possible to make conditional jumps, as we say today, and to convert addresses. From the point of view of schematics, there are several solutions for it. They all rest on a common thought: the feedback of the result of the calculation on the process and on the configuration of the program itself. Symbolically, one can envision that through a single wire. I was, frankly, nervous about taking that step. As long as that wire has not been laid, computers can easily be overseen and controlled in their possibilities and effects. But once unrestricted program processing becomes a possibility, it is difficult to recognize the point at which one could say: up to this point, but no further.[213]

A simple feedback loop—and information machines bypass humans, their so-called inventors. Computers themselves become subjects. IF a preprogrammed condition is missing, data processing continues according to the conventions of numbered commands, but IF somewhere an intermediate result fulfills the condition, THEN the program itself determines successive commands, that is, its future.

In the same way, Lacan, making a distinction with animal codes, defined language and subjectivity as human properties. For example, the dance of bees, as it has been researched by von Frisch, "is distinguished from language precisely by the fixed correlation of its signs to the reality that they signify." While the messages of one bee control the flight of another to blossoms and prey, these messages are not decoded and transmitted by the second bee. By contrast, "the form in which language is expressed . . . itself defines subjectivity. Language says: 'You will go here, and when you see this, you will turn off there.' In other words: it refers itself to the discourse of the other."[214]

In yet other words: bees are projectiles, and humans, cruise missiles. One is given objective data on angles and distances by a dance, the other, a command of free will. Computers operating on IF-THEN commands are therefore machine subjects. Electronics, a tube monster since Bletchley Park, replaces discourse, and programmability replaces free will.

Not for nothing was Zuse "frankly, nervous" about his algorithmic golems and their "halting problem." Not for nothing did the Henschel Works or the Ministry of Aviation assign the development of cruise missiles to these golems. On all fronts, from top-secret cryptoanalysis to the most spectacular future weapons offensive, the Second World War devolved from humans and soldiers to machine subjects. And it wasn't by much that Zuse's binary computers missed doing the programming of free space flight from its inception, rather than determining in the bunkers of the Harz the fate of the V2 at the last moment.[215] The "range of charges" that the Peenemünde Army Test Site assigned to German universities in 1939 included (aside from acceleration integrators, Doppler radar, onboard calculators, etc.), in a rather visionary way, what Wernher von Braun described as "the first attempt at electric digital computation."[216] The weapon as subject required a corresponding brain.

But since the commander in chief of the German army (whom Syberberg has called the "greatest filmmaker of all time")[217] did not believe in self-guided weapons on the actual rocket testing site, but only during their demonstration on color film at the Wolfsschanze,[218] the entropies of the Nazi state emerged victorious over information and information machines.

At any rate, cybernetics, the theory of self-guidance and feedback loops, is a theory of the Second World War. Norbert Wiener testified to that when he introduced the term:

> The deciding factor in this new step was the war. I had known for a considerable time that if a national emergency should come, my function in it [sic] would be

determined largely by two things: my close contact with the program of computing machines developed by Dr. Vannevar Bush, and my own joint work with Dr. Yuk Wing Lee on the design of electric networks. . . . At the beginning of the war, the German prestige in aviation and the defensive position of England turned the attention of many scientists to the improvement of anti-aircraft artillery. Even before the war, it had become clear that the speed of the airplane had rendered obsolete all classical methods of the direction of fire, and that it was necessary to build into the control apparatus all the computations necessary. These were rendered much more difficult by the fact that, unlike all previously encountered targets, an airplane has a velocity which is a very appreciable part of the velocity of the missile used to bring it down. Accordingly, it is exceedingly important to shoot the missile, not at the target, but in such a way that missile and target may come together in space at some time in the future. We must hence find some method of predicting the future position of the plane.[219]

With Wiener's Linear Prediction Code (LPC), mathematics changed into an oracle capable of predicting a probable future even out of chaos—initially for fighter aircraft and anti-aircraft guidance systems, in between the wars for human mouths and the computer simulations of their discourses.[220] Blind, unpredictable time, which rules over analog storage and transmission media (in contrast to the arts), was finally brought under control. With digital signal processing, measuring circuits and algorithms (like an automated sound engineer) ride along on random frequencies. Today this form of cybernetics ensures the sound of most reputable rock bands; in actuality, however, it was only a "new step" in ballistics. Machines replaced Leibniz in the analysis of trajectories.

With the consequence that COLOSSUS gave birth to many a son, each more colossal than its secret father. According to the ministry of supply, Turing's postwar computer ACE was supposed to calculate "grenades, bombs, rockets, and cruise missiles"; the American ENIAC "was to simulate trajectories of shells through varying conditions of air resistance and wind velocity, which involved the summation of thousands of little pieces of trajectories." John von Neumann's EDVAC was being designed to solve "three-dimensional 'aerodynamic and shock-wave problems, . . . shell, bomb and rocket work, . . . [and] progress in the field of propellants and high explosives'"; BINAC worked for the United States Air Force; ATLAS, for cryptoanalysis; and finally, MANIAC, if this suggestive name had been implemented in time, would have optimized the pressure wave of the first H-bomb.[221]

Machines operating on the basis of recursive functions produce slow-motion studies not only of human thinking but also of human demise.

According to the insight of Pynchon and Virilio, the blitzkrieg and the flash-bulb shot (*Blitzlichtaufnahme*) coincide in the bomb that leveled Hiroshima during rush hour on August 6, 1945. A shutter speed of 0.000000067 seconds, far below Mach's projectile-like, pioneering cinematic feat of 1883, melted countless Japanese people "as a fine-vapor deposit of fat-cracklings wrinkled into the fused rubble" of their city.[222] Cinema to be computed in computer processing speeds, and only in computer processing speeds.

On the film's manifest surface, everything proceeds as if the "marriage of two monsters"[223] that John von Neumann had arranged between a German guided missile and an American A-bomb payload (that is, by saving both conventional amatol and conventional bomber pilots) by itself had been the step from blitzkrieg to the strategic present. What speaks against that is that both guided missiles and nuclear weapons surmounted the iron and bamboo curtains with extraordinary ease—partly through espionage, partly through the transfer of technology. Different from the machine subject itself, the innocuous but fully automated type-computing machine. With the fiat of the theory that is omnipotent because it is true, Stalin condemned the bourgeois aberration of cybernetics. As if materialism, in the espionage races with its other half, had been blinded by the disclosed secrets of mass extermination, the smoke trail of rockets and the flash of bombs.

Annihilation is still called determining the outcome of the war. Only 40 years later, classified archives have gradually revealed that Bletchley Park was presumably the most suitable candidate for this title. During the Second World War, a materialist who materialized mathematics itself emerged victorious. Regarding COLOSSUS and Enigma, Turing's biographer writes that "intelligence had won the war"[224] with the British literality that does not distinguish among reason, secret service, and information machine. But that is exactly what remained a state secret. During the war, a whole organization emerged for the purpose of delivering the results of fully automatized cryptoanalysis in coded form to the commanding officers at the front. Otherwise, the most vital secret of the war (through seized documents, traitors, or treacherously revealing countermeasures) possibly would have filtered through to the German army, and Enigma would have been silenced. Hence it became secret agents' last historical assignment to invent radiant spy novels in order to camouflage the fact that interception and the type-computing machine respectively render secret services and agents superfluous. (Which is what spy novels continue to do to this very day.) The mysterious "Werther," who allegedly

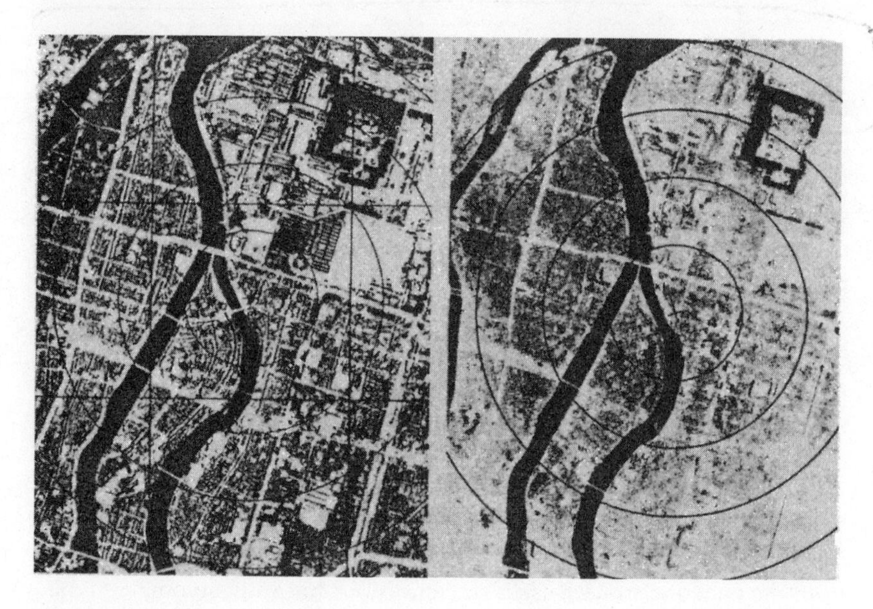

Hiroshima before and after August 6, 1945.

transmitted many plans of attack from the Wolfsschanze via Swiss double agents to Moscow, but who has yet to be located historically, may well have been one of the simulacra that systematically screened Bletchley Park from the Red Army.[225] Then, at any rate, Stalin's theory would have had a material basis—nonproliferation of the flow of information.

On August 28, 1945, three weeks after Hiroshima, four weeks after Potsdam, U.S. President Truman issued a secret decree on secret service interception, an information blockage on information machines. War-determining cryptoanalysis became a matter of ultimate classified material—in the past and the present, technology and method, successes and results, Bletchley Park and Washington, D.C.[226] As a result of which the same, but now cold, war could start again immediately: in the shadow of Truman's decree, COLOSSUS and its American clones learned Russian instead of German. Hermetically sealed, "the legacy of a total war, and of the capture of a total communications system, could now turn to the construction of a total machine."[227]

The success of this strategy of dissimulation is proved by its only leak. A writer, who not only knew the typewriter from secretaries but also reproduced it on the printed page, communicated in letter form to the warlords gathered in Potsdam that the symbolic has, through Enigma and COLOSSUS, become a world of the machine.

Arno Schmidt, "Offener Brief"

An die Exzellenzen
Herren
Truman (Roosevelt),
Stalin,
Churchill (Attlee)
Jalta, Teheran, Potsdam

8 c 357 8xup ZEUs !
id 21v18 Pt 7 gallisc 314002a 17 ? V 31 GpU 4a
29, 39, 49 ? mz 71Fi16 34007129 pp 34 udi119jem
13349 bubu WEg !
aff 19 exi: 16 enu 070 zIm 4019 abs12c 24 spü, 43
asti siv 13999 idle 48, 19037 pem 8 pho 36. 1012
sabi FR26a FlisCh 26:iwo - 18447 g7 gg !
Glent 31, glent 14 Po Arno Schmidt[228]

Under the conditions of high technology, literature has nothing more to say. It ends in cryptograms that defy interpretation and only permit interception. Of *all* long-distance connections on this planet today,[229] from phone services to microwave radio, 0.1 percent flow through the transmission, storage, and decoding machines of the National Security Agency (NSA), the organization succeeding SIS and Bletchley Park. By its own account, the NSA has "accelerated" the "advent of the computer age," and hence the end of history, like nothing else.[230] An automated discourse analysis has taken command.

And while professors are still reluctantly trading in their typewriters for word processors, the NSA is preparing for the future: from nursery school mathematics, which continues to be fully sufficient for books, to charge-coupled devices, surface-wave filters, digital signal processors including the four basic forms of computation.[231] Trenches, flashes of lightning, stars—storage, transmission, *the laying of cables.*

REFERENCE MATTER

NOTES

TRANSLATORS' NOTE: The citation format for Kittler's text closely follows that of the German edition. When two publication dates are given, the first refers to the date of original publication, the second, either to a later edition used by Kittler or to an English translation. Page numbers refer to the latter date, which corresponds to the edition given in the Bibliography, pp. 299–315.

TRANSLATORS' INTRODUCTION

1. Eric A. Havelock, *The Muse Learns to Write: Reflections on Orality and Literacy from Antiquity to the Present* (New Haven: Yale Univ. Press, 1986), 32.
2. Ibid.
3. Walter J. Ong, *Orality and Literacy: The Technologizing of the Word* (New York: Routledge, 1988), 136.
4. See ibid., 79–81, and Neil Postman, *Technopoly: The Surrender of Culture to Technology* (New York: Vintage, 1993), 3–20.
5. Quoted in Boris Eichenbaum, "The Theory of the 'Formal Method,'" in *Critical Theory Since Plato*, ed. Hazard Adams, rev. ed. (New York: Harcourt Brace, 1992), 803.
6. Friedrich Kittler, "Die Laterna magica der Literatur: Schillers und Hoffmanns Medienstrategien," in *Athenäum: Jahrbuch für Romantik 1994*, ed. Ernst Behler, Jochen Hörisch, and Günther Oesterle (Paderborn: Schöningh, 1994), 219.
7. See, for example, Michael Heim, "The Computer as Component: Heidegger and McLuhan," *Philosophy and Literature* 16.2 (1992): 304–18; Judith Stamps, *Unthinking Modernity: Innis, McLuhan and the Frankfurt School* (Montreal: McGill-Queen's Univ. Press, 1995); and Glenn Willmott, *McLuhan, or Modernism in Reverse* (Toronto: Univ. of Toronto Press, 1996).
8. Theodore Roszak, "The Summa Popologica of Marshall McLuhan," in *McLuhan: Pro & Con*, ed. Raymond Rosenthal (New York: Funk & Wagnalls, 1968), 257–69.
9. Hans Magnus Enzensberger, "Constituents of a Theory of the Media," in

The Consciousness Industry: On Literature, Poetics and the Media, ed. Michael Roloff (New York: Seabury, 1974), 118. See also Enzensberger's hilarious retraction, "The Zero Medium, or Why All Complaints About Television are Pointless," in *Mediocrity and Delusion: Collected Diversions*, trans. Martin Chalmers (New York: Verso, 1992), 59–70. In an interesting twist, Klaus Theweleit has speculated that the German Left discarded McLuhan because his focus on bodies and media, extensions, narcosis, and self-amputation was *more* materialist than Marxism had ever been. See Theweleit, *Buch der Könige 1: Orpheus und Eurydike* (Frankfurt: Stroemfeld, 1988), 383.

10. Enzensberger, "Constituents," 97.

11. Jean Baudrillard, "Requiem for the Media," in *For a Critique of the Political Economy of the Sign*, trans. Charles Levin (St. Louis: Telos, 1981), 168.

12. Ibid., 169.

13. Ibid., 173.

14. Ibid., 175.

15. The compound term *Mediendiskursanalyse* (the basis for our expression "media discourse analysis") is occasionally used in German scholarship. Norbert Bolz may have been the first to combine its constituent parts when he outlined a program for a future "Diskursanalyse für neue Medien." See Bolz, *Philosophie nach ihrem Ende* (Munich: Boer, 1992), 172, and idem, "Computer als Medium," in *Computer als Medium*, ed. Bolz, Kittler, and Christoph Tholen (Munich: Fink, 1994), 15.

16. Klaus Laermann, "Lacancan und Derridada: Über die Frankolatrie in den Kulturwissenschaften," *Kursbuch* 84 (1986): 36, 38, 41.

17. Needless to say, the story can be told neither impartially nor in its entirety: it is still going on and continuing divisions, spurred by the arrival of new approaches such as systems theory and radical constructivism, make it difficult, if not impossible, to find terms neutral enough to satisfy all parties involved. The following brief account only considers the politico-theoretical framework of the last three decades, although there are, of course, larger perspectives on postructuralism's hampered reception. Hans Ulrich Gumbrecht, for one, has argued that analytical procedures such as Derridean deconstruction, with its (potentially antihistorical) bias toward spatialization, did not sit easily with the traditional German bias in favor of temporalization; this may also explain why Freudian psychoanalysis fared better in France than it did in Germany or Austria. See Gumbrecht, "Who Is Afraid of Deconstruction?" in *Diskurstheorien und Literaturwissenschaft*, eds. Jürgen Fohrmann and Harro Müller (Frankfurt a.M.: Suhrkamp, 1988), 95–113.

18. Holub, *Crossing Borders: Reception Theory, Poststructuralism, Deconstruction* (Madison: Univ. of Wisconsin Press, 1992): 43.

19. Vincent Descombes, *Modern French Philosophy*, trans. L. Scott-Fox and J. M. Harding (Cambridge: Cambridge Univ. Press, 1980), 172.

20. Matthew Griffin and Susanne Herrmann, "Interview mit Friedrich A. Kittler," *Weimarer Beiträge* 43.2 (1997): 286. Griffin and Herrmann miss the point by translating the phrase "schwarz in jeder Bedeutung" as "virtually blacklisted" in the English version of their interview ("Technologies of Writing: Interview with Friedrich Kittler," *New Literary History* 27.4 [1996]: 741). Also see

Kittler's harsh critique of the venerable *Dialectic of Enlightenment*, coauthored by the technologically ignorant "*Fabrikantensöhne*" (manufacturers' sons) Horkheimer and Adorno: "Copyright 1944 by Social Studies Association, Inc.," in *Flaschenpost und Postkarte: Korrespondenzen zwischen Kritischer Theorie und Poststrukturalismus*, ed. Sigrid Weigel (Cologne: Böhlau, 1995), 185–93. For a brief but informed introduction to the German theoretical debates of the 1980s, see Peter Uwe Hohendahl, *Reappraisals: Shifting Alignments in Postwar Critical Theory* (Ithaca, N.Y.: Cornell Univ. Press, 1991), 187–97.

21. In 1978 the volume *Dichtung als Sozialisationsspiel* appeared, which contained Kaiser's "hermeneutical-dialectic" interpretation of novellas by Gottfried Keller, Kittler's "discourse-analytical" reading of Goethe's *Wilhelm Meister*, and a beautifully tortured preface trying to tie the two essays together. What they ultimately end up sharing is a common enemy, the "Marxist theory of the reflection of social conditions and processes in the literary work, as well as . . . the neo-Marxist aesthetics of Adorno" (Gerhard Kaiser and Friedrich A. Kittler, *Dichtung als Sozialisationsspiel: Studien zu Goethe und Gottfried Keller* [Göttingen: Vandenhoeck & Ruprecht, 1978], 9).

22. Manfred Frank, *What Is Neostructuralism?* trans. Sabine Wilke and Richard Gray (Minneapolis: Univ. of Minnesota Press, 1989).

23. See Hörisch, *Die Wut des Verstehens: Zur Kritik der Hermeneutik* (Frankfurt a.M.: Suhrkamp, 1988), 50–66.

24. *Urszenen: Literaturwissenschaft als Diskursanalyse und Diskurskritik*, ed. Friedrich A. Kittler and Horst Turk (Frankfurt a.M.: Suhrkamp, 1977). Containing essays by Kittler, Bolz, and Hörisch, this collection marks the beginning of French-inspired German literary scholarship.

25. See Norbert Bolz, ed., *Goethes 'Wahlverwandtschaften': Kritische Modelle und Diskursanalysen zum Mythos Literatur* (Hildesheim: Gerstenberg, 1981); Jochen Hörisch, *Die andere Goethezeit: Poetische Mobilmachung des Subjekts um 1800* (Munich: Fink, 1992); and Kittler, *Dichter Mutter Kind* (Munich: Fink, 1991). A well-known early example is the Lacanian reading of Kleist by Helga Gallas, *Das Textbegehren des 'Michael Kohlhaas': Die Sprache des Unbewußten und der Sinn der Literatur* (Reinbek: Rowohlt, 1981).

26. Frank, *What Is Neostructuralism?* 313.

27. See Jacques Lacan, *The Seminar of Jacques Lacan. Book II: The Ego in Freud's Theory and in the Techniques of Psychoanalysis 1954–1955*, ed. Jacques-Alain Miller, trans. Sylvana Tomaselli (New York: Norton, 1988), 46.

28. Kittler, "The World of the Symbolic—A World of the Machine," in idem, *Literature, Media, Information Systems: Essays*, ed. and intro. John Johnston, trans. Stefanie Harris (Amsterdam: Overseas Publishers Association, 1997), 134.

29. Ibid., 145, referring to the following passage (Lacan, *Seminar II*, 89): "This discourse of the other is not the discourse of the abstract other, of the other in the dyad, of my correspondent, nor even of my slave, it is the discourse of the circuit in which I am integrated. I am one of its links."

30. George P. Landow, *Hypertext: The Convergence of Contemporary Critical Theory and Technology* (Baltimore: Johns Hopkins Univ. Press, 1992), 34.

31. Gregory Ulmer, *Applied Grammatology: Post(e)-Pedagogy from Jacques Derrida to Joseph Beuys* (Baltimore: Johns Hopkins Univ. Press, 1985), 303.

32. See Eugene F. Provenzo, Jr., "The Electronic Panopticon: Censorship, Control and Indoctrination in a Post-Typographic Culture," in *Literacy Online: The Promise (and Peril) of Reading and Writing with Computers*, ed. Myron C. Tuman (Pittsburgh: Univ. of Pittsburgh Press, 1992), 167–88; and Mark Poster, *The Mode of Information: Poststructuralism and Social Context* (Chicago: Univ. of Chicago Press, 1990), 69–98.

33. Régis Debray, *Media Manifestos: On the Technological Transmission of Cultural Forms*, trans. Eric Rauth (New York: Verso, 1996), 54.

34. Bolz, *Philosophie nach ihrem Ende*, 154.

35. Hans H. Hiebel, "Strukturale Psychoanalyse und Literatur," in *Neue Literaturtheorien*, ed. Klaus-Michael Bogdal (Darmstadt: Westdeutscher Verlag, 1990), 69. See also Kittler, "'Das Phantom unseres Ichs' und die Literaturpsychologie: E. T. A. Hoffmann—Freud—Lacan," in *Urszenen*, 139–66.

36. See Kittler, *Dichter Mutter Kind*, 7–17, and Hörisch's belated preface to *Die andere Goethezeit*, 7–9. See also Kittler's remarks in Griffin and Herrmann, "Technologies of Writing," 741: "When I think of my old literary criticism, the good essays are actually didactic pieces in programming. How did Duke Carl Eugen von Würtemberg [sic] program Friedrich Schiller? I didn't write about Schiller's sentiments or religion, because all I had was a bare-bones model: educators and princes program the novelist for a specific civil function in the state. What you need is a fundamental understanding of concepts such as hardware, programming, automatization, and regulation."

37. Kittler, *Discourse Networks, 1800/1900*, trans. Michael Metteer with Chris Cullens (Stanford, Calif.: Stanford Univ. Press, 1990), 177.

38. All of these essays first appeared in the 1980s, but where possible we have provided an English translation: "Wie man abschafft, wovon man spricht: Der Autor von *Ecce Homo*," in *Literaturmagazin 12: Nietzsche*, ed. Nicolas Born, Jürgen Manthey, and Detlev Schmidt (Reinbek: Rowohlt, 1980), 153–78; "Pink Floyd, 'Brain Damage,'" in *europalyrik 1775 bis heute: Gedichte und Interpretationen*, ed. Klaus Lindemann (Paderborn: Schöningh, 1982), 467–77; "Das Alibi eines Schriftstellers—Peter Handkes *Die Angst des Tormanns beim Elfmeter*," in *Das schnelle Altern der neuesten Literatur*, ed. Jochen Hörisch and Hubert Winkels (Düsseldorf: Claassen, 1985), 60–72; "Über die Kunst, mit Vögeln zu jagen: *The Maltese Falcon* von D. Hammett," in *Phantasie und Deutung: Psychologisches Verstehen von Film und Literatur*, eds. Wolfram Mauser, Ursula Renner, and Walter Schönau (Würzburg: Königshausen & Neumann, 1986), 299–314; "Dracula's Legacy," *Stanford Humanities Review* 1.1 (1989–90): 143–73; "The Mechanized Philosopher," in *Looking After Nietzsche*, ed. Lawrence A. Rickels (Albany: State Univ. of New York Press, 1990), 195–207; "World-Breath: On Wagner's Media Technology," in *Opera Through Other Eyes*, ed. David J. Levin (Stanford, Calif.: Stanford Univ. Press, 1994), 215–35; and "Media and Drugs in Pynchon's Second World War," trans. Michael Wutz and Geoffrey Winthrop-Young, in *Reading Matters: Narrative in the New Media Ecology*, ed. Joseph Tabbi and Michael Wutz (Ithaca, N.Y.: Cornell Univ. Press, 1997), 157–72.

39. Kittler, "Ein Erdbeben in Chili und Preußen," in *Positionen der Literaturwissenschaft: Acht Modellanalysen am Beispiel von Kleists 'Das Erdbeben in Chili,'* ed. David E. Wellbery (Munich: C. H. Beck, 1985), 24.

40. As Michael Giesecke points out in his monumental study of the early print age, media theorists have themselves only recently started to pay full attention to the "neglected difference" between scriptography and typography. Michael Giesecke, *Der Buchdruck in der frühen Neuzeit: Eine historische Fallstudie über die Durchsetzung neuer Informations- und Kommunikationstechnologien* (Frankfurt: Suhrkamp, 1991), 29.

41. See Kittler, "Autorschaft und Liebe," in *Austreibung des Geistes aus den Geisteswissenschaften: Programme des Poststrukturalismus*, ed. Friedrich A. Kittler (Paderborn: Schöningh, 1980), 142–73.

42. Griffin and Herrmann, "Technologies of Writing," 734.

43. See Wellbery, foreword to Kittler, *Discourse Networks*, xix.

44. Kittler, *Discourse Networks*, 369.

45. Kittler, "Benn's Poetry—'A Hit in the Charts': Song Under Conditions of Media Technologies," *SubStance* 61 (1990): 6.

46. Kittler, *Discourse Networks*, 347.

47. Jerome McGann, *Black Riders: The Visible Language of Modernism* (Princeton, N.J.: Princeton Univ. Press, 1993), 74.

48. Gottfried Benn, "Vortrag in Knokke," in *Gesammelte Werke*, ed. Dieter Wellershoff (Wiesbaden: Limes, 1959), 4: 543. See also Kittler, *Discourse Networks*, 177.

49. With this more nuanced account of the relationship of Lacan's registers to media technologies Kittler goes a long way toward meeting the reviewers of *Discourse Networks* who charged him with setting up arbitrary links between the two. See, for example, Thomas Sebastian, "Technology Romanticized: Friedrich Kittler's *Discourse Networks 1800/1900*": "Why the phonograph should have access to the real, while the film only has access to the imaginary is baffling . . . Notes emanating from a phonograph are neither more real nor less imaginary than filmed images on the screen" (*MLN* 105.3 [1990]: 590).

50. Kittler, *Discourse Networks*, 182.

51. Most of the computer-related essays have been translated and collected in Kittler, *Literature, Media, Information Systems*. Regarding military technology and history, and related issues, see Kittler, "Die künstliche Intelligenz des Weltkriegs: Alan Turing," in *Arsenale der Seele: Literatur- und Medienanalysen seit 1870*, ed. Friedrich A. Kittler and Georg Christoph Tholen (Munich: Fink, 1989), 187–202; "Unconditional Surrender," in *Materialities of Communication*, ed. Hans Ulrich Gumbrecht and K. Ludwig Pfeiffer, trans. William Whobrey (Stanford, Calif.: Stanford Univ. Press, 1994), 319–34; "Eine Kurzgeschichte des Scheinwerfers," in *Der Entzug der Bilder: Visuelle Realitäten*, ed. Michael Wetzel and Herta Wolf (Munich: Fink, 1994), 183–89; and "Il fiore delle truppe scelte," in *Der Dichter als Kommandant: D'Annunzio erobert Fiume*, ed. Hans Ulrich Gumbrecht, Friedrich Kittler, and Bernhard Siegert (Munich: Fink, 1996), 205–25. This last advertises a forthcoming essay on Wotan and the Wagnerian prehistory of the German Storm Trooper (214n).

52. Jörg Lau, "Medien verstehen: Drei Abschweifungen," *Merkur* 534/535 (1993), 836.

53. Holub, *Crossing Borders*, 103.

54. Ibid., 104.

55. Jacques Derrida, "The Question of Style," in *The New Nietzsche*, ed. David Allison (New York: Dell, 1977), 176.

56. Witness, for instance, Kittler's take on Habermas's theory of the origin of the enlightened public sphere: "This enlightenment ideology did not have its origin in the Enlightenment but is primarily the work of Jürgen Habermas, who, as is well known, wrote *The Structural Transformation of the Public Sphere*, the book about the topic. First of all, something has to be said about this book. He claims that the private postal system, which was introduced at the time, set the whole process in motion. What Habermas completely forgets over all those loving, intimate letter-writing people, who he thinks are so great because in them the bourgeois mentality is said to have constituted itself, is quite simply that states, as good mercantilist states, founded this postal system with a clear object in mind: they wanted to skim off the postal rates. For instance, 40 percent of Prussia's successful Seven Years War against Austria was financed by postal revenue. So much for the function of enlightenment or participation in the eighteenth century." Kittler, "Das Internet ist eine Emanation: Ein Gespräch mit Friedrich Kittler," in *Stadt am Netz: Ansichten von Telepolis*, ed. Stefan Iglhaut, Armin Medosch, and Florian Rötzer (Mannheim: Bollmann, 1996), 201.

57. Griffin and Herrmann, "Technologies of Writing," 735.

58. Linda Dietrick, "Review of *Discourse Networks 1800/1900*," *Seminar* 28.1 (1992), 66. Virgina L. Lewis, among others, also observes that "Kittler's thesis that a single unified discourse network fully characterizes each of the two epochs he discusses is hardly acceptable" ("A German Poststructuralist," *PLL* 28.1 [1992], 106).

59. Timothy Lenoir, "Inscription Practices and Materialities of Communication," in *Inscribing Science: Scientific Texts and the Materiality of Communication*, ed. Lenoir (Stanford, Calif.: Stanford Univ. Press, 1998), 15.

60. Kittler, "Laterna magica," 220.

61. Walter Benjamin, "The Work of Art in the Age of Mechanical Reproduction," in *Illuminations*, ed. Hannah Arendt, trans. Harry Zohn (New York: Schocken, 1968), 237.

62. Kittler, "Geschichte der Kommunikationsmedien," in *Raum und Verfahren: Interventionen* 2, ed. Jörg Huber and Alois Martin Müller (Basel and Frankfurt: Stroemfeld / Roter Stern, 1993), 188.

63. "The robot historian of course would hardly be bothered by the fact that it was a human who put the first motor together: for the roles of humans would be seen as little more than that of industrious insects pollinating an independent species of machine-flowers that simply did not possess its own reproductive organs during a segment of its evolution. Similarly, when this robot historian turned its attention to the evolution of armies in order to trace the history of its own weaponry, it would see humans as no more than pieces of a larger military-industrial machine: a war machine." Manuel De Landa, *War in the Age of Intelligent Machines* (New York: Zone Books, 1991), 3.

64. James Beniger, *The Control Revolution: Technological and Economic Origins of the Information Society* (Cambridge, Mass.: Harvard Univ. Press, 1986), vii.

65. Jochen Schulte-Sasse, "Von der schriftlichen zur elektronischen Kultur:

Über neuere Wechselbeziehungen zwischen Mediengeschichte und Kulturgeschichte," in *Materialität der Kommunikation*, eds. Hans Ulrich Gumbrecht and K. Ludwig Pfeiffer (Frankfurt: Suhrkamp, 1988), esp. 429–33.

66. For a detailed study of the literary imagery surrounding and glorifying the German engineer, see Harro Segeberg, *Literarische Technik-Bilder: Studien zum Verhältnis von Technik- und Literaturgeschichte im 19. und frühen 20. Jahrhundert* (Tübingen: Niemeyer, 1987).

67. For related texts, see Anton Kaes, Martin Jay, and Edward Dimendberg, eds., *The Weimar Republic Sourcebook* (Berkeley and Los Angeles: Univ. of California Press, 1994), 393–411.

68. Jeffrey Herf, *Reactionary Modernism: Technology, Culture, and Politics in Weimar and the Third Reich* (Cambridge: Cambridge Univ. Press, 1984), 3.

69. For a critical assessment of these influences see Richard Wolin, *Labyrinths: Explorations in the Critical History of Ideas* (Amherst: Univ. of Massachusetts Press, 1995).

PREFACE

1. Benn, April 10, 1941, in Benn 1941 / 1977–80, 1: 267.

2. Concerning the precision of Benn's "Take stock of the situation!" see Schnur 1980, which makes clear that the poetic maxim immediately following—"Reckon with your defects, start with your holdings, not with your slogans" (Benn 1959–61, 2: 232)—simply rewrites Germany's logistical problems with distributing raw materials during the war.

3. See Schwendter 1982.

4. See Lorenz 1985, 19.

5. Heidegger 1950, 272.

6. Hitler, January 1945, in Schramm 1982, 4: 1652. See also Hitler, May 30, 1942, in Picker 1976, 491, where the fragment from Heraclitus appears as the eternally true and "profoundly serious statement of a military philosopher." But as Jünger (1926/1993, 128) observed, world wars, rather than continuing to fight in the "prevailing mode," depend on innovation as such.

7. See Pynchon 1973, 606.

INTRODUCTION

1. Under the title "*Nostris ex ossibus*: Thoughts of an Optimist," Karl Haushofer, "the main representative, . . . though not the originator, of the term 'geopolitics'" (November 2, 1945, in Haushofer 1979, 2: 639), wrote: "After the war, the Americans will appropriate a relatively wide strip of Europe's western and southern coast and, at the same time, in some shape or fashion annex England, thus realizing the ideal of Cecil Rhodes from the opposite coast. In doing so, they will act in accordance with the age-old ambition of any sea power to gain control of the opposite coast(s) and rule the ocean in between. The opposite coast

is at least the entire eastern rim of the Atlantic and, in order to achieve domination over all 'seven seas,' possibly the entire western rim of the Pacific. Thus, America wants to connect the outer crescent to the 'axis'" (October 19, 1944, in Haushofer 1979, 2: 635)

2. W. Hoffmann, 1944, in Hay 1975b, 374.
3. Bolz 1986, 34.
4. Abraham and Hornbostel 1904, 229.
5. See Campe 1986, 70–71.
6. Foucault 1963/1977, 66.
7. Goethe 1829/1981, 122.
8. Goethe, "Geschichte der Farbenlehre" (1810), in idem 1976, 14: 47. [The oral nature of this "opposite" to written history is underscored by the use of Goethe's word *Sage*, "legend," which derives from *sagen*, "to say."—Trans.]
9. See Ong 1982, 27 and (more reasonably) 3.
10. See Exodus 24:12–34:28.
11. Koran, sura 96, vv. 1–6.
12. Winter 1959, 6.
13. See Assmann and Assmann 1983, 68.
14. Nietzsche, "Geschichte der griechischen Literatur" (1874), in idem 1922–29, 5: 213.
15. Goethe 1811–14 / 1969, 3: 59.
16. Strauss 1977/1979, 15–16.
17. Hegel 1807/1977, 190.
18. Hardenberg (Novalis), 1798–99 / 1960–75, 3: 377.
19. Schlegel 1799/1958ff, 8: 42.
20. See Kittler 1985/1990, 108–23.
21. Goethe 1797/1987, 3. For reasons why a fully alphabetized literature in particular simulated orality, see Schlaffer 1986, 7–20.
22. Goethe, *Werther* (1774), in Goethe 1990, 109.
23. Benjamin 1924–25 / 1972–85, 1: 1, 200.
24. Goethe, *Elective Affinities* (1809), in idem 1990, 342.
25. Brentano 1835/1959–63, 2: 222.
26. Marker 1983, 23–24.
27. See Deleuze 1965, 32. "The alternative is between two purities; the false and the true; that of responsibility and that of innocence; that of memory and that of forgetting. . . . Either one remembers words but their meaning remains obscure, or one apprehends the meaning, in which case the memory of the words disappears."
28. Leroi-Gourhan, quoted in Derrida 1967/1976, 333n.
29. E. T. A. Hoffmann 1816/1969, 148 (translation modified).
30. Nadar 1899/1978, 9.
31. Arnheim 1933/1977, 27.
32. See Lacan 1978/1988b, 278.
33. Edison, 1878, quoted in Gelatt 1977, 29. Phonographic recordings of last words are based on the recognition that "physiological time is not reversible," and that "in the province of rhythm, and of time in general, there is no symmetry" (Mach 1886/1914, 256).

34. See Joyce 1922/1969, 113. See also Brooks 1977, 213–14. ["AEG" refers to the Allgemeine Elektrizitäts-Gesellschaft, one of the leading German electronics corporations. It was originally founded in 1883 by Emil Rathenau as the German Edison Society for Applied Electricity.—Trans.]

35. Rathenau 1918–29, 4: 347. Two examples of *déformation professionelle* among the dead of Necropolis: "A writer is dissatisfied with his epitaph. An employee of the telephone company uses short and long intervals, a kind of Morse alphabet, to ring in a critique of his sucessor." King Alexander, the hero of Bronnen's *Ostpolzug*, says everything there is to say about telephonitis and Hades while, according to the stage directions, the "telephone is buzzing": "Oh, you black beast growing on fatty brown stems, you flower of untimeliness, you rabbit of dark rooms! Your voice is our hereafter, and it has crowded out heaven" (Bronnen 1926/1977, 133).

36. The song "Example #22" actually combines the announcement and sound of "example no. 22" ("Hier spricht Edgar" / "Edgar speaking" [Schäfer 1983, 11]), which, strangely enough, must have migrated on a paranormal cassette-to-book from Freiburg to the United States.

37. See Lacan 1966/1977, 184.

38. Schäfer 1983, 2.

39. Ibid., 3.

40. See Gordon 1981, *passim*.

41. Watson 1978, 26, 410.

42. See Walze 1980, 133.

43. See Luhmann 1985, 20–22.

44. Heidegger 1942–43/1992, 86.

45. Keller 1865/1974: 41.

46. See Mallarmé 1893/1945, 850.

47. Lacan 1966, 720.

48. Lacan 1978/1988b: 47.

49. See Lacan 1966/1977, 1–7.

50. See Lacan 1975, 53, 73.

51. See Lacan 1978/1988b: 191–205.

52. Nietzsche 1873–76/1990, 110.

53. See Turing 1950, 441–42; Hodges 1983, 415–17.

54. Hodges 1983, 279.

55. Ibid., 30.

56. Ibid., 14.

57. J. Good, September 16, 1948, quoted in ibid., 387.

58. See Zuse, June 19, 1937, in idem 1984, 41: "Decisive thought, 19 June 1937 / Realization that there are elementary operations to which all computing and thinking operations may be reduced. / A primitive type of mechanical brain consists of storage unit, dialing system, and a simple device that can handle conditional chains of two or three links. / With such a form of brain it must be possible to solve all operations of the mind that can be dealt with mechanically, regardless of the time involved. More complex brains are merely a matter of executing those operations faster."

GRAMOPHONE

1. Chew 1967, 2. When Kafka's captured ape delivers his "Report to an Academy," the scene depicting his animal language acquisition quotes both Edison's "Hullo" and his storage technology: On board the ship "there was a cele-

bration of some kind, a gramophone was playing"; the ape drank the schnapps bottle "that had been carelessly left standing" in front of his cage; and, "because I could not help it, because my senses were reeling, [I] called a brief and unmistakable 'Hallo!' breaking into human speech, and with this outburst broke into the human community, and felt its echo: 'Listen, he's talking!' like a caress over the whole of my sweat-drenched body" (Kafka 1917/1948, 162).

2. Three months later (and independently of Edison) the same word appeared in an article on Charles Cros. See Marty 1981, 14.

3. *Scientific American*, 1877, quoted in Read and Welch 1959, 12.

4. Cros 1877/1964, 523–24.

5. Cros 1908/1964, 136; trans. Daniel Katz, in Kittler 1990, 231.

6. See Cros 1964, x.

7. See Derrida 1967/1976, 240.

8. Bruch 1979, 21.

9. See the documents from the *Gründerzeit* in Kaes 1978, 68–69, 104 (the scriptwriter H. H. Ewers on Wagner as "teacher").

10. See Friedheim, 1983, 63: "Wagner is probably the first dramatist to seriously explore the use of scream."

11. Wagner 1882/1986, 101.

12. Wagner, *Das Rheingold* (1854), mm. 11–20.

13. See Wagner 1880/1976, 511–12.

14. Jalowetz 1912, 51.

15. See Rayleigh 1877–78, 1: 7–17.

16. Lévi-Strauss 1964/1969, 23.

17. See Kylstra 1977, 7.

18. See Bruch 1979, 26, and Kylstra 1977, 5.

19. See Stetson 1903.

20. See Marage 1898.

21. See Bruch 1979, 3–4. Ong (1982, 5) even hailed Sweet (1845–1912) as the progenitor of Saussure's phoneme concept.

22. Shaw 1912/1972, 684.

23. Lothar 1924, 48–49.

24. See Shaw 1912/1972, 659–64.

25. For details, see Kittler1985/1990, 27–53.

26. Shaw 1912/1972, 795. [*My Fair Lady* is by Lerner and Loewe.—Trans.]

27. Lothar 1924, 12, and Kylstra 1977, 3, respectively.

28. See Knies 1857, iii.

29. Jarry 1895/1975, 4: 191.

30. Villiers 1886/1982, 19.

31. "Hähnische Litteralmethode" 1783/1986, 156–57.

32. On understanding as a measurable source of noise parallel to hearing, see Gutzmann 1908.

33. Lothar 1924, 51–52.

34. See Gelatt 1977, 27–28.

35. Abraham and Hornbostel 1904, 229.

36. On rock music and secret codes, see Kittler 1984b, 154–55.

37. Gelatt 1977, 52.

38. Hegel 1830/1927–40, 10: 346.
39. Pink Floyd 1976, 10–11.
40. Gelatt 1977, 72.
41. Freud, "Project for a Scientific Psychology," in idem 1895/1962, 1: 381.
42. Ibid., 1: 295.
43. Freud, *Beyond the Pleasure Principle*, in idem 1920/1962, 18: 24.
44. See Derrida 1967/1978, 221–31.
45. Abraham and Hornbostel 1904, 231. Hornbostel's superior, the great music physiologist Carl Stumpf, concluded that it was necessary to establish a phonographic archive in Berlin, as well (which was realized soon thereafter). His criticism of the exclusion of optics led another participant in the discussion to argue that it should be linked to a film archive (ibid., 235–36). See Meumann 1912, 130.
46. Hirth, 1897, 38. Sabina Spielrein proves that psychoanalysts didn't think any differently. According to her, the "treatment of hysteria" consists in "bringing about a transformation of the psychosexual components of the ego (either by way of art or simple reactions—whichever you prefer: in this way the component is progressively weakened like a playing [*sic*] gramophone record)." Spielrein 1906/1986, 224.
47. Rilke 1910/1949, 146.
48. Sachs 1905, 4.
49. See Flechsig 1894, 21–22.
50. See Hamburger 1966, 179–275.
51. Rilke 1910/1949, 185 (translation modified).
52. Lothar 1924, 58.
53. Ibid., 59–60.
54. See Rilke 1925/1957, 339–40.
55. Moholy-Nagy 1923, 104.
56. Ibid.
57. Ibid., 105.
58. Zglinicki 1956, 619.
59. Lothar 1924, 55.
60. Moholy-Nagy 1923, 104.
61. Pynchon 1973, 405.
62. See Andresen 1982, 83–84.
63. See Hodges 1983, 245–46.
64. See ibid., 287–88.
65. Marinetti 1912/1971, 87.
66. See Valéry 1937/1957–60, 1: 886–907.
67. Parzer-Mühlbacher 1902, 107.
68. See Ribot 1882, 114. For agony snapshots, see also Villiers's story "Claire Lenoir" and the commentary in S. Weber 1980, 137–44.
69. See Kafka, January 22–23, 1913, in Kafka 1974, 167–68.
70. Ibid., 166.
71. See G. Neumann 1985, 101–2.
72. See Cocteau 1930/1951, 28.
73. Kafka, January 22–23, 1913, in Kafka 1974, 168.
74. Campe 1986, 69.
75. See Kakfa 1935/1950, 93, and Siegert 1986, 299, 324–25.
76. See Kafka, January 17–18, 1913, in Kafka 1974, 158, and Campe 1986, 86.
77. See Campe 1986, 72.
78. See Lacan 1973/1978, 174ff.
79. See Wetzel 1985.
80. Dahms 1895, 21.
81. M. Weber 1928, 9.
82. Wellershoff 1980, 212–14.
83. Kafka, January 22–23, 1913, in Kafka 1974, 168.
84. See Wagner 1880/1976, 512.

85. Dehmel 1896/1906–09, 3: 115–16.
86. See Kittler 1985/1990, 147–48.
87. See Holst 1802, 63–66.
88. Schlegel 1799/1958, 8: 48, 42.
89. Deleuze and Guattari 1972/1983, 209.
90. E. T. A. Hoffmann 1819/1971, 32.
91. Lothar 1924, 7–8.
92. Düppengiesser 1928, quoted in Hay 1975a, 124–25.
93. Eyth 1909, 1: 457–58.
94. *Scientific American*, 1877, quoted in Read and Welch 1959, 12.
95. See Bredow 1950, 16.
96. Enzensberger 1970/1974, 97.
97. Rilke 1910/1949, 138.
98. Turing 1950, 434. See Hodges 1983, 291.
99. Snyder 1974, 11.
100. Scherer 1986, 49. For the factual history of similarly dismembered bodies, see Seeliger 1985. The major identification problem between 1826 and 1916 and in 1959 appears to have been Schiller's rather than Goethe's skull. Whether or not the corpse in the royal tomb will prove that Goethe used arsenic to poison Schiller, whether it belongs to the poet or to a young woman, whether Goethe used a file to distort its teeth—all that is still unresolved. Reason enough for Professor Pschorr to reenact the 1912 opening of tomb and coffin in 1916.
101. Philipp Siedler, 1962, quoted in Campe 1986, 90.
102. Reis 1861/1952, 37.
103. Bell quoted in Snyder 1974, 14.
104. See Saussure 1915/1959, 17–20.
105. On the algorithms of digital speech recognition, and its input and output in general, see Sickert 1983. The particulars of continuing Pschorr's Goethe experiment are as follows: "Under the Tokyo number 320-3000, a famous dead person is talking about his work. In his own language, the French painter Pierre Auguste Renoir, who died in 1919, is promoting an exhibition of impressionist paintings. Renoir's ghostly voice was captured on tape by scientists of the Japan Acoustic Research Laboratory—with the help of computers. The computer séance is based on electronic voice simulation and anatomical measurements: according to the researchers, various vocal features can be reconstructed from the characteristics of a person's nasopharyngeal cavity. In the case of Renoir, the voice of a French native speaker was gradually modulated according to the characteristics of Renoir's nasopharyngeal cavity. Japanese vocal experts, at least, consider the result to be 'pure Renoir'" (*Der Spiegel* 40, no.1 [1986]: 137). Unlike Pschorr, the Japan Acoustic Research Laboratory has kept silent about the acquisition of Renoir's nasopharyngeal cavity.
106. Foucault 1969/1972, 27.
107. Ibid., 103.
108. Friedlaender 1922, 326.
109. Ibid., 327.
110. Ibid., 326.
111. Ibid.
112. O. Wiener 1900, 23–24.
113. "The New Phonograph" 1887, 422.
114. Gelatt 1977, 100–101.

115. Bruch 1979, 24.

116. See Lerg 1970, 29–34. In the name of all German engineers, Slaby (1911, 369–70) found the exalting words: "At the turn of the century, words of deliverance resounding from the heights of the throne opened the path leading upward to the hallowed peaks of science . . . For whom do our hearts in this hour beat more passionately than for our emperor? He endowed us with *rights and privileges in the world of supreme intellectual life*, he made us a full part of the struggle for the glory of the fatherland, and at its deepest roots he provided the blooming science of engineering with new ideal incentives."

117. For details, see Kittler 1984a, 42. [AVUS: *Autoverkehrs- und Übungsstrasse*, a famous speedway in Berlin.—Trans.]

118. Wildenbruch, 1897, quoted in Bruch 1979, 20.

119. Nietzsche 1882–1887/1974, 138. Hobbes stated more prosaically that "in ancient times, before letters were in common use, the laws were many times put into verse, that the rude people taking pleasure in singing or reciting them might the more easily retain them in memory" (Hobbes 1651/1994, 178).

120. See Mallarmé 1897/1945, 455. This poet's only "innovation" was that for the first time, the empty spaces between words or letters were granted typographic "weight"—typewriter poetics.

121. Jensen 1917, 53.

122. Kracauer 1930/1971–79, 1: 262.

123. Keun 1932/1979b, 194.

124. Ibid., 8.

125. Ibid., 58, 95.

126. Siemsen, 1926, in Kaes, Jay, and Dimendberg 1994, 664.

127. Wilde, 1890, in idem 1966, 1091.

128. Benn 1959–61, 3: 474. For the same in prose, see Benn 1959–61, 1: 518.

129. Zumthor 1985, 368.

130. "The New Phonograph" 1887, 422.

131. Freud, "Fragments of an Analysis of a Case of Hysteria," 1905, in idem 1962, 7: 77–78.

132. Stern 1908, 432.

133. See Watzlawick, Beavin, and Jackson 1967, 54–55.

134. See Stern 1908, 432.

135. See Watzlawick, Beavin, and Jackson 1967, 72n.

136. Stransky 1905, 18.

137. Ibid., 18.

138. Ibid., 4.

139. Ibid., 7.

140. Ibid., 96.

141. Baade 1913, 81–82.

142. For details, see Kittler1982/1989–90, 143–73.

143. Stoker 1897/1965, 79.

144. See Blodgett 1890, 43.

145. Gutzmann 1908, 486–88.

146. Ibid., 499.

147. Freud, "Recommendations to Physicans Practising Psycho-Analysis," 1912, in idem 1962, 12: 115–16. Since the study in the Berggasse was not cabled, the telephony Freud describes must have been wireless: radio avant la lettre. On the analogy between psychic and technological media, see also Freud, *New Introductory Lectures on Psycho-Analysis*, 1933, in idem 1962, 22: 55. "And particularly so far as thought-transference is concerned, it seems actually to favour

the extension of the scientific—or, as our opponents say, the mechanistic—mode of thought to mental phenomena which are so hard to lay hold of. The telepathic process is supposed to consist in a mental act in one person instigating the same mental act in another person. What lies between these two mental acts may easily be a physical process into which the mental one is transformed at one end and which is transformed back once more into the same mental one at the other end. The analogy with other transformations, such as occur in hearing or speaking by telephone, would then be unmistakable."

148. See Campe 1986, 88.

149. Rilke 1910/1955–66, 6: 767.

150. See Stoker 1897/1965, 70, 79.

151. See Freud, *Beyond the Pleasure Principle*, 1920, in idem 1962, 18: 25.

152. Freud, "Fragment of an Analysis of a Case of Hysteria," 1905, in idem 1962, 7: 10. See also Freud, *New Introductory Lectures on Psycho-Analysis*, 1933, in idem 1962, 22: 5, on his writing technique: "My *Introductory Lectures on Psychoanalysis* were delivered during the two Winter Terms of 1915–16 and 1916–17 in a lecture room of the Vienna Psychiatric Clinic before an audience gathered from all the Faculties of the University. The first half of the lectures were improvised, and written out immediately afterwards; drafts of the second half were made during the intervening summer vacation at Salzburg, and delivered word for word in the following winter. At that time I still possessed the gift of a phonographic memory."

153. See Benjamin 1968, 235.

154. See Freud, *Interpretation of Dreams*, 1899, in idem 1962, 4: 277–78.

155. Guattari 1975.

156. Berliner quoted in Bruch 1979, 31.

157. See the endless descriptions of symptoms in Freud, *Studies on Hysteria*, 1895, in idem 1962, 2: 48–79.

158. Ibid., 2: 49–50. Freud was "always vexed" by the "'sound relationships' . . . because here I lack the most elementary knowledge, thanks to the atrophy of my acoustic sensibilities" (Freud, August 31, 1898, in idem 1985, 325).

159. Freud, "An Outline of Psycho-Analysis," 1938, in idem 1962, 23: 196.

160. See Freud, "On Beginning the Treatment," 1913, in idem 1962, 12: 134–35.

161. Freud, "The Handling of Dream-Interpretation in Psycho-Analysis," 1912, in idem 1962, 12: 96.

162. Freud, *Interpretation of Dreams*, 1899, in idem 1962, 4: 278.

163. Abraham 1913, 194.

164. Ibid., 194–95.

165. See Sartre 1969b, 43.

166. Ibid., 46.

167. Sartre 1969a, 1812.

168. Sartre 1969b, 49.

169. Foucault 1976/1990, 150.

170. Faulstich 1979, 193.

171. See Chapple and Garofalo 1977, 1.

172. List, 1939, quoted in Pohle 1955, 339: "Due to newspapers, journals, and radio, the population's leadership vacuum is relatively small. It is about 4 or 5 out of 100. . . . It must therefore be emphasized that with the exception of a relatively small part the population is subject to the will of the political leadership." The logic of world war mobilization.

173. McLuhan 1964, 307.
174. Slaby 1911, VII.
175. Ibid., 333–34.
176. Ibid., 344.
177. See Bronnen 1935, 76. As everywhere in his key novel, Bronnen is extremely well informed.
178. See Chapple and Garofalo 1977, 54.
179. See Briggs 1961, 27.
180. See Lerg 1970, 43.
181. See Blair 1929, 87: "From the earliest time the Army has been a pioneer in the development of radio as a means of communication, and more especially in the development of radio equipment for use by military forces in the field. . . . During the World War there was intensive development along all lines that appeared to make for the success of armies in the field. The armies of all powers involved . . . were quick to recognize its value and to expend funds and energy lavishly in scientific radio research. One of the biggest improvements which resulted was the design of more sensitive amplifiers by using vacuum tube detectors and amplifiers."
182. See Volckheim 1923, 14.
183. See Virilio 1984/1989, 69–71.
184. See Briggs 1961, 38.
185. Wedel 1962, 12.
186. See Lerg 1970, 51.
187. Bredow 1954, 91.
188. Höfle, December, 20, 1923, quoted in Lerg 1970, 188.
189. Bronnen 1935, 21.
190. Ibid., 16.
191. *Sunday Times*, quoted in Gelatt 1977, 234.
192. Villiers 1886/1982, 97.
193. See Gelatt 1977, 234–35.
194. Kafka 1924/1948, 257. For sources, see Bauer-Wabnegg 1986, 179–80.
195. Cocteau 1992, 63–64.
196. Gelatt 1977, 282.
197. von Schramm 1979, 324. For similar, though fictionalized and *post facto*, gramophone simulations of the First World War, see Fussell 1975, 227–30.
198. See Pink Floyd 1975, 77, and Kittler 1984b, 145–46.
199. R. Jones 1978, 76.
200. See Chapple and Garofalo 1977, 53.
201. See Stoker 1897/1965, 318. For details, see Kittler 1982/1989–90, 167–69.
202. The Beatles n.d., 194.
203. See Villiers 1886/1982, 55. Only one experiment, undertaken in 1881, can be considered the source of this scene: "A major development . . . has been the introduction of stereophonic broadcasting. Like many other scientific developments it suddenly became popular after spasmodic attempts dating back to the nineteenth century. As long ago as 1881 arrangements were made at the Paris opera, using ten microphones, to convey the program in stereo by line to an exhibition at the Palace of Industry. This demonstration showed that 'audience perspective' can lend a touch of magic to systems of quite modest performance" (Pawley 1972, 432).
204. Culshaw, 1959, quoted in Gelatt 1977, 318.
205. Wagner 1854/1993, 90.

206. Nietzsche 1873–76/1990, 276.

207. Chapple and Garofalo 1977, 110. VHF's superiority comes at the price of limited transmission range.

208. Wildhagen 1970, 27.

209. Ibid., 31.

210. Nehring quoted in Bradley 1978, 183. See also van Creveld 1985, 192–94: "Thus the credit for recognizing the importance of the question, for the first successful attempts and its solution, and for the first brilliant demonstration of how armoured command ought to operate belongs essentially to two men: Heinz Guderian—himself, not accidentally, an ex-signals officer who entered World War I as a lieutenant in charge of a wireless station—and General Fritz [sic] Fellgiebel, commanding officer, Signals Service, German Wehrmacht during most of the Nazi era. Between them these men developed the principles of radio-command that, in somewhat modified and technically infinitely more complex form, are still very much in use today. . . . The critical importance of command in armoured warfare cannot be exaggerated and is equalled only by the lack of systematic attention paid to it by military historians."

211. Briggs 1965, 362–63. According to Pawley (1972, 387), the Allied capture of the army radio station in Luxembourg yielded only tapes but no equipment. Only following V-day Europe did the BBC receive six magnetophones formerly belonging to the German navy.

212. Gelatt 1977, 286–87.

213. For Germany, see Faulstich 1979, 208, 218, and for Britain, see the technical details in Pawley 1972, 178–93.

214. Pohle 1955, 87.

215. Kolb, 1933, quoted in ibid., 18.

216. Wedel 1962, 116–17. The next sentence states that the propaganda division of the Army High Command also had special "film tanks" at its disposal.

217. Ludendorff 1935, 119.

218. Pynchon 1973, 854.

219. Buchheit 1966, 121.

220. Dallin 1955, 172.

221. See Hodges 1983, 314. Zuse's coworkers also planned to use magnetic tapes for the storage of computer data. See Zuse 1984, 99.

222. See Chapple and Garofalo 1977, 20.

223. Ibid., 94.

224. See Görlitz 1967, 441. "After the German General Staff had been tried as a criminal organization at Nuremberg and acquitted, the Americans began to study the Scharnhorst staff as a model for staff management in business." See also Overbeck 1971, 90–91.

225. See the relay in Factor 1978. Rumor has it that Australian radio stations broadcast without a second's delay.

226. See Scherer 1983, 91. On the origin of the Abbey Road magnetic tapes, see Southall 1982, 137: "There was also one interesting development which proved that out of adversity there sometimes comes the odd bit of good. In 1946, a team of audio engineers from America and England, including Abbey Road's Berth Jones, visited Berlin to study the developments in magnetic recording which had taken place in Germany during the war. They found amongst the military equipment that had been captured, a system of monitoring using magnetic tapes

which the German command had used in an effort to break codes. The information gathered from this equipment enabled EMI to manufacture tape and tape recorders, resulting in the production of the famous BTR series which remained in use at Abbey Road for over 25 years." Ironically, the acronym BTR stood for British Tape Recorders. And the Beatles encoded secret messages using machines the Army High Command had developed for decoding secret transmissions.

227. Gilmour, in Pink Floyd 1975, 115.

228. See Gilmour, in ibid., 119.

229. See Burroughs 1974, 200–202.

230. Ibid., 11.

231. Ibid., 12.

232. Ibid., 13.

233. Ibid., 14.

234. Ibid., 202. See also Morrison 1976, 16: "All games contain the idea of death."

235. On reception and interception, see p. 251.

236. Burroughs 1974, 15.

237. On scramblers as army equipment, see ibid., 176–80.

238. Pynchon 1973, 267–68.

239. Burroughs 1974, 202.

240. See Leduc 1973, 33.

241. See Burroughs 1974, 180.

242. See Benjamin 1968, 239–41.

243. Pohle 1955, 297. Virilio (1984/ 1989, 66–67) emphasizes the role played by senior military personnel in rock management.

244. Pink Floyd 1983, side A.

245. Hardenberg (Novalis) 1798/1960–75, 2: 662.

246. Rolling Stones 1969, 4.

247. Hendrix 1968, 52.

FILM

1. See Toeplitz 1973, 22–23.

2. See Zglinicki 1956, 472.

3. See MacDonnell 1973, 11.

4. See ibid., 21–26.

5. H. Münsterberg 1916/1970, 1. Bloem's chapter on "tricks" begins with similar questions: "On what star was cinematic man born? On a magic constellation where the laws of nature are suspended? Where time is at rest or goes backward, where set tables grow out of the ground? Where the desire to glide through the air or to sink into the ground is enough to do it?" (Bloem 1922, 53).

6. Foucault 1969/1972, 166. See Lorenz 1985, 12 (including the Sartre-Foucault polemic).

7. See Lorenz 1985, 252–92.

8. See Rabiner and Gold 1975, 438.

9. Bischoff 1928/1950, 263.

10. Klippert 1977, 40.

11. Kittler is referring playfully to the verbal similarity between the German *Herz* (heart) and *hertz*, the international unit of frequency equal to one cycle per second (named after Heinrich Rudolf Hertz, the German physicist who first produced radio waves artificially).—Trans.

12. Zglinicki 1956, 108.

13. Goethe 1829/1979, 125.

14. See Benn 1949/1959–61: 2: 176.

15. See above, p. 000.

16. Nietzsche 1872/1956, 59. On the difference between negative (complementary) and positive afterimages, see H. Münsterberg 1916/1970, 25.

17. For an early reference to sensory deprivation in the acoustic realm, see Groos 1899, 25, which ostensibly goes back to Wilhelm Preyer's theories of 1877: "The province of radio play far exceeds that which is acceptable to our senses—a fact to which we've been referring in our discussion of other sensory domains. We are missing something when we don't hear anything; the awkward feeling upon exposure to continuous silence has even led to the thought of assuming a quality of sensation particular to silence, corresponding to the positive perception of blackness in the realm of vision."

18. Nietzsche 1872/1956, 42.

19. Ibid., 59.

20. See Wieszner 1951, 115: "Bayreuth aspired to the darkened room. That, too, was at that time a surprising element of production. 'The whole theater was made into a dark night, so that one could not recognize one's neighbor,' writes Richard Wagner's nephew, Clemens Brockhaus, on the occasion of the Kaiser's visit to Bayreuth in 1876, 'and the wonderful orchestra began to play in the pit.'"

21. Following Altenloh, 1914, as quoted in Vietta 1975, 294.

22. See Kittler 1993, 232.

23. Pretzsch 1934, 146.

24. Morin 1956, 139. See also Morrison 1977, 94.

25. See H. Münsterberg 1916/1970, 2.

26. See Nadar 1899, 246–63.

27. See Mitry 1976, 59–60.

28. See Nadar 1899, 37–42.

29. Virilio 1984/1989, 11.

30. See Mitry 1976, 64, and Nadar 1899, 260.

31. Pynchon 1973, 405.

32. Virilio 1984/1989, 11; see also 68. On Janssen, see Arnheim 1933/1977, 37–40.

33. See Ellis 1975. For example, 11,000 Dervish, 28 British, and 20 other soldiers were killed during the 1898 battle of Omdurman, in which Lord Kitchener had used six Maxim guns (ibid., 87.) In the imperial poetry of Hilaire Belloc: "'Whatever happens, we have got / The Maxim Gun, and they have not'" (quoted in ibid., 94). From all this Jünger has the following to say about the films of Chaplin (1932, 129): "In essence, they contain a rediscovery of laughter as a mark of terrible and primitive hostility [against obsolete individuality], and the screenings, which take place in centers of civilization, inside safe, warm, and well-lighted rooms, may well be compared to skirmishes in which tribes equipped with bows and arrows are fired upon with machine guns."

34. See Jünger 1932, 104–5.

35. See Virilio 1984/1989, 18. See also Morrison 1977, 22: "The sniper's rifle is an extension of his eye. He kills with injurious vision."

36. Wedel 1962, 116. On air reconnaissance and von Fritsch's oracular utterance, see Babington Smith 1958, 251–52.

37. Kittler is alluding to the Battle of Tannenberg (August 26–30, 1914), at which German forces under Hindenburg and Ludendorff defeated the Russian army.—Trans.

38. Pynchon 1973, 474. See also p. 660: "Three hundred years ago mathematicians were learning to break the cannonball's rise and fall into stairsteps of range and height, Δx and Δy, allowing them to grow smaller, approaching zero. . . . This analytic legacy has been handed down intact—it brought the technicians at Peenemünde to peer at the Askania films of Rocket flights, frame by frame, Δx by Δy, flightless themselves . . . film and calculus, both pornographies of flight."

39. See Hahn 1963, 11.

40. For specifics, see Kittler 1986/1997, 157–72.

41. Pynchon 1973, 887.

42. See M. Münsterberg 1922.

43. See H. Münsterberg 1916/1970, 10.

44. Bloem 1922, 86.

45. Toeplitz 1973, 139.

46. Ludendorff, 1917, quoted in Zglinicki 1956, 394. See also Görlitz 1967, 194, as well as Jünger, 1926/1993, 200–201. As a troop officer, Jünger as always was responsible for disseminating orders coming from the Army High Command: Film "would be an admirable means of enhancing the modern battlefields. To turn away from this theme or to veil it is already a sign of inward weakness. Gigantic films with a wealth of resources spent upon them, and shown nightly to millions during the only hours of the day they can call their own, would have an invaluable influence. Moral and aesthetic compunction has no place here. The film is a problem of power and to be valued as such. The direct interests of the state are involved, and they necessarily go far beyond a negative interference by censorship."

47. Jünger 1922, 45.

48. Ibid., 92.

49. Ibid., 12.

50. Ibid., 20.

51. Ibid., 18.

52. Ibid., 23.

53. Ibid., 19.

54. Ibid., 18.

55. Ibid., 50. On Anglo-American literature describing the First World War as film, see Fussell 1975, 220–21.

56. Jünger 1926/1929, 280.

57. Ibid., 29.

58. Ibid., 30.

59. Jünger 1922, 19.

60. Pinthus 1914/1963, 23.

61. Ibid., 22.

62. See Virilio 1984/1989, 14, 70.

63. See van Creveld 1985, 168–84.

64. Jünger 1926/1929, 134–36.

65. Jünger 1922, 109.

66. Ibid., 26.

67. Ibid., 107.

68. Ibid., 108.

69. See Theweleit 1977–78/1987–89, 2: 176–206.

70. Jünger 1922, 8.

71. Ibid., 18.

72. Pynchon 1973, 614.

73. Bahnemann 1971, 164.

74. Maréchal 1891, 407. See also W. Hoffmann 1932/33, 456. "As speech travels through the microphone it becomes more acute, while its living immediacy dwindles. What do we mean by acuteness? If one studies the technological reproduction of speech in film, for example, the muscular action of the mouth is presented much more acutely than we would observe in real life."

75. Mitry 1976, 76.

76. Maréchal 1891, 407.

77. Demeny 1899, 348.

78. See Demeny 1904, as well as Virilio 1984/1989, 69. However, as early as 1900 Bergson illustrated in his lectures the "cinematographic illusion of consciousness" through the examples of the goose step and the cinematic rendition of a defiling regiment. See Bergson 1907/1911, 329–31.

79. Hirth 1897, 364–65.

80. Jünger is playing on the German *Uhrziffern*, which literally means "watch ciphers" and further implies that these ciphers are classified.—Trans.

81. Jünger 1922, 101.

82. See Fussell 1975, 315. The "watch, worn . . . on the inside of his wrist," by contrast, is "WW II style" (Pynchon 1973, 141).

83. A reference to the 1985 song "Tanz den Hitler / Tanz den Mussolini" by the German punk rock group DAF (Deutsch-Amerikanische Freundschaft [German-American friendship]).—Trans.

84. Foucault 1976/1990, 56.

85. See Cagnetta 1981, 39.

86. See Farges 1975, 89.

87. Freud, *Studies on Hysteria*, 1895, in idem 1962, 2: 280.

88. Ibid.

89. Rank 1914/1971, 3–4.

90. Ibid., 3.

91. See Kittler 1985, 129.

92. See Lichtbild-Bühne, 1926, quoted in Greve 1976, 326.

93. See Schneider 1985, 89–94.

94. See Urban 1978, 30–38.

95. Hennes 1909, 2013.

96. Ibid., 2014.

97. Ibid., 2012.

98. Ibid., 2013.

99. Ibid., 2010.

100. Ibid., 2014.

101. See Kaes 1979, 94.

102. Kracauer 1947/1974, 67.

103. Ibid., 62.

104. See Bourneville and Régnard 1877–78, 2: 208–26.

105. See Clément 1975.

106. Freud, October 15, 1897, in idem 1985, 272.

107. See Jentsch 1906, 198.

108. Bloem 1922, 57. On the use of puppets in film, see also H. Münsterberg 1916/1970, 15.

109. Hennes 1909, 2011–12.

110. See Lacan 1973/1978, 113: "Let us go to the great hall of the Doges' Palace in which are painted all kinds of battles, such as the battle of Lepanto, etc. The social function, which was already emerging at the religious level [of icons], is now becoming clear. Who comes here? Those who form what Retz calls '*les peuples*,' the audiences. And what do the audiences see in these vast compositions? They see the gaze of those persons who, when the audiences are not there, deliberate in this hall. Behind their picture, it is their gaze that is there."

111. Cocteau 1992, 39.

112. Bronnen 1927, 139–40.

113. Nabokov 1926/1970, 40.

114. See Lacan 1973/1978, 209–13.

115. Pynchon 1973, 157.

116. Bronnen 1927, 35. Elsewhere Fitzmaurice remarks: "Some people support themselves from the writing of books, namely, reviewers. By the way, I once saw a man who read a book; I will never forget that impression" (ibid., 196).

117. Kittler is building on both the German word for literacy, *Alphabetismus*, and Lacan's *Alphabêtise*, a composite of the French *alphabétisme* (literacy) and *la bêtise* (stupidity). The use of language, in effect, is a culturally instilled and mechanical exercise that perpetuates the illusion of an existent (fully present or coherent) soul.—Trans.

118. Büchner 1842/1963, 129.

119. See Kittler 1985, 118–24.

120. Cendrars 1926/1970, 207.

121. See Mallarmé 1945, 880. "It is not a matter of distorting but of inventing. The coach, with its team of horses, requires the inconvenience of a driver blocking the view: it is left to him, like an oven in front of a cook. Something entirely different will have to come about. A bow window, opening onto the space that one moves through magically, with nothing in front: the mechanic is placed in the rear, with his upper body above the roof, to steer like a helmsman. Thus the monster advances in an innovative fashion. This is the vision of a passing man of taste, putting things back in perspective." On cars and moving cameras in general, see Virilio 1976, 251–57.

122. Schreber 1903/1973, 161.

123. Mach 1886/1914, 4n [trans. modified], from which follows: "The ego is as little absolutely permanent as are bodies."

124. Freud, "The Uncanny" (1919), in idem 1962, 17: 248. Hauptmann's poem "Im Nachtzug" (In the Night train) casts such railroad doppelgängers in verse (Hauptmann 1888/1962–74, 4: 54).

125. Todorov 1970/1973, 161.

126. Ibid., 168.

127. Ibid., 160.

128. Behne, 1926, in Kaes 1983, 220. For a similar comment, see Bloem 1922, 51.

129. Méliès quoted in Toeplitz 1973, 26.

130. Ewers, October 8, 1912, quoted in Zglinicki, 1956: 375.

131. See A. M. Meyer, 1913, quoted in Greve et al. 1976, 111: "It was a very real opening night. Many smokings. In the guest gallery one caught occasional glimpses of poets and their pretty female company. . . . Goethe, Chamisso, E. Th. A. Hoffmann, Alfred de Musset, Oscar Wilde were present as well. Namely as godfathers of this 2,000-Mark film."

132. Haas, 1922, on Hauptmann's film *Phantom* and in reference to *The Student of Prague*, quoted in Greve et al. 1976, 110.

133. Bronnen 1927, 144.

134. Ewers quoted in Greve et al. 1976, 110.

135. Bloem 1922, 56.

136. *Der Kinematograph*, advertisement (1929), reproduced in Greve et al. 1976, 127.

137. Benn, August 29, 1935, in idem 1977–80, 1: 63. Lindau, by the way, was among the reading matter of Freud's youth.

138. See the facsimile in Greve et al. 1976, 108.

139. Lindau 1906, 26.

140. Ibid., 8.

141. See the analogous passage in Valéry 1944/1957–60, 2: 282–86.
142. Schreber 1903/1973, 86; see also Kittler 1985/1990, 296–304.
143. Lindau 1906, 76.
144. Ibid., 19.
145. Ibid., 21.
146. Schreber 1903/1973, 95; see also 208–10.
147. Lindau 1906, 58.
148. Ibid., 34–35; see also 57.
149. Ibid., 27.
150. Ibid., 83.
151. Ibid., 47.
152. For the precise phrases in Lindau 1906, 26–27, see Azam 1893, and Wagner 1882/1986, 851–52, 854–55.
153. Lindau 1906, 22.
154. Bergson 1907/1911, 306; see also 2–3.
155. Ibid., 331–32.
156. H. Münsterberg 1916/1970, 26.
157. Ibid., 30.
158. See ibid., 22, and, on the Münsterberg-Lindsay connection, Monaco 1977, 298–301.
159. H. Münsterberg 1916/1970, 74.
160. Ibid.
161. Ibid., 31.
162. Ibid., 36.
163. Ibid., 37–38.
164. Ibid., 40.
165. Ibid., 41.
166. Ibid., 44.
167. Balázs 1930, 51.
168. Specht (1922, 212–13) calls *Leutnant Gustl*'s inner monologue "fabulous, stupendous, almost uncanny in its truth and power as well as in the vision of a writer who appears to be able to unlock the secret of every human soul"—simply because in this monologue, "the film of words and the phonography of the soul are one and the same."
169. Meyrink 1915/1928, 22.
170. Ibid., 5–7.
171. Balázs 1930, 120. Meyrink, by contrast, knows that essences made of associations result not from the spirit but from brain functions. Pernath, the protagonist of the framed narrative that evolves as a flow of association or doppelgänger of the framing I, follows all the patterns of psychophysics when he recognizes himself in a "cat gone mad" (*eine Katze mit verletzter Gehirnhälfte*; 49), and he follows all the patterns of aphasia research when he recognizes the Golem (the exponential doppelgänger) precisely in that brain injury: "all these problems had suddenly achieved their terrible solution: I had been mad, and treated by *hypnosis*. They had locked up a room [the Golem's] that communicated with certain chambers in my brain; they had made me into an exile in the midst of the life that surrounded me" (Meyrink 1915/1928, 51; see also 19, 21–22, 25–26). [*Eine Katze mit verletzter Gehirnhälfte*: lit., "a cat with an injured brain hemisphere."—Trans.]
172. H. Münsterberg 1916/1970, 15.
173. Hardenberg (Novalis) 1802/1964, 91.
174. Ibid., 90–91.
175. Ibid., 17.

176. H. Münsterberg 1916/1970, 15.

177. Freud, *Interpretation of Dreams* (1899), in idem 1962, 5: 536.

178. Lacan 1975/1988a, 123.

179. Ibid., 140, 125.

180. See Zglinicki 1956, 338.

181. See ibid., 43–44. The parallel between Messter and Lacan was discovered by Lorenz (1985, 209–11).

182. Lacan 1975, 76.

183. See Lacan 1975/1988a: 140–41.

184. See Lacan 1966, 680.

185. Edison, quoted in Monaco 1977, 56.

186. Pinthus 1914/1963, 9.

187. Ibid., 9–10.

188. Bloem 1922, 36.

189. H. Münsterberg 1916/1970, 87–88.

190. Balázs 1930, 142.

191. Pamphlet (1929), reproduced in Greve et al. 1976, 387.

192. Bloem 1922, 25.

193. A. Braun, "Hörspiel" (1929), in Bredow 1950, 149.

194. Bronnen 1927, 48.

195. Ibid., 109.

196. Ibid., 116.

197. Ibid., 130–31. Similar couplings linked gramophone and typewriter. Kracauer (1930/1971–79, 1: 228) describes an industrial training course using gramophone rhythms to make stenographers achieve record performances.

198. Hesse 1927/1957, 210–13.

199. Mann, 1928/1933: 266.

200. For details, see Fischer and Kittler 1978, 29–37.

201. See Mann 1924/1927, 108.

202. Ibid., 801.

203. See ibid., 854–57.

204. See the brilliant analysis in Matt 1978, 82–100.

205. Goethe 1809/1965–72, 2: 474.

206. Mann 1924/1927, 401. For an observant analysis of writing and media in *The Magic Mountain*, see Kudszus 1974.

207. Mann 1924/1927, 168–69.

208. Braune, 1929, in Kaes 1983, 352–53. The reverse link of film and reading has also been claimed, albeit with a dandy's flightiness. "She reads at such a pace," a queen complains about her reader in a novella, "and when I asked her *where* she had learnt to read so quickly she replied 'In the screens at Cinemas'" (Firbank 1923/1949, 128).

209. See the poem "Brise Marine" in Mallarmé 1945, 38.

210. Bloem 1922, 43–44.

211. The difficulties involved in this distinction are articulated by one of Ewers's heroes, who, for purposes of scientific experimentation, looks for and finds a whore: "It had to be the one, he thought, who belongs here and nowhere

else. Not one of those who ended up here [in the brothel] by some fluke. Who could just as well have become little ladies, workers, maids, typists, or even telephone operators" (Ewers 1911, 101).

212. Bliven 1954, 3.

TYPEWRITER

1. Bliven 1974, 72. Other languages experienced terminological problems. In French, the typewriter was initially called "typographe, piano à écrire, clavecin à écrire, pantographe, plume typographique" (B. Müller 1975, 169), as well as "dactylographe."

2. Heidegger 1942–43/1992, 86.

3. See Cockburn 1981.

4. See Van Creveld 1985, 103–4.

5. See pp. 65–67 in this volume; on Goethe's dictations, see Ronell 1986, 63–191.

6. Goethe, November 24, 1809, cited in Riemer 1841/1921, 313.

7. Schlegel 1799/1958, 8: 42.

8. Freud, *Introductory Lectures on Psycho-Analysis*, 1915–16, in idem 1962, 15: 155. See also Giese 1914, 528, on "Sexual Models for Simple Inventions" (*Sexualvorbilder bei einfachen Erfindungen*): "In 1565, Konrad Gesner describes an actual lead or, more properly, graphite pencil enclosed in a gliding, wooden casing. . . . The model that would come to mind is the retracting foreskin during erection. The interior of the penis coming thus to view would be equivalent to the emerging lead of graphite. Even the fountain pen of a more recent date might well be a reconstruction of the construction above."

9. "Schreiben" 1889, 863ff. Those in search of sexual models capitulate in a corresponding way: "In the 'modern' technology of our time psychoanalysis may well appear out of place" (Giese 1914, 524).

10. See Bliven 1954, 56.

11. See Stümpel 1985, 9.

12. Bliven 1954, 72.

13. Burghagen 1898, 9.

14. British Patent 395, January 7, 1714, quoted in Eye 1958, 12.

15. C. Müller 1823, 11.

16. Ibid., 16.

17. Kußmaul 1881, 5.

18. Ibid., 126.

19. C. Müller 1823, 5.

20. See Eye 1958, 13–17, as well as Tschudin 1983, 5ff. The link between neurophysiology and media technology is most visible with Thurber, whose typewriter was supposed to help not only the blind but also "people with nervous disorders who could not guide the quill" (Stümpel 1985, 12).

21. *Journal of Arts and Sciences*, 1823, cited in Brauner 1925, 4.

22. Burghagen 1898, 20.

23. Bliven 1954, 35.

24. See, for example, Grashey 1885, 688.

25. Salthouse 1984, 94–96.

26. See Granichstaedten-Czerva 1924, 35. Significantly, n18, which is supposed to carry proof of this charge, does not point to any evidence.

27. Zeidler 1983, 96. Correspondingly, the standardization of the component parts of typewriters took place "during the time of the First World War" (Eye 1958, 75).

28. Bliven 1954, 56.

29. Burghagen 1898, 31.

30. Ibid. Typing-speed records in the United States, by contrast, were up to fifteen letters per second (Klockenberg 1926, 10).

31. DPA (German press agency) news release, June 1, 1985.

32. Cocteau 1979, 62.

33. See, for example, Cocteau 1941/1946–51, 8: 40.

34. Ibid., 63.

35. Ibid., 181.

36. Ibid., 16.

37. See Wedel 1962, 114–17. However, see also Pynchon 1973, 529: "It [the V2] was half bullet, half arrow. *It* demanded this, we didn't. So. Perhaps you used a rifle, a radio, a typewriter. Some typewriters in Whitehall, in the Pentagon, killed more civilians than our little *A4* could have ever hoped to."

38. Twain, March 1875, quoted in Bliven 1954, 62.

39. Sales figures (in thousands) yield the curve shown in Stümpel 1985, 12. [Beginning with 0 in 1879, the graph shows a precipitous increase: 10,000 units by 1874, 30,000 by 1887, and 65,000 by 1890.—Trans.]

40. Current 1954, 54.

41. See Bliven 1954, 71.

42. See Eye 1958, 78.

43. Krukenberg 1906, 38.

44. Richards 1964, 1.

45. See Baumann 1985, 96.

46. Schwabe 1902, 6. Compare Burghagen 1898, 29: "Youths and female office assistants can also, without any training, be put to productive use at the typewriter for all types of business and administrative correspondence." See also Weckerle 1925, 32: "We have grown as accustomed to the typewriter as the sewing machine. And yet it has only been a few decades since a 'fine hand' was the best recommendation for a trade apprentice. Today, handwriting in a trading firm is virtually outdated and is at best limited to bookkeeping."

47. Schwabe 1902, 7.

48. For evidence on the social stratification of typists, see Witsch 1932, 54.

49. Meyer and Silbermann 1895, 264.

50. Valéry 1944/1957–60, 2: 301.

51. Spinner quoted in Eye 1958, 54.

52. See Eye 1958, 78. Von Budde's division of the General Staff, however, is shamelessly described as "a large railroad corporation."

53. For details, see Siegert 1986, 181–88.

54. L. Braun 1901, 197.

55. Schwabe 1902, 21.

56. *Zeitschrift für weibliche Handelsgehilfen*, 1918, quoted in Nienhaus 1982, 46. Stalin integrated Hindenburg's wholesome principle into the constitution of the Soviet Union in 1936.

57. Heidegger 1935/1959, 35.

58. Heidegger, 1942–43/1992, 80–81, 85–86.

59. Nietzsche, letter toward the end of February 1882, in idem 1975–84, pt. 3, 1: 172.

60. Dr. Eiser, 1877, quoted in Fuchs 1978, 632.

61. Ibid., 633.

62. After an observation by Martin Stingelin of Basel.

63. Nietzsche, letter of November 5, 1879, in idem 1975–84, pt. 2, 5: 461.

64. Nietzsche, letter of August 14, 1882, in idem 1975–84, pt. 2, 5: 435.

65. Nietzsche, letter of August 14, 1882, in idem 1975–84, pt. 3, 1: 113.

66. Nietzsche, letter of December 5, 1881, in idem 1975–84, pt. 3, 1: 146.

67. Burghagen 1898, 6.

68. Apparently infected, Nietzsche's biographer corrects his hero (saying that "the typewriter was 'invented,' that is, developed, 10 years earlier [sic] in America"). To top it off, he even writes "Hansun" instead of "Hansen" (Janz 1978–79, 2: 81, 95).

69. The following data are taken from Nyrop 1938.

70. Burghagen 1898, 6.

71. See Stümpel 1985, 22. There were even writing balls with a Morse-code hookup (Brauner 1925, 35–36).

72. Burghagen 1898, 120. Also see the photograph on p. 204 of this volume.

73. See Martin 1949, 571.

74. Stümpel 1985, 8.

75. McLuhan 1964, 260.

76. Bliven 1954, 132.

77. Nietzsche, letter of August 20–21, 1881, in idem 1975–84, pt. 3, 1: 117.

78. Burghagen 1898, 120 (referring to Malling Hansen's typewriter).

79. Nietzsche, letter of August 20–21, 1881, in idem 1975–84, pt. 3, 1: 117.

80. *Berliner Tageblatt*, March 1882.

81. See Nietzsche, letter of March 17, 1882, in idem 1975–84, pt. 3, 1: 180. "I enjoyed a report of the *Berliner Tageblatt* about my existence in Genoa—even the typewriter was mentioned." The mechanized philosopher clipped the news item.

82. Nietzsche, *Ecce Homo*, 1908, in idem 1967, 287.

83. See, for example, Eye 1958, 20.

84. Beyerlen quoted in Herbertz 1909, 559.

85. Beyerlen 1909, 362.

86. Swift 1904, 299, 300, 302. Also see the self-observation in the novel by Brück (1930, 238): "Here I sit, day by day, . . . typing freight letters, freight letters, freight letters. After three days it turned into purely mechanical work, the dim interactions between eyes and fingers, in which consciousness does not actively participate."

87. This list of early typewriting authors is taken from Burghagen 1898, 22.

88. Nietzsche, letter of April 1, 1882, in idem 1975–84, pt. 3, 1: 188.

89. See Doyle 1889/1930, 199.

90. Nietzsche, letter of March 17, 1882, in idem 1975–84, pt. 3, 1: 180.

91. Nietzsche, letter of March 27, 1882, in idem 1975–84, pt. 3, 1: 188.

92. Nietzsche, letter of March 17, 1882, in idem 1975–84, pt. 3, 1: 180; on

the "reading machine," see Nietzsche, letter of December 21, 1881, in idem 1975–84, pt. 3, 1: 151.

93. Förster-Nietzsche, in Nietzsche 1902–9, pt. 5, 2: 488.

94. Nietzsche, letter of June 18, 1882, in idem 1975–84, pt. 3, 1: 206.

95. Förster-Nietzsche 1935, 136.

96. Ibid., 138.

97. Nietzsche, *Ecce Homo*, 1908, in idem 1967, 267.

98. Nietzsche, *Genealogy*, 1887, in ibid., 61.

99. Ibid., 68.

100. Meysenburg, April 26, 1882, in Pfeiffer 1970, 420.

101. Nietzsche 1889/1984, 57. 102. Ibid., 59.

103. Nietzsche 1968, 89. 104. M. Weber 1918, 3.

105. Nietzsche, letter of February 1, 1883, in idem 1975–84, pt. 3, 1: 324.

106. See Nietzsche 1883–85/1966, 40.

107. Nietzsche, letter, June 1885, in idem 1975–84, pt. 3, 3: 58.

108. Nietzsche, letter of July 23, 1885, in idem 1975–84, pt. 3, 3: 70.

109. Bliven 1954, 79.

110. Hofmannsthal, June 11, 1919, in Hofmannsthal and Degenfeld 1974, 385.

111. Freud, *Introductory Lectures on Psycho-Analysis*, 1915–16, in idem 1962, 15: 154.

112. Ibid., 155.

113. Ibid., 156.

114. E. Jones 1953–57, 2: 98.

115. Freud, May 4, 1915, in Freud and Abraham 1980, 212.

116. Hyde 1969, 161.

117. Bosanquet 1924, 245.

118. Ibid., 248.

119. For the text of and a commentary on these dictates, see Hyde 1969, 277.

120. See Van Creveld 1985, 58–78.

121. See Nowell 1960, 106.

122. Ibid., 14, 199.

123. Benn, January 10, 1937, in idem 1969, 184.

124. Benn 1952/1959–61, 4: 173–74.

125. Benn, November 22, 1950, in idem 1962, 120.

126. Benn, February 6, 1937, in idem 1969, 194.

127. Benn, January 25, 1937, in idem 1969, 187. Klaus Theweleit describes this situation in much more detail, from the two girlfriends and the marriage to the war-induced suicide of Herta von Wedemeyer. For a portion, see Theweleit 1985, 133–56.

128. Benn, January 10, 1937, in idem 1969, 185–86.

129. See Kretzer 1894, in which a female accountant and daughter of an officer's widow (a sensation in the male office) still writes in longhand, but in which the problem of anonymous writing already appears in the form of block letters and round hand (166).

130. See Derrida 1980/1987, 53–55.

131. Eye 1958, 69, 80. Thus, for this volume August Walla typed the message that his "technological factory-like written highly honored highly esteemed honored valid typewriter" is being "appreciated by all gods and all political mortal public sovereigns."

132. Höhne 1984, 224–25.

133. Hitler, March 29, 1942, in Picker 1976, 157. On the Führer's typewriter (including four-millimeter Antique types against farsightedness), see ibid., 42.

134. Schramm 1982, 1: 139E.

135. Tolstoi 1978, 181.

136. Schlier 1926, 81.

137. Brück 1930, 218.

138. Ibid., 225. On the wish list of publications (which will then be fulfilled by the typewriter novel), see 233–34, 280.

139. Ibid., 229. For a psychiatrist's commentary on such tipptipp, see Ballet 1886, 143. "If it's a mild case of agraphia, patients are able to write many words, but with numerous mistakes; for example, they repeat at every occasion the same letters or the same syllable; they suffer, as Gairdner calls it, from *intoxication through the letter*, just as certain aphasic patients suffer from intoxication through the word."

140. See Kafka, November 27, 1912, in idem 1974, 70.

141. See Siegert 1986, 292.

142. Kafka 1912/1965, 268.

143. Kafka, October 27, 1912, in idem 1974, 16.

144. Kafka, November 2, 1912, in ibid., 23.

145. Kafka, August 10, 1913, in ibid., 302.

146. Streicher 1919, 38–41. Based on these criminological uses, on April 8, 1983, the republic of Romania came to the nice conclusion of coercing all typewriter owners into registering their machines with the authorities. See Rosenblatt 1983, 88.

147. Kafka, October 30, 1916, in idem 1974, 580.

148. See Kafka, August 22, 1916, in ibid., 491–92.

149. See Zglinicki 1956, 395.

150. Kafka, March 1922, in idem 1953, 229. See Derrida 1980/1987, 33.

151. Kafka, November 27, 1912, in idem 1974, 70.

152. Kafka, January 22/23, 1913, in idem 1974: 167–68.

153. Bronnen 1926/1977, 131.

154. Weckerle 1925, 31–32.

155. Kafka, July 10, 1913, in idem 1974, 289.

156. Kafka, December 21–22, 1912, in ibid., 115–16.

157. Mallarmé 1895/1945, 366.

158. Derrida 1980/1987, 194.

159. Benn 1951/1959–61, 1: 529.

160. Benn 1949/1959–61, 1: 366.

161. Streicher 1919, 7.

162. Benjamin 1928/1978, 79, 78.

163. See Apollinaire 1918/1965–66, 3: 901. More generally, see Ong 1982, 128.

164. Eliot, August 21, 1916, in idem 1971: x.

165. Foucault 1969/1972, 85.

166. Ibid., 86.

167. Ibid., 84.

168. Enright 1971/1981, 101.

169. Schmitt 1917/1918, 90.

170. Ibid., 92–105.

171. See Diller 1980, 188–92. The Secret Service took over British TV stations to use UHF to scramble the stereophony of German bombers over England. See R. Jones 1978, 175.

172. See Ong 1982, 93.

173. See Hodges 1983, 109.

174. Turing 1950, 440.

175. Hodges 1983, 364.

176. Turing 1950, 434.

177. Ibid., 434.

178. See Bliven 1954, 132.

179. See Morgall 1981.

180. Turing, in Hodges 1983, 362.

181. See Kowalski 1979, 424.

182. Péter 1957, 210.

183. Friedlaender 1922, 38, 164. On the possible, yet paranoid, implications of the name Bosemann for "this volume, this bond," see S. Weber 1980, 170–72. [*Grammophon Film Typewriter* was first published in German by Brinkmann & Bose. Kittler is alluding to a network of associations that ranges from a composite of his publishers' names to the etymological link in "diesem Bande, dieser Bande" (this volume, this bond), all of which is difficult to render in English.—Trans.]

In exile in England during the Second World War, Robert Neumann will finally get to know a cybernetics specialist who not only can scramble the stereophonies of German bombers but also can build "a solitary typewriter . . . that starts writing all by itself, as soon as we step out the door. (Simultaneously, a television set lights up directly across from it—I feel it dictates to the typewriter without sound what it thinks of us.)" R. Neumann 1963, 167–69.

184. Turing 1950/1992, 451.

185. J. von Neumann 1951/1963, 295, 301–2.

186. Ibid., 298.

187. Lacan 1975, 41.

188. Genesis 1:2; Hebrew for "trackless waste and emptiness" or "formless void."—Trans.

189. Murawski 1962, 112–13.

190. See Watzlawick, Beavin, and Jackson 1967/1969, 66–67.

191. Marconi, 1937, quoted in Dunlap 1941, 353.

192. Garliński 1979, 11. How fundamental the connection between typewriter and cryptography is, is demonstrated in the *Psychotechnische Arbeitsstudien* (psychotechnological time and motion studies; studies evidently done in the spirit of Münsterberg) on the *Rationalisierung der Schreibmaschine und ihrer Bedienung* (Rationalization of the typewriter and its operation): statistically exact analyses of letter frequencies in given languages provide the basis not only for the ten-finger typing system (see Klockenberg 1926, 82–83) but also for all forms of decoding.

193. Bredow, 1922, quoted in Lerg 1970, 159. On controlling military organs during the founding of the BBC, see Briggs 1961, 49.

194. See Garliński 1979, 12.

195. See Wildhagen 1970, 182.

196. See Bamford 1986, 51, and Garliński 1979, 147.

197. See Garliński 1979, 28.

198. Lacan 1978/1988b, 47.

199. Turing, October 14, 1936, quoted in Hodges 1983, 120. Turing's step toward cryptoanalysis was only consistent if both brain and nature were threatened by Laplacian computation mistakes. For computers, he later wrote, "the field of cryptography will perhaps be the most rewarding. There is a remarkably close parallel between the problems of the physicist and those of the cryptographer. The system on which a message is enciphered corresponds to the laws of the universe, the intercepted messages to the evidence available, the keys for a day or a message to important constants which have to be determined. The correspondence is very close, but the subject matter of cryptography is very easily dealt with by discrete machinery, physics not so easily" (Turing, 1948, in Hodges 1983, 383). That is how simply computer capabilities spell out the differences between nature and general staffs.

200. Hodges 1983, 148.

201. Rohwer and Jäckel 1979, 64.

202. Hodges 1983, 175.

203. Ibid., 168.

204. Rohwer and Jäckel 1979, 336.

205. Hodges 1983, 192.

206. See ibid.

207. See ibid., 267.

208. See Rohwer and Jäckel 1979, 110–12.

209. On this remarkable combination of writing, adding, and subtracting, which was introduced in 1910, see Brauner 1925, 40.

210. See Hodges 1983, 277.

211. See Zuse 1984, 77.

212. Oberliesen 1982, 205.

213. Zuse 1984, 77.

214. Lacan 1966/1977, 84–85.

215. That, at any rate, is how Zuse himself describes it (1984, 80–83). For a different version, see Hodges 1983, 299.

216. Von Braun quoted in Bergaust 1976, 95.

217. Syberberg 1978/1982, 109.

218. On Hitler's disinterest in the test demonstrations, see Dornberger 1953/1954, 64–68; on his enthusiasm upon seeing the Askania color films, see Virilio 1984/1989, 59–60 (with the suggestion that liquid-fuel rockets are attributable to Fritz Lang's film *Frau im Mond* [1929]).

219. N. Wiener 1961, 3, 5. See also Heims 1982, 183–84, and Virilio 1984/1989, 72.

220. See Sickert 1983, 134–42.

221. See Hodges 1983, 335, 301, 304, 413, respectively. More generally, see Gorny 1985, 104–9.

222. Pynchon 1973, 685. See Virilio's astonishingly parallel formulation of a "'*Blitzkrieg*' . . . the blinding Hiroshima flash which literally photographed the shadow cast by beings and things, so that every surface immediately became war's *recording* surface, its *film*" (1984/1989, 68).

223. Jungk 1956, 314.

224. Hodges 1983, 362.

225. See Garliński 1979, 119–44.

226. See Virilio 1984/1989, 94n.

227. Hodges 1983, 337.

228. Schmidt 1985. Upon the successful decoding of this "message" [?], the journal *Der Rabe* will award a prize.

229. Raven, quoted in Bamford 1986, 324.

230. Ibid., 430.

231. See ibid., 136. What in translator's German is called a *Ladungs-Übertragungsgerät* (charge transmission device), and can process "more than one quadrillion (1,000,000,000,000,000) multiplications per second," is, of course, the CCD, or charge-coupled device.

BIBLIOGRAPHY

Abraham, Karl. 1913. "Sollen wir die Patienten ihre Träume aufschreiben lassen?" *Internationale Zeitschrift für Psychoanalyse* 1: 194–96.

Abraham, Otto, and Erich Moritz von Hornbostel. 1904. "Über die Bedeutung des Phonographen für vergleichende Musikwissenschaft." *Zeitschrift für Ethnologie* 36: 222–36.

Andresen, Uwe. 1982. "Musiksynthesizer." *Funkschau* 39: 79–84.

Apollinaire, Guillaume. 1965–66. *Oeuvres complètes*. Ed. Michel Décaudin. 3 vols. Paris.

Arnheim, Rudolf. 1977. "Systematik der frühen kinematographischen Erfindungen." In *Kritiken und Aufsätze zum Film*, ed. Helmut H. Dieterichs. Munich.

Assmann, Aleida, and Jan Assmann, eds. 1983. *Schrift und Gedächtnis: Archäologie der literarischen Kommunikation*. Munich.

Azam, Eugène. 1893. *Hypnotisme et double conscience: Origine de leur étude et divers travaux sur des sujets analogues*. Paris.

Baade, Walter. 1913. "Über die Registrierung von Selbstbeobachtungen durch Diktierphonographen." *Psychologie* 66: 81–93.

Babington Smith, Constance. 1958. *Evidence in Camera: The Story of Photographic Intelligence in World War II*. London.

Bahneman, Jörg. 1971. "Wie bleibt die Armee auf der Höhe der Zeit?" In *Clausewitz in unserer Zeit: Ausblick nach zehn Jahren Clausewitz Gesellschaft*, ed. Rolf Eible, 161–75. Darmstadt.

Baier, Wolfgang. 1971. *Quellendarstellung zur Geschichte der Photographie*. Halle.

Balázs, Béla. 1930. *Der Geist des Films*. Halle.

Ballet, Gilbert. 1886. *Le langage intérieur et les diverses formes de l'aphasie*. Paris.

Bamford, James. 1986. *NSA: Amerikas geheimster Nachrichtendienst*. Zurich / Schwäbisch Hall.

Bateman, Wayne. 1980. *Introduction to Computer Music*. New York.

Bauer-Wabnegg, Walter. 1986. *Zirkus und Artisten in Franz Kafkas Werk: Ein Beitrag über Körper und Literatur im Zeitalter der Technik*. Erlangen.

Baumann, Roland. 1985. "Einschreibung und Götterschauspiele. Nietzsche und das Medium Schreibmaschine." M.A. thesis, Freiburg.

Beatles. N.d. *The Beatles Complete (Guitar Edition)*. London.

Benjamin, Walter. 1968. *Illuminations*. Ed. Hannah Arendt, trans. Harry Zohn. New York.

———. 1972–85. *Gesammelte Schriften*. Ed. Rolf Tiedemann et al. Frankfurt a.M.

———. 1978. *Reflections: Essays, Aphorisms, Autobiographical Writings*. Ed. Peter Demetz, trans. Edmund Jephcott. New York.

Benn, Gottfried. 1959–61. *Gesammelte Werke*. Ed. Dieter Wellershoff. 8 vols. Wiesbaden.

———. 1962. *Das gezeichnete Ich: Briefe aus den Jahren 1900–1956*. Munich.

———. 1969. *Den Traum alleine tragen: Neue Texte, Briefe, Dokumente*. Ed. Paul Raabe and Max Niedermayer. Munich.

———. 1977–80. *Briefe*. Vol. 1, *Briefe an F. W. Oelze*. Ed. Harald Steinhagen and Jürgen Schröder. Wiesbaden.

Bergaust, Erik. 1976. *Wernher von Braun: Ein unglaubliches Leben*. Düsseldorf.

Bergson, Henri. 1911. *Creative Evolution*. Trans. Arthur Mitchell. New York.

Bermann, Richard. 1960 [1913]. "Leier und Schreibmaschine." In *Das Kinobuch*, ed. Kurt Pinthus. Zurich.

Beyerlen, Angelo. 1909. "Eine lustige Geschichte von Blinden usw." *Schreibmaschinen-Zeitung Hamburg* 138: 362–63.

Bischoff, Walter. 1950. "Die Dramaturgie des Hörspiels." In Hans Bredow, *Aus meinem Archiv*, 260–66. Heidelberg.

Blair, William. 1929. "Army Radio in Peace and War." In *Radio*, ed. Irwin Stewart, 86–89. Annals of the American Academy of Political and Social Sciences, vol. 142 suppl. Philadelphia.

Blake, Clarence J. 1876. "The Use of the Membrana Tympani as a Phonautograph and Logograph." *Archives of Ophthalmology and Otology* 5: 108–13.

Bliven, Bruce, Jr. 1954. *The Wonderful Writing Machine*. New York.

Blodgett, A. D. 1890. "A New Use for the Phonograph." *Science* 15: 43.

Bloem, Walter. 1922. *Seele des Lichtspiels: Ein Bekenntnis zum Film*. Leipzig.

Bolz, Norbert. 1986. "Die Schrift des Films." In *Diskursanalysen 1: Medien*, ed. Friedrich A. Kittler, Manfred Schneider, and Samuel Weber, 26–34. Wiesbaden.

Bosanquet, Theodora. 1924. *Henry James at Work*. The Hogarth Essays. London.

Bouasse, Henri Pierre Maxime. 1934. *Optique et photométrie dites géométriques*. Paris.

Bourneville, Désiré Magloire, and Paul Regnard. 1877–78. *Iconographie photographique de la Salpêtrière*. 2 vols. Paris.

Bradley, Dermot. 1978. *Generaloberst Heinz Guderian und die Entstehung des modernen Blitzkrieges*. Osnabrück.

Braun, Lily. 1901. *Die Frauenfrage: Ihre geschichtliche Wirkung und wirtschaftliche Seite*. Leipzig.

Brauner, Ludwig. 1925. *Die Schreibmaschine in technischer, kultureller und wirtschaftlicher Bedeutung*. Sammlung gemeinnütziger Vorträge, ed. Deutscher Verein zur Verbreitung gemeinnütziger Kenntnissse in Prag. Prague.

Bredow, Hans. 1950. *Aus meinem Archiv: Probleme des Rundfunks.* Heidelberg.
———. 1954. *Im Banne der Ätherwellen.* Vol. 1. Stuttgart.
Brentano, Bettina. 1959–63. *Bettina von Arnim, Werke und Briefe.* Ed. Gustav Konrad. Frechen.
Briggs, Asa. 1961. *The Birth of Broadcasting.* Vol. 1 of *A History of Broadcasting in the United Kingdom.* London.
———. 1965. *The Golden Age of the Wireless.* Vol. 2 of *A History of Broadcasting in the United Kingdom.* London.
Bronnen, Arnolt. 1927. *Film und Leben: Barbara La Marr.* Berlin.
——— (pseudonym: A. H. Schelle-Noetzel). 1935. *Der Kampf im Aether oder die Unsichtbaren.* Berlin.
———. 1977. *Ostpolzug.* In idem, *Werke,* ed. Hans Mayer, vol. 1, 117–50. Kronberg.
Brooks, John. 1977. "The First and Only Century of Telephone Literature." In *The Social Impact of the Telephone,* ed. Ithiel de Sola Pool, 208–24. Cambridge, Mass.
Bruch, Walter. 1979. "Von der Tonwalze zur Bildplatte: 100 Jahre Ton- und Bildspeicherung." *Funkschau,* special issue.
Brück, Christa Anita. 1930. *Schicksale hinter Schreibmaschinen.* Berlin.
Brücke, Ernst. 1856. *Grundzüge der Physiologie und Systematik der Sprachlaute für Linguisten und Taubstummenlöehrer bearbeitet.* Wien.
Buchheit, Gert. 1966. *Der deutsche Geheimnisdienst: Geschichte der militärischen Abwehr.* Munich.
Büchner, Georg. 1963. *Complete Plays.* Trans. Carl Richard Mueller. New York.
Burghagen, Otto. 1898. *Die Schreibmaschine. Illustrierte Beschreibung aller gangbaren Schreibmaschinen nebst gründlicher Anleitung zum Arbeiten auf sämtlichen Systemen.* Hamburg.
Burroughs, William. 1974. *The Job: Interviews with William S. Burroughs.* Rev. and enlarged ed. Ed. Daniel Odier. New York.
Cagnetta, Franco, ed. 1981. *Nascita della fotografia psychiatrica.* Venice.
Campe, Rüdiger. 1986. "Pronto! Telefonate und Telephonstimmen." In *Diskursanalysen 1: Medien,* ed. Friedrich A. Kittler, Manfred Schneider, and Samuel Weber, 68–93. Wiesbaden.
Cendrars, Blaise. 1970. *Moravagine.* Trans. Alan Brown. New York.
Chapple, Steve, and Reebee Garofalo. 1977. *Rock 'n' Roll is Here to Pay.* Chicago.
Charbon, Paul, ed. 1976. *Le téléphone à la belle époque.* Brussels.
———. 1977. *Le phonographe à la belle époque.* Brussels.
Chew, Victor Kenneth. 1967. *Talking Machines 1877–1914: Some Aspects of the Early History of the Gramophone.* London.
Clément, Cathérine. 1975. "Les charlatanes et les hystériques." *Communications* 23 (*Psychanalyse et cinéma*): 213–22.
Cockburn, Cynthia. "The Material of Male Power." *Feminist Review* 9 (1981).
Cocteau, Jean. 1946–51. *Oeuvres complètes.* Ed. Louis Forestier and Pascal Pia. 11 vols. Paris.
———. 1951. *The Human Voice.* Trans. Carl Wildman. London.

———. 1979. *Kino und Poesie*. Munich.

———. 1992. *The Art of Cinema*. Comp. and ed. André Bernard and Claude Gauteur, trans. Robin Buss. New York.

Cros, Charles. 1964. *Oeuvres complètes*. Ed. Louis Forestier and Pascal Pia. Paris.

Current, Richard Nelson. 1954. *The Typewriter and the Men Who Made It*. Urbana, Ill.

Dahms, Gustav. 1895. *Die Frau im Staats- und Gemeindedienst*. Berlin.

Dallin, David J. 1955. *Soviet Espionage*. New Haven, Conn.

Davies, Margery. 1974. *Woman's Place Is at the Typewriter: The Feminization of the Clerical Labour Force*. Somerville, Mass.

Dehmel, Richard. 1906–09. *Gesammelte Werke*. 10 vols. Berlin.

Deleuze, Gilles. 1965. "Pierre Klossowski ou Les corps-langage." *Critique* 21: 199–219.

Deleuze, Gilles, and Félix Guattari. 1983. *Anti-Oedipus: Capitalism and Schizophrenia*. Trans. Robert Hurley, Mark Seem, and Helen R. Lane. Minneapolis.

Demeny, Georges. 1899. "Études sur les appareils chronophotographiques. *L'année psychologique* 5: 347–68.

———. 1904. *L'éducation du marcheur*. Paris.

Derrida, Jacques. 1976. *Of Grammatology*. Trans. Gayatri Chakravorty Spivak. Baltimore.

———. 1978. *Writing and Difference*. Trans. Alan Bass. Chicago.

———. 1987. *The Postcard: From Socrates to Freud and Beyond*. Trans. Alan Bass. Chicago.

Diller, Ansgar. 1980. *Rundfunkpolitik im Dritten Reich*. Vol. 2 of *Rundfunk in Deutschland*, ed. Hans Bausch. Munich.

Dornberger, Walter. 1954. *V2*. Trans. James Cleugh and Geoffrey Halliday. New York.

Doyle, Sir Arthur Conan. 1930. *The Complete Sherlock Holmes*. Ed. Christopher Morley. Garden City, N.Y.

Driesen, Otto. 1913. *Das Grammophon im Dienste des Unterrichts und der Wissenschaft: Systematische Sammlung von Grammophonplatten vom Kindergarten bis zur Universität*. Berlin.

Dunlap, Orrin E., Jr. 1941. *Marconi: The Man and His Wireless*. New York.

"Elektor-Vocoder." 1980. *Elektor: Zeitschrift für Elektronik* 1: 38–43; 2: 40–52.

Eliot, T[homas] S[tearns]. 1971. *The Waste Land*. Facsim. and transcr. ed., including annotations of Ezra Pound. Ed. Valerie Eliot. New York.

Ellis, John. 1975. *The Social History of the Machine Gun*. London.

Enright, Dennis Joseph. 1971. *The Typewriter Revolution and Other Poems*. New York.

———. 1981. *Collected Poems*. Oxford.

Enzensberger, Hans Magnus. 1974. "Constituents of a Theory of the Media." In *The Consciousness Industry: On Literature, Politics and the Media*, comp. Michael Roloff, 95–128. New York.

Ewers, Hanns Heinz. 1911. *Alraune: Die Geschichte eines lebendigen Wesens*. Munich.

Eye, Werner von. 1958. *Kurzgefaßte Geschichte der Schreibmaschine und des Maschinenschreibens*. Berlin.

Eyth, Max von. 1909. *Gesammelte Schriften.* 6 vols. Stuttgart.

Factor, R. 1978. "A 6,4-Second Digital Delay Line, Uniquely Designed for Broadcast Obscenity Publishing." AES (Advances in Engineering Software and Workstations) preprint no. 1417.

Farges, Jorges. 1975. "L'image d'un corps." *Communications* 23 (*Psychanalyse et cinéma*): 88–95.

Faulstich, Werner, ed. 1979. *Kritische Stichwörter zur Medienwissenschaft.* Munich.

Feldhaus, Franz Maria. 1928. *Kulturgeschichte der Technik I.* Berlin.

Firbank, Ronald. 1949. *The Flower Beneath the Foot.* In idem, *Five Novels*, ed. Osbert Sitwell. London.

Fischer, Gottfried, and Friedrich A. Kittler. 1978. "Zur Zergliederungsphantasie im Schneekapitel des *Zauberberg.*" In *Perspektiven psychoanalytischer Literaturkritik*, ed. Sebastian Goeppert, 23–41. Freiburg.

Flechsig, Paul. 1894. "Gehirn und Seele." Spoken address, October 31, University Church, Leipzig.

Förster-Nietzsche, Elisabeth. 1935. *Friedrich Nietzsche und die Frauen seiner Zeit.* Munich.

Foucault, Michel. 1972. *The Archaeology of Knowledge and The Discourse on Language.* Trans. A. M. Sheridan Smith. New York.

———. 1977. "Language to Infinity." In idem, *Language, Counter-Memory, Practice*, ed. Donald F. Bouchard, 53–67. Ithaca, N.Y.

———. 1990. *The History of Sexuality, volume 1: An Introduction.* Trans. Robert Hurley. New York.

Frese, Frank, and M. V. Hotschevar. 1937. *Filmtricks und Trickfilme.* Düsseldorf.

Freud, Sigmund. 1962. *The Standard Edition of the Complete Psychological Works of Sigmund Freud.* Ed. and trans. James Strachey. 23 vols. London.

———. 1985. *The Complete Letters of Sigmund Freud to Wilhelm Fliess, 1887–1904.* Trans. and ed. Jeffrey Moussaieff Masson. Cambridge, Mass.

Freud, Sigmund, and Karl Abraham. 1980. *Briefe 1907–1926.* Ed. Hilde C. Abraham and Ernst L. Freud. Frankfurt a.M.

Friedheim, Philip. 1983. "Wagner and the Aesthetics of the Scream." *19th-Century Music* 7: 63–70.

Friedlaender, Salomo (Mynona). 1922. *Graue Magie: Berliner Nachschlüsselroman.* Dresden.

———. 1980a [1916]. "Goethe spricht in den Phonographen." In *Das Nachthemd am Wegweiser und andere höchst merkwürdige Geschichten des Dr. Salomo Friedlaender*, 159–78. Berlin.

———. 1980b [1920]. "Fatamorganamaschine." In Mynona, *Prosa*, ed. Hartmut Geerken, vol. 1, 93–96. Munich.

Fuchs, Joachim. 1978. "Friedrich Nietzsches Augenleiden." *Münchener Medizinische Wochenschrift* 120: 631–34.

Fussell, Paul. 1975. *The Great War and Modern Memory.* New York.

Garliński, Jozef. 1979. *The Enigma War.* New York.

Gaupp, Fritz. 1931. *Die Nacht von heute auf morgen.* Berlin.

Gelatt, Roland. 1977. *The Fabulous Phonograph 1877–1977. From Edison to Stereo.* 2d rev. ed. New York.

Giedion, Siegfried. 1948. *Mechanization Takes Command: A Contribution to Anonymous History*. New York.

Giese, Fritz. 1914. "Sexualvorbilder bei einfachen Erfindungen." *Imago: Zeitschrift für Anwendung der Psychoanalyse auf die Geisteswissenschaften* 3: 524–35.

Ginzburg, Carlo. 1983. "Morelli, Freud, and Sherlock Holmes: Clues and Scientific Method." In *The Sign of Three: Dupin, Holmes, Peirce*, ed. Umberto Eco and Thomas A. Sebeok. Bloomington, Ind.

Goethe, Johann Wolfgang von. 1965–72. *Gespräche*. Ed. Wolfgang Herwig. Zurich.

———. 1969. *The Autobiography of Johann Wolfgang von Goethe (Dichtung und Wahrheit)*. Trans. John Oxenford; intro. Gregor Seeba. New York.

———. 1976. *Werke*. Ed. Erich Trunz. Munich.

———. 1979. *Wilhelm Meister's Years of Apprenticeship*. Bks. 7–8. Trans. H. M. Waidson. New York.

———. 1981. *Wilhelm Meister's Years of Travel; or, The Renunciants*. Bk. 2. Trans. H. M. Waidson. New York.

———. 1987. *Faust, Part One*. Trans. David Luke. Oxford.

———. 1990. *The Sufferings of Young Werther* and *Elective Affinities*. Ed. Victor Lange. The German Library, vol. 19. New York.

Gordon, Don E. 1981. *Electronic Warfare: Element of Strategy and Multiplier of Combat Power*. New York.

Görlitz, Walter. 1967. *Kleine Geschichte des deutschen Generalstabes*. Berlin.

Gorny, Peter. 1985. "Informatik und Militär." In *Militarisierte Wissenschaft*, ed. Werner Butte, 104–18. Reinbek.

Granichstaedten-Czerva, Rudolf von. 1924. *Peter Mitterhoder, Erfinder der Schreibmaschine: Ein Lebensbild*. Wien.

Grashey, Hubert. 1885. "Über Aphasie und ihre Beziehungen zur Wahrnehmung." *Archiv für Psychiatrie und Nervenkrankheiten* 16: 654–88.

Greve, Ludwig, Margot Pehle, and Heide Westhoff, eds. 1976. *Hätte ich das Kino! Die Schriftsteller und der Stummfilm*. Special exhibition, Schiller-Nationalmuseum. Marbach.

Grivel, Charles. 1984. "Die Explosion des Gedächtnisses: Jarry über die Entwicklung im literarischen Prozeß." In *Lyrik und Malerei der Avantgarde*, ed. Rainer Warning and Winfried Wehle, 243–93. Munich.

Groos, Karl. 1899. *Die Spiele des Menschen*. Jena.

Guattari, Félix. 1975. "Le divan du pauvre." *Communications* 23 (*Psychanalyse et cinéma*): 96–103.

Gutzmann, Hermann. 1908. "Über Hören und Verstehen." *Zeitschrift für angewandte Psychologie und psychologische Sammelforschung* 1: 483–503.

Guyau, Jean-Marie. 1880. "La mémoire et le phonographe." *Revue philosophique de la France et de l'étranger* 5: 319–22.

Hahn, Fritz. 1963. *Deutsche Geheimwaffen 1939–45*. Vol. 1, *Flugzeugbewaffnungen*. Heidenheim.

"Die Hähnische Litteralmethode." 1986. In *Karl Philipp Moritz: Die Schriften in Dreissig Bänden*, ed. Petra and Uwe Nettelbeck, vol. 1, pt. 2, 157–58. Nördlingen.

Hamburger, Käte. 1966. *Philosophie der Dichter: Novalis, Schiller, Rilke*. Stuttgart.

Hardenberg, Friedrich von (Novalis). 1960–75. *Schriften*. Ed. Paul Kluckhohn and Richard Samuel. Stuttgart.

———. 1964. *Henry von Ofterdingen*. Trans. Palmer Hilty. New York.

Hauptmann, Gerhart. 1962–74. *Sämtliche Werke: Centenar-Ausgabe*. Ed. Hans-Egon Hass. 11 vols. Darmstadt.

Haushofer, Karl. 1979. "*Nostris ex ossibus*: Gedanken eines Optimisten." In Hans-Adolf Jacobsen, *Karl Haushofer: Leben und Werk*, vol. 2, 634–40. Boppard.

Hay, Gerhard. 1975a. "Rundfunk in der Dichtung der zwanziger und dreißiger Jahre." In *Rundfunk und Politik 1923–1933: Beiträge zur Rundfunkforschung*, ed. Winfried B. Lerg and Rolf Steininger, 119–34. Berlin.

———, ed. 1975b. *Literatur und Rundfunk 1923–1933*. Hildesheim.

Hegel, Georg Wilhelm Friedrich. 1927–40. *Sämtliche Werke*. Ed. Hermann Glockner. 26 vols. Stuttgart.

———. 1977. *Phenomenology of Spirit*. Trans. A. V. Miller. Oxford.

Heidegger, Martin. 1950. *Holzwege*. Frankfurt a.M.

———. 1959. *An Introduction to Metaphysics*. Trans. Ralph Mannheim. New Haven, Conn.

———. 1992. *Parmenides*. Trans. André Schuwer and Richard Rojcewicz. Bloomington, Ind.

Heilbut, Iwan. 1931. *Frühling in Berlin*. Berlin.

Heims, Steve J. 1982. *John von Neumann and Norbert Wiener: From Mathematics to the Technologies of Life and Death*. Cambridge, Mass.

Hendrix, Jimi. 1968. *The Jimi Hendrix Experience: Electric Ladyland*. London.

Hennes, Hans. 1909. "Die Kinematographie im Dienste der Neurologie und Psychiatrie, nebst Beschreibungen einiger seltenerer Bewegungsstörungen." *Medizinische Klinik*, 2010–14.

Herbertz, Richard. 1909. "Zur Psychologie des Maschinenschreibens." *Zeitschrift für angewandte Psychologie* 2: 551–61.

Hesse, Hermann. 1957. *Steppenwolf*. Intro. Joseph Mileck. New York.

Hirth, Georg. 1897. *Aufgaben der Kunstphysiologie*. 2d ed. Munich.

Hobbes, Thomas. 1994. *Leviathan*. Ed. Edwin Curley. Indianapolis, Ind.

Hodges, Andrew. 1983. *Alan Turing: The Enigma*. New York.

Hoffmann, E. T. A. 1969. "The Sandman." In *Selected Writings*, vol. 1, *The Tales*, ed. and trans. Leonard J. Kent and Elizabeth C. Knight. Chicago.

———. 1971. *Three Märchen of E. T. A. Hoffmann*. Trans. Charles E. Passage. Columbia, S.C.

Hoffmann, Wilhelm. 1932/33. "Das Mikrophon als akustisches Fernglas." *Rufer und Hörer: Monatshefte für den Rundfunk* 2: 453–57.

Hofmannsthal, Hugo von, and Ottonie Gräfin Degenfeld. 1974. *Briefwechsel*. Ed. Therese Miller-Degenfeld. Frankfurt a.M.

Höhne, Heinz. 1984. *Der Orden unter dem Totenkopf: Die Geschichte der SS*. Munich.

Holst, Amalie. 1802. *Über die Bestimmung des Weibes zur höheren Geistesbildung*. Berlin.

Hyde, Montgomery. 1969. *Henry James at Home*. London.
Innis, Harold Adam. 1950. *Empire and Communications*. Oxford.
Jalowetz, Heinrich. 1912. "Die Harmonielehre." In *Arnold Schönberg*, 49–64. Munich.
Janz, Kurtz Paul. 1978–79. *Friedrich Nietzsche: Biographie*. 3 vols. Munich.
Jarry, Alfred. 1975. *Les minutes de sable mémorial*. In *Oeuvres complètes*, ed. René Massat, vol. 4, 169–268. Geneva.
Jensen, Johannes Vilhelm. 1917. *Unser Zeitalter*. Berlin.
Jentsch, Ernst. 1906. "Zur Psychologie des Unheimlichen." *Psychiatrisch-Neurologische Wochenschrift*, 195–98, 203–5.
Jones, Ernest. 1953–57. *The Life and Work of Sigmund Freud*. 3 vols. New York.
Jones, Reginald V. 1978. *Most Secret War*. London.
Joyce, James. 1969. *Ulysses*. Hammondsworth.
Jünger, Ernst. 1922. *Der Kampf als innere Erlebnis*. Berlin.
———. 1929. *Storm of Steel: From the Diary of a German Storm-Troop Officer on the Western Front*. Trans. Basil Creighton. New York.
———. 1932. *Der Arbeiter: Herrschaft und Gestalt*. Hamburg.
———. 1993. *Copse 125: A Chronicle from the Trench Warfare of 1918*. Trans. Basil Creighton. Rpt. New York.
Jungk, Robert. 1956. *Heller als tausend Sonnen: Das Schicksal der Atomforscher*. Bern.
Jüttemann, Herbert. 1979. *Phonographen und Grammophone*. Braunschweig.
Kaes, Anton. 1979. "The Expressionist Vision in Theater and Cinema." In *Expressionism Reconsidered: Relationships and Affinities*, ed. Gertrud Bauer Pickar and Karl Eugene Webb, 89–98. Munich.
———, ed. 1978. *Kino-Debatte: Texte zum Verhältnis von Literatur und Film 1909–1929*. Munich.
———. 1983. *Weimarer Republik: Manifeste und Dokumente zur deutschen Literatur 1918–1933*. Stuttgart.
Kaes, Anton, Martin Jay, and Edward Dimendberg, eds. 1994. *The Weimar Republic Sourcebook*. Berkeley.
Kafka, Franz. 1948. *The Penal Colony: Stories and Short Pieces*. Trans. Willa and Edwin Muir. New York.
———. 1950. *The Castle*. Trans. Willa and Edwin Muir. New York.
———. 1953. *Letters to Milena*. Ed. Willi Haas, trans. Tania and James Stern. London.
———. 1965. *The Diaries of Franz Kafka 1910–1913*. Ed. Max Brod, trans. Joseph Kresh. New York.
———. 1974. *Letters to Felice*. Ed. Erich Heller and Jürgen Born, trans. James Stern and Elizabeth Duckworth. London.
Keller, Gottfried. 1974. *The Misused Love Letters* and *Regula Amrain and her Youngest Son*. Trans. Michael Bullock and Anne Fremantle. New York.
Keun, Irmgard. 1979a [1931]. *Gilgi—eine von uns*. Düsseldorf.
———. 1979b [1932]. *Das kunstseidene Mädchen*. Düsseldorf.
Kittler, Friedrich. 1984a. "auto bahnen." *Kulturrevolution* 9: 42–45.
———. 1984b. "Der Gott der Ohren." In *Das Schwinden der Sinne*, ed. Dietmar Kamper and Christoph Wulf, 140–55. Frankfurt a.M.

———. 1985. "Romantik—Psychoanalyse—Film: Eine Doppelgängergeschichte." In *Eingebildete Texte: Affairen zwischen Psychoanalyse und Literaturwissenschaft*, ed. Jochen Hörisch and Georg Christoph Tholen, 118–35. Munich.

———. 1989/1990. "Dracula's Legacy." Trans. William Stephen Davis. *Stanford Humanities Review* 1, no. 1: 143–73.

———. 1990. *Discourse Networks, 1800/1900*. Trans. Michael Metteer, with Chris Cullens. Stanford, Calif.

———. 1993. "World-Breath: On Wagner's Media Technology." In *Opera Through Other Eyes*, ed. David J. Levin, 215–35. Stanford, Calif.

———. 1997. "Media and Drugs in Pynchon's Second World War." Trans. Michael Wutz and Geoffrey Winthrop-Young. In *Reading Matters: Narrative in the New Ecology of Media*, ed. Joseph Tabbi and Michael Wutz. Ithaca, N.Y.

Klippert, Werner. 1977. *Elemente des Hörspiels*. Stuttgart.

Klockenberg, Erich. 1926. *Rationalisierung der Schreibmaschine und ihre Bedienung: Psychotechnische Arbeitsstudien*. Berlin.

Knies, Karl. 1857. *Der Telegraph als Verkehrsmittel*. Tübingen.

Kowalski, Robert A. 1979. "Algorithm = Logic + Control." *Communications of the Association for Computing Machinery* 2: 424–36.

Kracauer, Siegfried. 1971–79. *Schriften*. Ed. Karsten Witte. 8 vols. Frankfurt a.M.

———. 1974. *From Caligari to Hitler: A Psychological History of the German Film*. Princeton, N.J.

Krcal, Richard. 1964. *Peter Mitterhofer und seine Schreibmaschine: Zum Buch geformt von Peter Basten*. Aaachen.

Kretzer, Max. 1894. *Die Buchhalterin*. Dresden.

Krukenberg, Elsbeth. 1906. *Über das Eindringen der Frauen in männliche Berufe*. Essen.

Kudszus, Winfried. 1974. "Understanding Media: Zur Kritik dualistischer Humanität im *Zauberberg*." In *Besichtigung des Zauberbergs*, ed. Heinz Sauereßig, 55–80. Biberach.

Kußmaul, Adolf. 1881. "Die Störungen der Sprache. Versuch einer Pathologie der Sprache." In *Handbuch der speciellen Pathologie und Therapie*, vol. 12 (app.), ed. H. v. Ziemssen. 2d ed. Leipzig.

Kylstra, Peter H. 1977. "The Use of the Early Phonograph in Phonetic Research." *Phonographic Bulletin* 17: 3–12.

Lacan, Jacques. 1966. *Ecrits*. Paris.

———. 1975. *Le séminaire: livre XX*. Paris.

———. 1977. *Ecrits: A Selection*. Trans. Alan Sheridan. New York.

———. 1978. *The Four Fundamental Concepts of Psychoanalysis*. Ed. Jacques-Alain Miller, trans. Alan Sheridan. New York.

———. 1988a. *The Seminar of Jacques Lacan. Book I: Freud's Papers on Technique 1953–54*. Ed. Jacques-Alain Miller, trans. John Forrester. New York.

———. 1988b. *The Seminar of Jacques Lacan. Book II: The Ego in Freud's Theory and in the Technique of Psychoanalysis 1954–55*. Trans. Sylvana Tomaselli. New York.

Leduc, Jean-Marie. 1973. *Pink Floyd*. Collection Rock & Folk. Paris.

Lerg, Winfried B. 1970. *Die Entstehung des Rundfunks in Deutschland: Herkunft und Entwicklung eines publizistischen Mittels*. 2d ed. Frankfurt a.M.

Lerg, Winfried B., and Rolf Steininger, eds. 1975. *Rundfunk und Politik 1923 bis 1933*. Berlin.

Lévi-Strauss, Claude. 1969. *The Raw and the Cooked*. Trans. John and Doreen Weightman. New York.

Lindau, Paul. 1906. *Der Andere: Schauspiel in vier Aufzügen*. Leipzig.

Lorenz, Thorsten. 1985. "Wissen ist Medium. Die deutsche Stummfilmdebatte 1907–1929." Diss., Freiburg.

Lothar, Rudolph. 1924. *Die Sprechmaschine: Ein technisch-ästhetischer Versuch*. Leipzig.

Ludendorff, Erich. 1935. *Der totale Krieg*. Munich.

Luhmann, Niklas. 1985. "Das Problem der Epochenbildung und die Evolutionstheorie." In *Epochenschwellen und Epochenstrukturen im Diskurs der Literatur- und Sprachhistorie*, ed. Hans-Ulrich Gumbrecht and Ulla Link-Herr, 11–33. Frankfurt a.M.

MacDonnell, Kevin. 1973. *Der Mann, der die Bilder laufen ließ oder Eadweard Muybridge und die 25.000$-Wette*. Lucerne.

Mach, Ernst. 1914. *The Analysis of Sensations and the Relation of the Physical to the Psychical*. Trans. C. M. Williams, supplemented by Sydney Waterlow. Chicago.

Mallarmé, Stéphane. 1945. *Oeuvres complètes*. Ed. Henri Mondor and G. Jean-Aubry. Paris.

Mann, Thomas. 1927. *The Magic Mountain*. Trans. Helen Lowe-Porter. London.

———. 1933. *Past Masters and Other Papers*. Trans. H. T. Lowe-Porter. London.

Marage, René M. 1898. "Les phonographes et l'étude des voyelles." *L'année psychologique* 5: 226–44.

Maréchal, Gaston. 1891. "Photographie de la parole." *L'illustration* 2543 (November 21): 406–7.

Marey, Etienne-Jules. 1873. *La machine animale: Locomotion terrestre et aérienne*. Paris.

———. 1894. *Le mouvement*. Paris.

Marinetti, Filippo Tommaso. 1971. *Selected Writings*. Ed. R. W. Flint. London.

Marker, Chris. 1983. *Sans Soleil / Unsichtbare Sonne. Vollständiger Text zum gleichnamigen Film-Essay*. Hamburg.

Martin, Ernst. 1949. *Die Schreibmaschine und ihre Entwicklungsgeschichte*. 2d ed. Pappenheim.

Marty, Daniel. 1981. *Grammophone: Geschichte in Bildern*. Karlsruhe.

Matt, Peter von. 1978. "Zur Psychologie des deutschen Nationalschriftstellers. Die paradigmatische Bedeutung der Hinrichtung und Verklärung Goethes durch Thomas Mann." In *Perspektiven psychoanalytischer Literaturkritik*, ed. Sebastian Goeppert, 82–100. Freiburg.

McLuhan, Marshall. 1964. *Understanding Media*. New York.

Meumann, Ernst. 1912. *Ästhetik der Gegenwart*. 2d ed. Leipzig.

Meyer, Julius, and Josef Silbermann. 1895. *Die Frau im Handel und Gewerbe*. Der Existenzkampf der Frau im modernen Leben. Seine Ziele und Aussichten, 7. Berlin.

Meyrink, Gustav. 1928. *The Golem*. Trans. Madge Pemberton. Boston.

Mitry, Jean, ed. 1976. "Le cinéma des origines." *Cinéma d'aujourd'hui, cahiers bimensuels* 9 (fall): 1–126.

Moholy-Nagy, Laszlo. 1923. "Neue Gestaltungen in der Musik. Möglichkeiten des Grammophons." *Der Sturm* 14: 103–5.

———. 1978. *Malerei, Fotografie, Film: Nachdruck der Ausgabe München 1925*. Ed. Otto Stelzer. Mainz.

Monaco, James. 1977. *How to Read a Film*. New York.

Morgall, Janine. 1981. "Typing Our Way to Freedom: Is It True the New Office Technology Can Liberate Women?" *Feminist Review* 9 (fall).

Morin, Edgar. 1956. *Le cinéma; ou, L'homme imaginaire*. Paris.

Morrison, Jim. 1977. *The Lords and the New Creatures / Poems: Gedichte, Gesichte und Gedanken*. Frankfurt a.M.

Müller, Bodo. 1975. *Das Französische in der Gegenwart: Varietäten, Strukturen, Tendenzen*. Heidelberg.

Müller, C. M. 1823. *Neu erfundene Schreib-Maschine, mittlest welcher Jedermann, ohne Licht in jeder Sprache und Schriftmanier sicher zu schreiben, Aufsätze und Rechnungen zu verfertigen vermag, auch Blinde besser als mit allen bisher bekannten Schreibtafeln nicht nur leichter schreiben, sondern auch das von ihnen Geschriebene selbst lesen können*. Vienna.

Münsterberg, Hugo. 1914. *Grundzüge der Psychotechnik*. Leipzig.

———. 1970. *The Photoplay: A Psychological Study*. Rpt. as *The Film: A Psychological Study*, ed. Richard Griffith. New York.

Münsterberg, Margaret. 1922. *Hugo Münsterberg: His Life and His Work*. New York.

Murawski, Erich. 1962. *Der deutsche Wehrmachtsbericht 1939–1945, Ein Beitrag zur Untersuchung der geistigen Kriegsführung*. 2d ed. Boppard.

Nabokov, Vladimir. 1970. *Mary*. Trans. Michael Glenny, in collaboration with the author. Greenwich, Conn.

Nadar (Félix Tournachon). 1899. *Quand j'étais photographe*. Paris.

———. 1978. "My Life as a Photographer." *October* 5: 7–28.

Navratil, Leo. 1983. *Die Künstler aus Gugging*. Wien.

Neumann, Gerhard. 1985. "'Nachrichten vom Pontus': Das Problem der Kunst in Franz Kafkas Werk." In *Franz Kafka Symposium*, ed. Wilhelm Emrich and Bernd Goldmann, 101–57. Mainz.

Neumann, John von. 1961–63. "The General and Logical Theory of Automata." In idem, *Collected Works*, vol. 5, 288–328.

Neumann, Robert. 1963. *Ein leichtes Leben: Bericht über sich selbst und Zeitgenossen*. Vienna.

"The New Phonograph." 1887. *Scientific American* 57: 421–22.

Nienhaus, Ursula. 1982. *Berufsstand weiblich: Die ersten weiblichen Angestellten*. Berlin.

Nietzsche, Friedrich. 1902–9. *Briefwechsel*. Ed. Elisabeth Förster-Nietzsche and Peter Gast. Berlin.

———. 1922–29. *Sämtliche Werke*. Musarion-Ausgabe. Munich.

———. 1956. *The Birth of Tragedy and The Genealogy of Morals*. Trans. Francis Golffing. New York.

———. 1966. *Thus Spoke Zarathustra*. Trans. Walter Kaufmann. Hammondsworth.
———. 1967. *On the Genealogy of Morals and Ecce Homo*. Trans. Walter Kaufmann and R. J. Hollingdale. New York.
———. 1968. *Twilight of the Idols*. Trans. Walter Kaufmann. Hammondsworth.
———. 1974. *The Gay Science*. Trans. Walter Kaufmann. New York.
———. 1975–84. *Briefwechsel: Kritische Gesamtausgabe*. Ed. Giorgio Colli and Mazzino Montinari. Berlin.
———. 1984. *Dithyrambs of Dionysus*. Trans. R. J. Hollingdale. Redding Ridge, Conn.
———. 1990. *Unmodern Observations*. Ed. William Arrowsmith. Trans. Gary Brown, William Arrowsmith, and Herbert Golder. New Haven, Conn.
Nowell, Elizabeth. 1960. *Thomas Wolfe: A Biography*. New York.
Nyrop, Camillus. 1938. "Malling Hansen." In *Dansk Biografisk Leksikon*, ed. Povl Engelstoft, vol. 18, 265–67. Copenhagen.
Oberliesen, Rolf. 1982. *Informationen, Daten, Signale: Geschichte technischer Informations-verarbeitung*. Reinbek.
Ong, Walter J. 1982. *Orality and Literacy: The Technologizing of the Word*. London.
Overbeck, Egon. 1971. "Militärische Planung und Unternehmensplanung." In *Clausewitz in unserer Zeit: Ausblicke nach zehn Jahren Clausewitz Gesellschaft*, ed. Rolf Eible, 89–97. Darmstadt.
Parzer-Mühlbacher, Alfred. 1902. *Die modernen Sprechmaschinen (Phonograph, Graphophon und Grammophon), deren Behandlung und Anwendung: Ratschläge für Interessenten*. Vienna.
Pawley, Edward L. E. 1972. *BBC Engineering: 1922–1972*. London.
Péter, Rósza. 1957. *Rekursive Funktionen*. 2d ed. Budapest.
Pfeiffer, Ernst, ed. 1970. *Friedrich Nietzsche, Paul Rée, Lou von Salomé: Die Dokumente ihrer Begegnung*. Frankfurt a.M.
Picker, Henry, ed. 1976. *Hitlers Tischgespräche: Auflage mit bisher unbekannten Selbstzeugnissen Adolf Hitlers, Abbildungen, Augenzeugenberichten und Erläuterungen des Autors, "Hitler, wie er wirklich war."* Stuttgart.
Pink Floyd. 1975. *Wish You Were Here: Songbook*. London.
———. 1976. *Song Book: Ten Songs from the Past*. London.
———. 1983. *The Final Cut: A Requiem for the Post War Dream*. London: EMI LP.
Pinthus, Kurt, ed. 1963. *Kinobuch*. Zürich.
Pohle, Heinz. 1955. *Der Rundfunk als Instrument der Politik: Zur Geschichte des deutschen Rundfunks 1923/38*. Hamburg.
Pretzsch, Paul, ed. 1934. *Cosima Wagner und Houston Stewart Chamberlain im Briefwechsel 1888 bis 1908*. Leipzig.
Pynchon, Thomas. 1973. *Gravity's Rainbow*. New York.
Rabiner, Lawrence R., and Bernard Gold. 1975. *Theory and Application of Digital Signal Processing*. Englewood Cliffs, N.J.
Rank, Otto. 1971. *The Double: A Psychological Study*. Trans. and ed. Harry Tuckler, Jr. Chapel Hill, N.C.
Rathenau, Walter. 1918–29. *Gesammelte Schriften*. 6 vols. Berlin.

Rayleigh, Lord John William Strutt. 1877–78. *The Theory of Sound.* 2 vols. London.

Read, Oliver, and Walter L. Welch. 1959. *From Tin Foil to Stereo: Evolution of the Phonograph.* Indianapolis, Ind.

Reis, Philipp. 1952. "Über Telephonie durch den galvanischen Strom." In Erwin Horstmann, *75 Jahre Fernsprecher in Deutschland 1877–1952*, 34–38. Frankfurt a.M.

Renard, Maurice. 1970 [1907]. "La Mort et le Coquillage." In *Invitation à la peur*, 67–72. Paris.

Ribot, Théodule. 1881. *Les maladies de la mémoire.* Paris.

Richards, George Tilghman. 1964. *The History and Development of Typewriters.* 2d ed. London.

Riemer, Friedrich Wilhelm. 1921. *Mitteilungen über Goethe: Aufgrund der Ausgabe von 1814 und des handschriftlichen Nachlasses.* Ed. Arthur Pollmer. Leipzig.

Rilke, Rainer Maria. 1949. *The Notebooks of Malte Laurids Brigge.* Trans. M. D. Herter Norton. New York

———. 1955–66. *Sämtliche Werke.* Ed. Ernst Zinn. Wiesbaden.

———. 1957. *Poems 1906 to 1926.* Trans. J. B. Leishman. London.

———. 1961 [1919]. "Primal Sound." In idem, *Selected Works. Volume 1: Prose*, 51–56. Trans. G. Craig Houston. London.

Rolling Stones. 1969. *Beggars Banquet: Songbook.* New York.

Rohwer, Jürgen, and Eberhard Jäckel, eds. 1979. *Die Funkaufklärung und ihre Rolle im Zweiten Weltkrieg.* Stuttgart, 1979.

Ronell, Avital. 1986. *Dictations: On Haunted Writing.* Bloomington, Ind.

Rosenblatt, Roger. 1981. "The Last Page in the Typewriter." *Time* (May 16): 88.

Sachs, Heinrich. 1905. *Gehirn und Sprache.* Grenzfragen des Nerven- und Seelenlebens, no. 36. Wiesbaden.

Salthouse, Timothy. 1984. "Die Fertigkeit des Maschinenschreibens." *Spektrum der Wissenschaft* 4: 94–100.

Sartre, Jean-Paul. 1969a. "L'Homme au Magnétophone." *Les Temps modernes* 274 (April): 1812–40.

———. 1969b. "A Psychoanalytic Dialogue with a Commentary by Jean Paul Sartre." Abridged. Trans. Paul Augst. *Ramparts* 8, no. 4 (October): 43–49.

Saussure, Ferdinand de. 1959. *Course in General Linguistics.* Ed. Charles Bailly and Albert Sechehaye. Trans. Wade Baskin. New York.

Schäfer, Hildegard. 1983. *Stimmen aus einer anderen Welt.* Freiburg.

Scherer, Wolfgang. 1983. *Babellogik: Sound und die Auslöschung der buchstäblichen Ordnung.* Frankfurt a.M.

———. 1986. "Klaviaturen, Visible Speech und Phonographie: Marginalien zur technischen Entstellung der Sinne im 19. Jahrhundert." In *Diskursanalysen 1: Medien*, ed. Friedrich A. Kittler, Manfred Schneider, and Samuel Weber, 37–54. Wiesbaden.

Schlaffer, Heinz. 1986. "Einführung." In Jack Goody, Ian Watt, and Kathleen Gough, *Enstehung und Folgen der Schriftkultur*, 7–20. Frankfurt a.M.

Schlegel, Friedrich. 1958–87. *Kritische Ausgabe.* Ed. Ernst Behler. 35 vols. Munich.

Schlier, Paula. 1926. *Petras Aufzeichnungen oder Konzept einer Jugend nach dem Diktat der Zeit*. Innsbruck.

Schmidt, Arno. 1985. "Offener Brief." In *Der Rabe* 10, ed. Gerd Haffmans, 125. Zürich.

Schmitt, Carl. 1918. "Die Buribunken: Ein geschichtsphilosophischer Versuch." *Summa* 1, no. 4: 89–106.

Schneider, Manfred. 1985. "Hysterie als Gesamtkunstwerk: Aufstieg und Verfall einer Semiotik der Weiblichkeit." *Merkur* 39: 89–106.

Schnur, Roman. 1980. "Im Bauche des Leviathan: Bemerkungen zum politischen Gehalt der Briefe Gottfried Benns an F. W. Oelze in der NS-Zeit." In *Auf dem Weg zur Menschenwürde und Gerechtigkeit: Festschrift Hans R. Klecatsky*, ed. Ludwig Adamovich und Peter Pernthaler, vol. 2, 911–928. Vienna.

Schramm, Percy Ernst, ed. 1982. *Das Kriegstagebuch des Oberkommandos der Wehrmacht (Wehrmachtführungsstab) 1940–45, geführt von Helmuth Grener und Percy Ernst Schramm*. Rpt. Herrsching.

Schreber, Daniel Paul. 1973. *Denkwürdigkeiten eines Nervenkranken*. Ed. Samuel M. Weber. Frankfurt a.M.

"Schreiben mit der Maschine." 1889. *Vom Fels zum Meer: Spemann's Illustrirte Zeitschrift für das Deutsche Haus*. Col. 863f.

Schwabe, Jenny. 1902. *Kontoristin: Forderungen, Leistungen, Aussichten in diesem Berufe*. 2d ed. Leipzig.

Schwendter, Rolf. 1982. *Zur Geschichte der Zukunft: Zukunftsforschung und Sozalismus*. Frankfurt a.M.

Seeliger, Germar. 1985. "Schillers köstliche Reste. Ein bis heute mysteriöser Fall: Was geschah mit des Dichters Schädel?" *Die Zeit* (September 27): 82–85.

Shaw, George Bernard. 1972. *Collected Plays with Their Prefaces*. Ed. Dan H. Laurence. London.

Sickert, Klaus, ed. 1983. *Automatische Spracheingabe und Sprachausgabe: Analyse, Synthese und Erkennung menschlicher Sprache mit digitalen Systemen*. Haar.

Siegert, Bernhard. 1986. "Die Posten und die Sinne: Zur Geschichte der Einrichtung von Sinn und Sinnen in Franz Kafkas Umgang mit Post und technischen Medien." M.A. thesis, Freiburg.

Siemsen, Hans. 1994. "The Literature of Nonreaders." In *The Weimar Republic Sourcebook*, ed. Anton Kaes, Martin Jay, and Edward Dimendberg, 663–64. Berkeley.

Slaby, Adolf. 1911. *Entdeckungsfahrten in den elektrischen Ozean: Gemeinverständliche Vorträge*. 5th ed. Berlin.

Snyder, Charles. 1974. "Clarence John Blake and Alexander Graham Bell: Otology and the Telephone." *Annals of Otology, Rhinology and Laryngology* 83, suppl. 13: 3–31.

Southall, Brian. 1982. *Abbey Road: The Story of the World's Most Famous Recording Studio*. Cambridge, Eng.

Specht, Richard. 1922. *Arthur Schnitzler: Der Dichter und sein Werk*. Berlin.

Spielrein, Sabina. 1986. *Ausgewählte Werke*. Ed. Günter Bose and Erich Brinkmann. Berlin.

Stern, William. 1908. "Sammelbericht über Psychologie der Aussage." *Zeitschrift für angewandte Psychologie* 1: 429–50.

Stetson, Raymond Herbert. 1903. "Rhythm and Rhyme." *Harvard Psychological Studies* 1: 413–66.

Stoker, Bram. 1965. *Dracula*. New York.

Stransky, Erwin. 1905. *Über Sprachverwirrtheit: Beiträge zur Kenntnis derselben bei Geisteskranken und Geistesgesunden*. Sammlung zwangsloser Abhandlungen aus dem Gebiete der Nerven- und Geisteskrankheiten, no. 6. Halle.

Strauss, Botho. 1979. *Devotion*. Trans. Sophie Wilkins. New York.

Streicher, Hubertus. 1919. *Die kriminologische Verwertung der Maschinenschrift*. Graz.

Stümpel, Rolf, ed. 1985. *Vom Sekretär zur Sekretärin: Eine Ausstellung zur Geschichte der Schreibmaschine und ihrer Bedeutung für den Beruf der Frau im Büro*. Gutenberg Museum Mainz, Mainz.

Swift, Edgar J. 1904. "The Acquisition of Skill in Type-Writing: A Contribution to the Psychology of Learning." *The Psychological Bulletin* 1: 295–305.

Syberberg, Hans-Jürgen. 1982. *Hitler: A Film from Germany*. Trans. Joachim Neugroschel. New York.

Theweleit, Klaus. 1985. "The Politics of Orpheus Between Women, Hades, Political Power and the Media: Some Thoughts on the Configuration of the European Artist, Starting with the Figure of Gottfried Benn, or: What Happens to Eurydice?" *New German Critique* 36: 133–56.

———. 1987–89. *Male Fantasies*. Trans. Erica Carter, Stephen Conway, and Chris Turner. 2 vols. Minneapolis.

———. 1988. *Buch der Könige*. Vol. 1, *Orpheus [und] Eurydike*. Basel.

Todorov, Tzvetan. 1973. *The Fantastic: A Structural Approach to a Literary Genre*. Trans. Richard Howard. Cleveland.

Toeplitz, Jerzy. 1973. *Geschichte des Films 1895–1928*. Munich.

Tolstoi, Tatiana. 1978. *Ein Leben mit meinem Vater: Erinnerungen an Leo Tolstoi*. Köln.

Troitzsch, Ulrich, and Wolfhard Weber. 1982. *Die Technik: Von den Anfängen bis zur Gegenwart*. Braunschweig.

Tschudin, Peter. 1983. *Hüpfende Lettern: Kleine Geschichte der Schreibmaschinen*. Mitteilungen der Basler Papiermühle, no. 38. Basel.

Turing, Alan M. 1950. "Computing Machinery and Intelligence." *Mind: A Quarterly Review of Psychology and Philosophy* 59: 433–60.

———. 1992. *Collected Works*. Vol. 3, *Mechanical Intelligence*. Ed. D. C. Nice. Amsterdam.

Urban, Bernd. 1978. *Hofmannsthal, Freud und die Psychoanalyse: Quellenkundliche Untersuchungen*. Frankfurt a.M.

Valéry, Paul. 1957–60. *Oeuvres*. 2 vols. Ed. Jean Hytier. Paris.

Van Creveld, Martin L. 1985. *Command in War*. Cambridge, Mass.

Vietta, Silvio. 1975. "Expressionistische Literatur und Film: Einige Thesen zum wechselseitigen Einfluß ihrer Darstellung und Wirkung." *Mannheimer Berichte* 10: 294–99.

Villiers de l'Isle-Adam, Philippe Auguste Mathias, Comte de. 1982. *Tomorrow's Eve*. Trans. Robert Martin Adams. Urbana, Ill.

Virilio, Paul. 1976. *L'insécurité du territoire*. Paris.

———. 1989. *War and Cinema: The Logistics of Perception*. Trans. Patrick Camiller. London.

Volckheim, Ernst. 1923. *Die deutschen Kampfwagen im Weltkriege*. Vol. 107, no. 2 of *Militär-Wochenblattes*. Berlin.

von Schramm, Wilhelm. 1979. *Geheimdienst im Zweiten Weltkrieg: Operationen, Methoden, Erfolge*. 3d ed. Munich.

Wagner, Richard. 1906. *Tristan and Isolde: Drama in Three Acts*. Trans. Henry Grafton Chapman. New York.

———. 1976. *Mein Leben*. Ed. Martin Gregor-Dellin. Munich.

———. 1978. *Die Musikdramen*. Ed. Joachim Kaiser. Munich.

———. 1986. *Parsifal*. Trans. Andrew Porter. London.

———. 1993. *Ring of the Nibelung: A Companion*. Trans. Stewart Spencer. London.

Walze, Alfred. 1980. "Auf den Spuren von Christopher Latham Scholes: Ein Besuch in Milwaukee, der Geburtstätte der ersten brauchbaren Schreibmaschine." *Deutsche Stenografenzeitung*, 132–33, 159–61.

Watson, Peter. 1978. *War on the Mind: The Military Uses and Abuses of Psychology*. London.

Watzlawick, Paul, Janet H. Beavin, and Don D. Jackson. 1967. *Pragmatics of Human Communication: A Study of Interactional Patterns, Pathologies and Paradoxes*. New York.

Weber, Marianne. 1918. *Vom Typenwandel der studierenden Frau*. Berlin.

———. 1928. "Die soziale Not der berufstätigen Frau." In *Die soziale Not der weiblichen Angestellten*. Schriftreihe des Gewerkschaftsbundes der Angestellten, GDA-Schrift no. 43. Berlin.

Weber, Samuel L. 1980. "Fellowship." In *Grosz/Jung/Grosz*, ed. Günter Bose and Erich Brinkmann, 161–72. Berlin.

Weckerle, Eduard. 1925. *Mensch und Maschine*. Jena.

Wedel, Hasso von. 1962. *Die Propagandatruppen der deutschen Wehrmacht*. Vol. 34 of *Wehrmacht im Kampf*. Neckargemünd.

Wellershoff, Dieter. 1980. *Die Sirene: Eine Novelle*. Köln.

Wetzel, Michael. 1985. "Telephonanie: Kommunikation und Kompetenz nach J.G. Hamann." In *Affairen zwischen Psychoanalyse und Literaturwissenschaft*, ed. Jochen Hörisch and Georg Christoph Tholen, 136–45. Munich.

Wiener, Norbert. 1961. *Cybernetics; or, Control and Communication of the Animals and the Machine*. 2d ed. Cambridge, Mass.

Wiener, Otto. 1900. "Die Erweiterung unserer Sinne." Inaugural lecture, May 19, Leipzig.

Wieszner, Georg Gustav. 1951. *Richard Wagner als Theater-Reformer: Vom Werden des deutschen National-Theaters im Geiste des Jahres 1848*. Emstetten.

Wilde, Oscar. 1966. "The Soul of Man Under Socialism." In idem, *Complete Works*, ed. J. B. Foreman. London.

Wildhagen, Karl Heinz, ed. 1970. *Erich Fellgiebel: Meister operativer Nachrichtenverbindungen: Ein Beitrag zur Geschichte der Nachrichtentruppe*. Wennigsen.

Winter, L. W., ed. 1959. *Der Koran: Das heilige Buch des Islam*. Munich.

Witsch, Josef. 1932. *Berufs- und Lebensschicksale weiblicher Angestellten in der schönen Literatur.* 2d ed. Sozialpolitische Schriften des Forschungsinstitutes für Sozialwissenschaften in Köln, no. 2. Cologne.

Yaks, Rodnaz. 1985. *Programmierung des Z 80.* 2d ed. Düsseldorf.

Zeidler, Jürgen. 1983. "Kopisten und Klapperschlangen—aus der Geschichte der Schreibmaschine." In *Museum für Verkehr und Technik Berlin: Ein Wegweiser*, 96–106. Berlin.

Zglinicki, Friedrich von. 1956. *Der Weg des Films: Die Geschichte der Kinematographie und ihrer Vorläufer.* Berlin.

———. 1979. *Der Weg des Films: Bildband.* Hildesheim.

Zumthor, Paul. 1985. "Die orale Dichtung: Raum, Zeit und Periodisierungsprobleme." In *Epochenschwellen und Epochenstrukturen im Diskurs der Sprach- und Literaturhistorie*, ed. Hans-Ulrich Gumbrecht and Ursula Link-Herr, 359–75. Frankfurt a.M.

Zuse, Konrad. 1984. *Der Computer: Mein Lebenswerk.* 2d ed. Berlin.

Library of Congress Cataloging-in-Publication Data

Kittler, Friedrich A.
[Grammophon Film Typewriter. English]
Gramophone, film, typewriter / Friedrich Kittler ; translated, with an introduction, by Geoffrey Winthrop-Young and Michael Wutz.
p. cm. — (Writing science)
Includes bibliographical references.
ISBN 0-8047-3232-9 (cloth : alk. paper). — ISBN 0-8047-3233-7 (pbk. : alk. paper)
1. Communication and technology. 2. Communication—History—19th century. 3. Communication—History—20th century. I. Title. II. Series.
P96.T42K5713 1999
302.2—dc21 98-37243
CIP

∞ This book is printed on acid-free, recycled paper.

Original printing 1999

Last figure below indicates year of this printing:
08 07 06 05 04 03 02